2021 International Conference on Intelligent Power and Systems

Journal of Physics: Conference Series Volume 2215

Online
24 – 26 September 2021

ISBN: 978-1-7138-5178-3
ISSN: 1742-6588

Printed from e-media with permission by:

Curran Associates, Inc.
57 Morehouse Lane
Red Hook, NY 12571

CURRAN ASSOCIATES INC.
proceedings
.com

Some format issues inherent in the e-media version may also appear in this print version.

Institute of Physics
Dirac House, Temple Back
Bristol BS1 6BE UK

Phone: 44 1 17 929 7481
Fax: 44 1 17 920 0979

techtracking@iop.org

Additional copies of this publication are available from:

Curran Associates, Inc.
57 Morehouse Lane
Red Hook, NY 12571 USA
Phone: 845-758-0400
Fax: 845-758-2633
Email: curran@proceedings.com
Web: www.proceedings.com

TABLE OF CONTENTS

Journal of Physics: Conference Series **2215** (2022) 011001 doi:10.1088/1742-6596/2215/1/011001

Preface

Due to recent pandemic, 2021 International Conference on Intelligent Power and Systems (ICIPS 2021) was held virtually online during September 24-26, 2021. The decision to hold the virtual conference was made in compliance with many restrictions and regulations that were imposed by countries around the globe. Such restrictions were made to minimize the risk of people contracting or spreading the COVID-19 through physical contact. There were 50 individuals who attended this on-line conference, represented many countries.

ICIPS 2021 brings together top professionals from industry, government, and academia from around the world. ICIPS 2021 includes invited talks, oral presentations, and poster presentations of refereed papers. We invite submissions of papers and abstracts on all topics related to power electronics and power system. The conference provides networking opportunities for participants to share ideas, designs, and experiences on the state-of-the-art and future direction of power electronics and power system.

During the conference, the conference model was divided into three sessions, including oral presentations, keynote speeches, and online Q&A discussion. In the first part, some scholars, whose submissions were selected as the excellent papers, were given about 5-10 minutes to perform their oral presentations one by one. Then in the second part, keynote speakers were each allocated 30-45 minutes to hold their speeches. In the second part, we invited three professors as our keynote speakers. Prof. Shunchun Yao, South China University of Technology School of Electric Power Engineering, China. He was the first one to perform a thought-provoking speech: *On-line detection techniques for low-carbon and clean energy utilization.* And then we had Prof. Eryu Zhu, Beijing Jiaotong University, China. He presented an insightful speech: *Report of the Next Generation of Civil Engineering.* Our finale keynote speaker, Prof. Changbin Hu, North China University of Technology, China. He made a wonderful speech: *Research on multi-VSG fast pre-synchronization grid connection with dual power loop.* Their insightful speeches had triggered heated discussion in the third session of the conference. Every participant praised this conference for disseminating useful and insightful knowledge.

The proceedings are a compilation of the accepted papers and represent an interesting outcome of the conference. Topics include but are not limited to the following areas: Power Electronics, Power System and Intelligent System and other related topics. All the papers have been through rigorous review and process to meet the requirements of international publication standard.

We would like to acknowledge all of those who supported ICIPS 2021. The help and contribution of each individual and institution was instrumental in the success of the conference. In particular, we would like to thank the organizing committee for its valuable inputs in shaping the conference program and reviewing the submitted papers.

We sincerely hope that the ICIPS 2021 turned out to be a forum for excellent discussions that enabled new ideas to come about, promoting collaborative research. We are sure that the proceedings will serve as an important research source of references and knowledge, which will lead to not only scientific and engineering findings but also new products and technologies.

The Committee of ICIPS 2021

Journal of Physics: Conference Series **2215** (2022) 011001 doi:10.1088/1742-6596/2215/1/011001

Committee member

Conference Chairman
Prof. Steivan Defilla, Tianjin University, China
Prof. SHAHID HUSSAIN, Jiangsu University, China
Prof. Wei Qin, Chongqing University, China

Program Committees
Prof. Noreddine Gherabi, Sultan Moulay Slimane University, Morocco
Prof. Sandeep Saxena, Galgotias College of Engineering and Technology, Greater Noida
Prof. Said Fathy EL-Zoghdy, Department of Mathematics & Computer Science, Faculty of Science, Menoufia University, Egypt
Dr. Attlee Munyaradzi Gamundani, Namibia University of Science and Technology, Namibia
Dr. Dimitrios Kollias, School of Computing & Mathematical Sciences, University of Greenwich, UK
Dr. Azim Zaliha Abd Aziz, Faculty of Informatics & Computing, Universiti Sultan Zainal Abidin, Malaysia

Technical Program Committees
Dr. Omar DIB, College of Science and Technology, Wenzhou-Kean University, China
Dr. Aslina Baharum, User Experience Research Lab, Universiti Malaysia Sabah, Malaysia
Dr. Muhammad Aslam, School of Cyber Science and Engineering, Wuhan University, China
Dr. Ariffin NOR HAPIZA, INFORMATION SYSTEM DEPARTMENT, UNIVERSITI TEKNOLOGI MARA, MALAYSIA

Journal of Physics: Conference Series **2215** (2022) 011002 doi:10.1088/1742-6596/2215/1/011002

Peer Review Statement

All papers published in this volume have been reviewed through processes administered by the Editors. Reviews were conducted by expert referees to the professional and scientific standards expected of a proceedings journal published by IOP Publishing.

- **Type of peer review:** Single Anonymous
- **Conference submission management system:** Morressier
- **Number of submissions received:** 57
- **Number of submissions sent for review:** 28
- **Number of submissions accepted:** 28
- **Acceptance Rate (Submissions Accepted / Submissions Received × 100):** 49.1
- **Average number of reviews per paper:** 2
- **Total number of reviewers involved:** 9
- **Contact person for queries:**
  Name: Xuexia Ye
  Email: xx.ye@keoaeic.org
  Affiliation: AEIC Academic Exchange Information Centre

Robust optimal dispatch of integrated energy system considering with coupled wind and hydrogen system

Liang Ding[1,a], Jie Gao[1,b], Guangnan Shi[1,c], Zihua Ni[2,d]

[1]State Grid Zhejiang Shaoxing Shangyu District Power Supply Co.,Ltd , China,Zhejiang

[2]Zhejiang Univeresity , China,Zhejiang

[a]Email:21910142@zju.edu.cn, [b]Email:15397038935@163.com, [c]Email:3140102814@zju.edu.cn, [d]Email:2814701822@qq.com

Abstract—Hydrogen production from renewable energy will become one of the important features of the integrated energy system in the future. By considering the coupled wind and hydrogen system, it can not only take the advantages of fast power regulation of hydrogen energy, but improve the economy of wind and hydrogen system. In view of the high volatility of the wind power, this paper proposes a robust optimal scheduling model for integrated energy system considering with coupled wind and hydrogen system. Firstly, it establishes a comprehensive energy system model with wind and hydrogen system, forming an efficient electric-gas-electric energy closed-loop flow model with fuel cells. Then, based on improved multi-interval uncertainty sets, the model of wind turbines, electrical load and hydrogen load are described respectively, which can reduce the conservatism compared with the traditional robust optimization. And the two-stage robust optimal scheduling model is established and it can be solved by adopting Column and Constraint Generation method (C&CG). Finally, an example is given to verify the advantages of the proposed method.

1. INTRODUCTION

Low-carbon energy system has drawn considerable attention to address the challenges caused by climate change. And developing renewable energy generation and reducing traditional fossil energy are identified as vital measures to achieve the goal [1][2]. Wind power, one of the absolute main energy of the renewable energy installation, has formed a completely industrial chain and supply chain in China [3]. By the end of 2020, the wind power capacity in China has reached 280 million kilowatts. However, the large-scale wind power integration will pose a great challenge to the grid, caused by its high volatility and seasonal variation of power generation [4]. Hydrogen energy is generally acknowledged as clean energy with high energy density, which is considered to be one of the most promising energy in the 21st century [5]. The coupled wind and hydrogen system can regulate the power rapidly and make full use of the abandoned wind power to improve the economic efficiency of wind power [6].

Research institutes at home and abroad have carried out extensive research for the combination of wind power and hydrogen energy. In [7], it makes water-electrolytic hydrogen making equipment as a flexible and controllable load to use the wasted wind power during the peak of the energy system efficiently. A coordinated optimization model of the thermal-electric-hydrogen system is presented in [8], which contains the fuel cell power generation system to improves the efficiency of the wind power. The wind power hydrogen-energy coupling system is proved that it can shorten the investment payback period from the perspective of life-cycle cost evaluation [9]. It further combined the wind-hydrogen energy storage system with ultracapacitors and make the simulation of the model in [10]. However, the above studies mainly focus on the economy of the wind power and hydrogen system. The above studies consider the impact of the uncertainty of wind power and load, which will lead to the infeasibility of system optimization results and damage the whole system severely.

Several studies are conducted in research work on the methods with the uncertainty of wind power, which mainly includes random optimization [11]and robust optimization [12]. Compared to the random optimization, the robust optimization can meet all the constraints and achieve the optimal solution with the greatest toughness. Besides it does not require accurate probability distribution function for the uncertain set. In [13], the study presents a typical structure of two-stage robust optimization model, and provides the solution of the system operation strategy under the worst scenarios. However, the probability of extreme scenarios is extremely small, which make the solution the solution relatively conservative when pursue to operate stably under the extreme scenarios. A flexible optimization model is proposed to improve the economy and robustness of the energy system by considering the uncertainty of wind power [14]. The actual output of wind power is the sum of the predicted power and prediction error. However, robust optimization often too conservative to ignore the actual probability distribution function. In [15], it adopts an uncertain set about moment uncertainty that it can to obtain statistical information from the samples of random data, match the actual output of renewable energy and establish a data-driven optimization model to reduce the conservativeness of traditional robust optimization and random data and obtain more accurate scheduling results. In [16], it combines kernel density estimation and autoregressive moving average model to establish the probability distribution model of wind power, which is used to solve the cooperative optimization problem of long-term operation and short-term dispatching of units.

Based on the aforementioned discussions, this paper proposes a new robust optimization model of integrated energy system with coupled wind power and hydrogen. It takes electric energy and hydrogen energy as the main input energy and models the hydrogen electric energy conversion equipment. Then, it constructs the robust optimization model of the integrated energy system based on multi-interval uncertain sets of the wind power. Finally, the model is solved by the column-and-constraints Generation (C&CG) algorithm [17].

2. The Optimization Model With Wind-Coupled And Hudrogrn System

Coupled wind and hydrogen system can be divided into three parts, which include electrolytic-hydrogen unit, fuel cell unit and hydrogen storage unit. Electrolytic-hydrogen unit is an electrolytic water process that produces hydrogen and oxygen without any contamination of other production. When the power is charged too much, the electrical energy can flow into the hydrogen-making equipment to produce hydrogen and store it in hydrogen storage equipment such as hydrogen storage tanks or directly supply hydrogen loads such as fuel cells. When the system is short of charge, the fuel cell is activated for discharge and the chemical energy stored by hydrogen is converted into electrical energy to into the power grid.

The integrated energy system is mainly composed of wind power, Coupled wind and hydrogen system, fuel cell, hydrogen storage tank and battery. The scheduling of the system is globally optimized, and all the equipment in the system is controllable, and the minimum operating cost of the system is realized by adjusting the contribution of each equipment in the system and the load response from the outside world.

2.1. Objective Function

The optimization model is formulated as a linear model that identifies the least cost of the system to meet the power balance and the restriction on system operation.

$$C = C_e + C_{ope} + C_{su} + C_{cur} \tag{1}$$

$$0 \le q_{sell}^t \le Q_{sell}^{\max},\ 0 \le q_{buy}^t \le Q_{buy}^{\max}, \forall t \in N_t \tag{2}$$

Where C reflects the whole operation cost of the system. C_e is the interaction cost between the main grid and the system and C_{sur} is the start-stop cost of the electrolytic-hydrogen unit. C_{ope} is the operating cost and Ccur is the punishment penalty of wind abandonment.

1) The interacting cost between the main grid and the system contains the expense of buying and selling electricity from the main grid.

$$C_e = \sum_{t\in N_t} \left(p^t_{buy} q^t_{buy} - p^t_{sell} q^t_{sell} \right) \tag{3}$$

In (3), p^t_{buy}, p^t_{sell} represent the price of electricity purchased and sold by the main grid respectively. And q^t_{buy} , q^t_{sell} represents the power bought and sold by the main grid. Q^{max}_{buy} , Q^{max}_{sell} are the maximum power for buying and selling electricity.

2) It needs extra expense when the equipment is operating and starting up.

$$C^{ope}_{t} = C^{op} \delta_i(t) \tag{4}$$

$$C^{SU}_{t} = C^{SU} \left(\delta_i(t) - \delta_i(t-1) \right) \tag{5}$$

C^{op}, C^{SU} are the operating cost and start-up cost of hydrogen-making equipment, respectively. C^{ope}_t ,C^{SU}_t represent the actual operating cost and start-stop cost of the equipment at the time *t*. $\delta(t)$ is the binary variable and reflects the operation status of equipment at the time *t* .

3) It needs extra cost when the wind power is abandoned for the power balance.

$$C_{cur} = \sum_{t\in N^t} c_{cur} p_{cur,t} \tag{6}$$

In (6), $p_{cur,t}$ is the abandoned power at the time *t* and c_{cur} is the punishment factor.

2.2. Operational Constraints of the Integrated Energy System

1) The model of electrolytic-hydrogen equipment is established as a linearized model to reduce the calculation time without significant loss of model accuracy within the normal operating range [18].

$$\mathbf{F}^{ptH}_t = \Pi^{ptH} \cdot \eta^{ptH} \cdot \mathbf{P}^{PtH}_t \tag{7}$$

$$F^{\min} \le \mathbf{F}^{ptH}_t \le F^{\max} \quad t \in N_{\mathrm{t}} \tag{8}$$

In (7) and (8), F^{ptH}_t, P^{ptH}_t are the flow of the hydrogen produced and the electricity power consumed by the equipment at the time *t*. η^{ptH}, Π^{ptH} are the efficiency of the equipment and the conversion efficiency of the process making-hydrogen, respectively. F^{min}, F^{max} are the max and min hydrogen flow. N^t is the number of hydrogen-making equipment.

2) Considering the safety of the tank, the capacity of the tank must be limited to a certain extent..

$$\mathrm{SOC}_{h,t} = \mathrm{SOC}_{h,(t-1)} + \left(F^{hs,ch}_{h,t} - F^{hs,dis}_{h,t} - \gamma^{Dsp}.\mathrm{SOC}_{h,t} \right) \times \Delta t \tag{9}$$

$$0 \le F^{hs,ch}_{h,t} \le F^{\max,\mathrm{ch}}_{h,t}, \quad 0 \le F^{hs,dis}_{h,t} \le F^{\max,dis}_{h,t} \tag{10}$$

$$SOC^{\min}_{\mathrm{h}} \le SOC_{\mathrm{h},t} \le SOC^{\max}_{\mathrm{h}}, \quad t \in N_{\mathrm{t}}, \quad h \in N_h \tag{11}$$

The equation (9) reflects the variation of the hydrogen storage capacity between the time *t-1* and the time *t*. $SOC_{h,t}$, $SOC_{h,t-1}$ represent the storage capacity of hydrogen in the tank at the time *t* and at the time *t-1*. $F^{max,ch}_{h,t}$, $F^{min,dis}_{h,t}$ represent the rate of hydrogen charge and hydrogen discharge; γ^{dsp} is the rate of hydrogen dissipation; $F^{max,ch}_{h,t}$, $F^{min,dis}_{h,t}$ represents the maximum and minimum charge rate of hydrogen, SOC^{max}_h ,SOC^{min}_h represents the maximum and minimum storage of hydrogen and N_h is the number of tanks..

3) The fuel cell can produce electricity by consuming hydrogen, which is the critical equipment of the system [19].

$$\mathbf{F}^{ptH}_t = \Pi^{ptH} \cdot \eta^{ptH} \cdot \mathbf{P}^{PtH}_t \tag{12}$$

$$F^{\min} \le \mathbf{F}^{ptH}_t \le F^{\max} \quad t \in N_{\mathrm{t}} \tag{13}$$

The equation (12) presents the conversion function and physical restrictions of the fuel cell. F_t^{ptH}, P_t^{ptH} are the consumed hydrogen and the generated electricity. $P^{fuel,max}$, $P^{fuel,min}$ are the max and min flow of hydrogen and η^{fuel} is the efficiency of the hydrogen production.

4) In order to extend the life of the battery, over-charging and over-discharging is not allowed.

$$SOC_{i,t} = SOC_{i,t-1} + \eta_i^{\mathrm{ES,ch}} P_{i,t}^{\mathrm{ES,ch}} - \eta_i^{\mathrm{ES,dis}} P_{i,t}^{\mathrm{ES,dis}} \tag{14}$$

$$0 \le P_{i,t}^{\mathrm{ES,dis}} \le P_{i,\max}^{\mathrm{ES,dis}}, 0 \le P_{i,t}^{\mathrm{ES,ch}} \le P_{i,\max}^{\mathrm{ES,ch}} \tag{15}$$

$$SOC_i^{\min} \le SOC_{i,t} \le SOC_i^{\max} \tag{16}$$

In (14), $SOC_{i,t}$, $SOC_{i,t-1}$ represents the battery's storage during the t-period; $P_{i,t}^{Es,ch}$, $P_{i,t}^{Es,dis}$ represents the charge and discharge rate of the battery; $P_{i,t}^{max,ch}$, $P_{i,t}^{min,dis}$ represents the maximum charge and discharge rate of the battery; $\eta_i^{Es,ch}$, $\eta_i^{Es,dis}$ represents the charge and discharge efficiency of the battery *and* SOC_h^{max}, SOC_h^{min} represents the maximum and minimum storage of the battery, respectively.

5) Power balance and hydrogen flow balance

$$\begin{aligned} &P_{buy}^{\mathrm{t}} - P_{sell}^{\mathrm{t}} + \sum_{i=1}^{n_g} g_i^{\,t} - \sum_{i=1}^{n_{es}} P_{i,t}^{\mathrm{ES,ch}} + \sum_{i=1}^{n_{es}} P_{i,t}^{\mathrm{ES},dis} \\ &- \sum_{i=1}^{n_h} P_{i,t}^{PtH} + \sum_{i=1}^{n_f} P_{i,t}^{\mathrm{fuel}} - \sum_{i=1}^{n_h} x^t = 0 \end{aligned} \tag{17}$$

$$\sum_{i=1}^{n_{es}} F_{i,t}^{\mathrm{HS,ch}} + \sum_{i=1}^{n_{es}} F_{i,t}^{\mathrm{HS},dis} - \sum_{i=1}^{n_h} F_{i,t}^{PtH} - \sum_{i=1}^{n_h} F_{i,t}^{\mathrm{dem}} - \sum_{i=1}^{n_f} F_{i,t}^{\mathrm{fuel}} = 0 \tag{18}$$

Where (17) represents the system's dynamic power balance and (18) presents the system's hydrogen balance.

3. Robust Optimize The Scheduling Model And Solvitation

3.1. Classic Robust Optimization Scheduling

The basic requirement of a classic robust optimization model is the toughness that the system is able to combat interruptions under the most extreme scenarios [19]. The model can be described as (19).

$$\begin{gathered} \min_x \left[c^t x + \max_{w \in W} \min_{\Delta x \in (x,w)} b^t \Delta x \right] \\ \text{s. t.} \begin{cases} Dx..d \\ Gx = 0 \\ D(x+\Delta x)..d \\ G(x+\Delta x) = 0 \end{cases} \end{gathered} \tag{19}$$

In (11) , c^t, b^t are cost factor vectors associated with x , Δx, respectively; d is the constant column vector; x is the decision variable, which is the system power under the deterministic constraint, D, G is the coefficient matrix to meet the constraint of the variable.

The uncertain set of the robust model is normally described as a predictive deviation model of the power generation.

$$\mathbf{W}_\partial^* = \left\{ w_{\partial,t} = w_{\partial,t}^* + w_{\partial,t}^+ \varepsilon_{\partial,t}^+ - w_{\partial,t}^- \varepsilon_{\partial,t}^-, 0 \le \varepsilon_{\partial,t}^+, \varepsilon_{\partial,t}^- \le 1 \right\} \tag{20}$$

$w_{\partial,t}$, $w_{\partial,t}^*$ for the actual power generation load, forecast force, $w_{\partial,t}^+$,$w_{\partial,t}^-$ is the maximum up-and-down deviation of the actual power generation load, $\varepsilon_{\partial,t}^+$, $\varepsilon_{\partial,t}^-$ The control factor for predicting the deviation for the power generation load makes $w_{\partial,t}$ in the intervals *[*$w_{\partial,t}^*+ w_{\partial,t}^+$*,* $w_{\partial,t}^*- w_{\partial,t}^-$*]*.

3.2. The improved robust optimization scheduling model

The classical robust optimization model has a extremely robustness, but it often ignores the probability information of the data itself, which makes the results too conservative. As shown in the Fig.1, the classic robust optimization often produces a strategy that makes the strategic result on the boundary. However, the probability that the result occurs is tiny in practice, which often results in a greater loss on the economics of the power system.

Therefore, this paper proposes an improved robust optimization model with the multi-interval uncertainty set based on the actual probability of power generation to analyze the characteristics of the data. As shown in the Fig.2, the basics of the model is that the system power generation output basically falls in the place where the prediction deviation is small and only a small part falls at the larger deviation according to the probability of the actual power, so as to improve the practicality of the scheduling strategy.

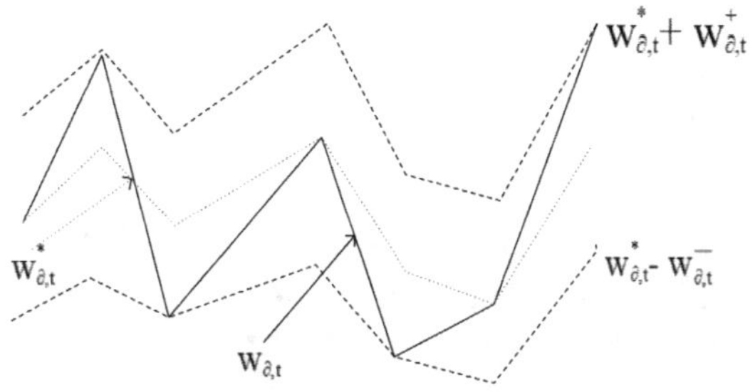

Fig.1: The traditional robust optimization

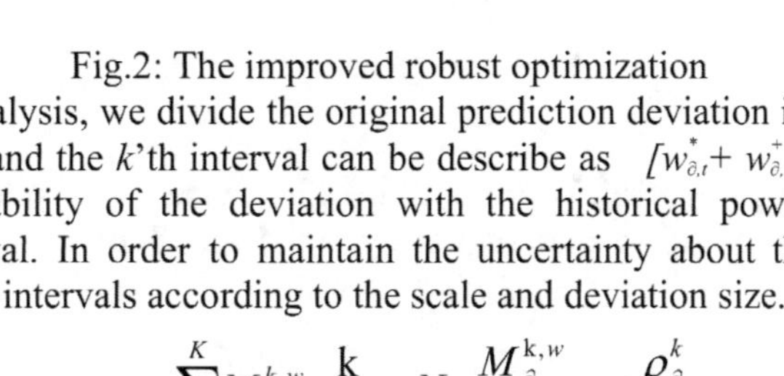

Fig.2: The improved robust optimization

Based on the above analysis, we divide the original prediction deviation intervals of $[w^*_{\partial,t}+w^+_{\partial,t}, w^*_{\partial,t}-w^-_{\partial,t}]$ into the Intervals K and the k'th interval can be describe as $[w^*_{\partial,t}+w^+_{\partial,t}*k/K, w^*_{\partial,t}-w^-_{\partial,t}*k/K]$. Then we can obtain the probability of the deviation with the historical power data by Monte Carlo Simulation in each interval. In order to maintain the uncertainty about the power generation, the uncertainty is divided into intervals according to the scale and deviation size.

$$\sum_{k=1}^{K} M_{\partial}^{k,w}\cdot\frac{k}{K}=N,\quad \frac{M_{\partial}^{k,w}}{M_{\partial}^{k+1,w}}=\frac{\rho_{\partial}^{k}}{\rho_{\partial}^{k+1}} \tag{21}$$

In (21), $M_{\partial}^{k,w}$, $M+1_{\partial}^{k,w}$ are the uncertainty allocated by the k'th interval and the $k+1$'th interval. ρ^k, ρ^{k+1} the probability of deviation in the k'th interval and the $k+1$'th interval.

As a result, an improved uncertainty set can be obtained:

$$w_{\partial,t}=w^*_{\partial,t}+\sum_{k=1}^{K}\left(w^{k+}_{\partial,t}\varepsilon^{k+}_{\partial,t}-w^{k-}_{\partial,t}\varepsilon^{k-}_{\partial,t}\right) \tag{22}$$

$$0 \leq \varepsilon_{\partial,t}^{k\pm} \leq 1 \tag{23}$$

$$0 \leq \sum_{k=1}^{K}\left(\varepsilon_{\partial,t}^{k+} + \varepsilon_{\partial,t}^{k-}\right) \leq 1 \tag{24}$$

$$\sum_{t=1}^{N_t}\left(\varepsilon_{\partial,t}^{k+} + \varepsilon_{\partial,t}^{k-}\right) \leq M_{\partial}^{k,w} \tag{25}$$

$$\sum_{t=1}^{N_t}\sum_{k=1}^{K}\varepsilon_{\partial,t}^{k+} = \sum_{t=1}^{N_t}\sum_{k=1}^{K}\varepsilon_{\partial,t}^{k-} \tag{26}$$

Where (22) to (26) describe the entire uncertainty set. $w_{\partial,t}^{k+}$, $w_{\partial,t}^{k-}$ represent the maximum upper and lower limits of power generation in the interval *k*. ε^{k+} and ε^{k} represent the control factors for the deviation of power generation in the interval *k* respectively. Equation (22) and (23) reflect the relationship between the predicted power and the actual power. Equation (24) and (25) ensure that the deviation size will be in any interval and in any period and (26) indicates that the upper deviation is basically equal to the lower deviation.

3.3. Solve the algorithm

In view of the robust optimization model proposed above, this paper solves it by column and constraint generation algorithm(C&CG). The algorithm decomposes the original problem as the main problem and sub-problem, the main problem solves the optimal solution to meet the constraint conditions under the known probability to realize the system's daily scheduling, where x^m is the optimal solution of the *m*th iteration, as shown in [20].

$$\min\ C^T \mathrm{x}+\eta \tag{27}$$

$$\eta > \mathrm{b}^T \mathrm{x}^{\mathrm{m}}$$

The sub-problem is to look for the worst scenario under given the variable x given by the main problem and then return to the main problem for the next iteration

$$\eta(*)=\max_{w\in W}\ \min_{\Delta x\in(x,w)} b'\Delta x \tag{28}$$

According to the theory of pairing, the inner minimum function is transformed to the maximum and its upper bound can be calculated. In (28), λ, γ are the pair variables of each constraint; x is the solution of the main problem, Δx is the variation of power.

$$\begin{aligned} &\max_{u\in U,\lambda,\gamma} d^{\mathrm{T}}\lambda \\ &\text{s.t. } D^{\mathrm{T}}\lambda + G^{\mathrm{T}}\gamma \leqslant b \\ &\lambda \geqslant 0, \gamma \geqslant 0 \end{aligned} \tag{29}$$

The equation (17) and (29) can be iterated until the difference between the upper and lower bound values reaches the desired precision ε, then stop the iteration and return the optimal solution.

4. Case Study

To verify the validity of the proposed model and method, this paper analyzes the study in the integrated energy system .

The historical predication data and the actual power data of a wind farm in northwest China are selected. The proportion of each error interval is obtained by calculating the prediction error of wind power transmission. The wind power forecast curve is shown in Fig.3. The forecast curve for electrical and hydrogen load is shown in Fig.4. The time of use electricity price is shown in Fig.5.

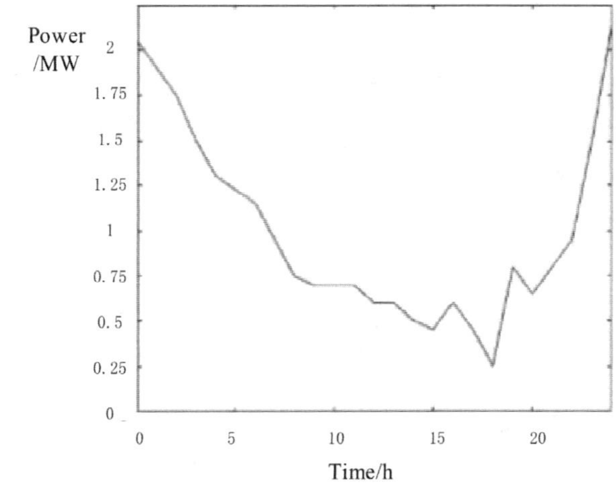

Fig.3: The wind power prediction

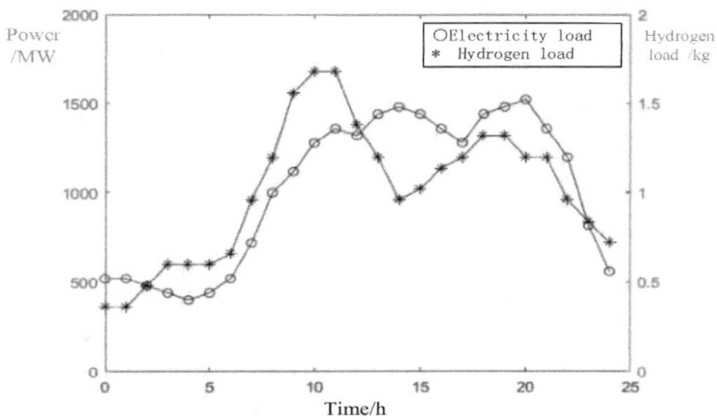

Fig. 4:The electric load and hydrogen load prediction

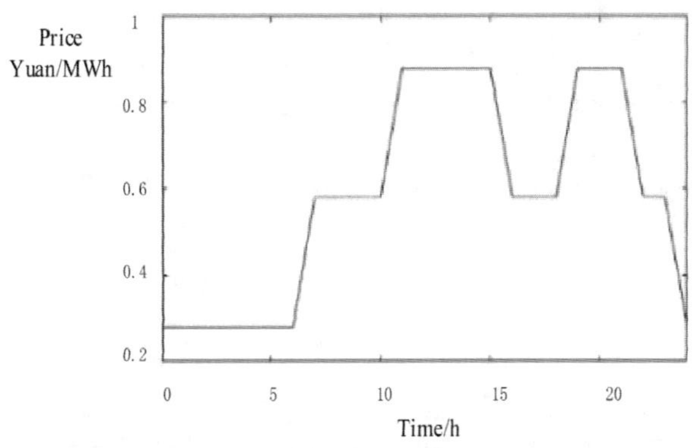

Fig.5: The peak and valley time price

4.1. Study Optimization Results

As Fig.6 and Fig.7 show, wind power is higher than the electrical load, the system has residual power and starts to charge and the hydrogen-making equipment also begin to work from 1h to 6h and from 23h to 24h. Besides, the system also starts to buy electricity on account of the relatively low electricity price. From 7h to 9h, the time-of-use electricity price is lower than fuel cell power generation cost, so it will take priority over purchasing electricity from the main grid to meet the power gap. From 9h to 13h, the power price gets higher and the cost of generating electricity is lower than buying price from outside, it is still necessary to continue to purchase electricity from outside the network to meet the power shortage due to the ramping-rate constraint and the high demand for hydrogen load at this time. From 13h to 17h and 20h to 23h, time-of-use electricity price is higher than the cost of power generation, so the system will give priority to generate electricity by itself and at this time hydrogen production is surplus, the fuel cell will start to work to consume hydrogen and generate electricity to make up the power gap.

4.2. Model Economic Analysis

In order to verify the validity of the model proposed by this paper, a comparative analysis is made with the classical robust optimization model. Fig.8 shows the abandoned wind power in the classical robust algorithm and the improved algorithm, it can be concluded that the improved robust optimization reduces the amount of abandoned wind power in the system and promotes wind power consumption.

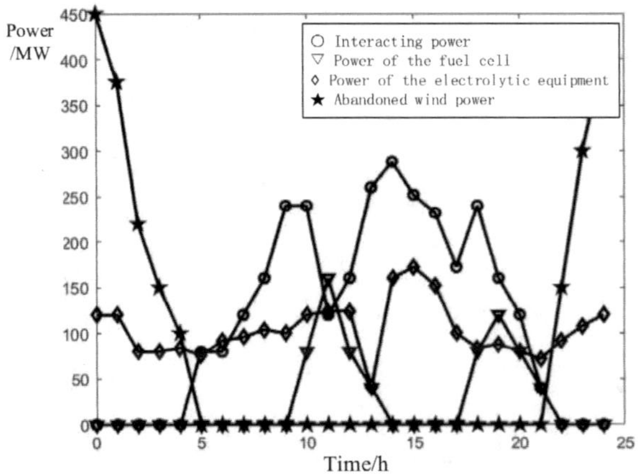

Fig.6 The power optimization results of the model

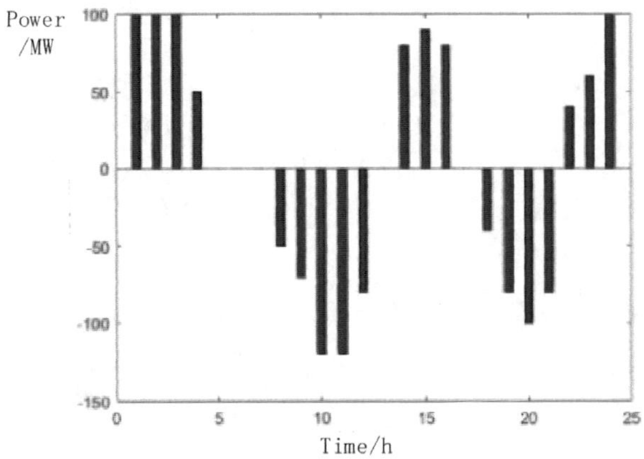

Fig.7 The optimization results of the charge and discharge power of energy storage

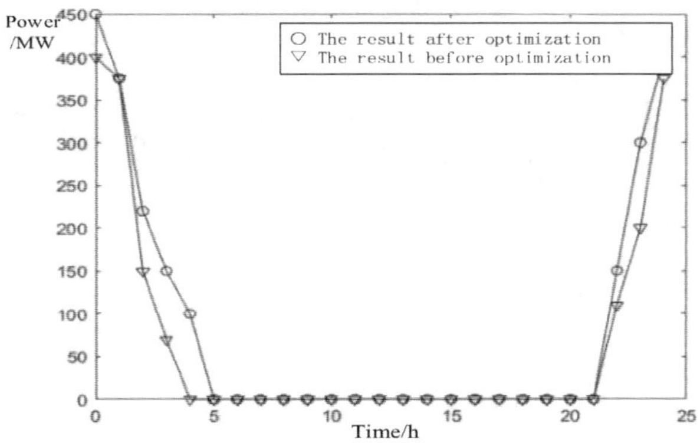

Fig.8 The optimization results of wind abandoned power

As Tab.1, the improved robust optimization model not only reduces the abandoned wind power, but also reduces the scheduling costs. The classic robust optimization aims to find the operation strategy under the worst scenario, so the regulation scheme should set aside enough spare capacity for the equipment to meet the maximum possible fluctuation of the system. But the improved robust optimization can control the spare more accurately to obtain a better economy. In the real-time operation, the classic two-stage robust optimization may make the conservative scheduling plan, which can lead to the increase of wind power disposal with selling excess power at a lower price and buying the missing power at a higher price.

Table.1:Operation costs under different optimizations models

scheme	Classic robust optimization	Improve robust optimization
Total cost	1970.16 yuan	1748.14 yuan
The amount of abandoned power	2.176 MW*h	1.817 MW*h

5. CONCLUSION

Based on the improved robust optimization model, this paper constructs a recent scheduling model of the integrated energy system that takes into account wind power production and the uncertainty of the user's electrical load, optimizes the energy scheduling of the system, and discusses the economics and environmental protection of hydrogen in the integrated energy system, and its broad application scenario. Finally, the feasibility of the proposed model is verified by the study, and it can be found that compared with the classical robust optimization, the scheduling scheme proposed to improve the two-stage robust optimization, without affecting robustness, reduces the conservativeness of the scheme, is more in line with the practical engineering application, and is more economical.

ACKNOWLEDGMENT

This work was supported by Technology Project of Stategrid Zhejiang Co. Ltd. (Grant ID: SGTYHT/19-JS-217).

REFERENCES

[1] Meng Ming, Chen Shichao, Zhao Shujun,Overview on Research of Rnewable Eenergy Microgrid[J]. Modern Elextric Power, 2017,34(01):1-7

[2] Huang Guodong, Li Weigang , Li Zhenbing . A review on generation dispatching of power system with large-scale wind power[J]. Electrical Measurement & Instrumentation, 2015,52(09)

[3] CHEN Guoping,DONG Yu,LIANG Zhifeng. Analysis and Reflection on High-quality Development of New Energy With Chinese Characteristics in EnergyTransition[J]. Proceedings of the CSEE, 2020: 1-13.
[4] WANG Zhenhao,YANG Lu,TIAN Chunguang, et al. Energy Optimization for Combined System of Wind-electric Energy Storage-regenerative Electric Boiler Considering Wind Consumption[J]. ZheJiang Electric Power 2020,39(05):55-60.
[5] Proceedings of the CSEE, 2017. 37(S1): 137-143. MORIARTY P, HONNERY D, Intermittent renewable energy: The only future source of hydrogen[J].International Journal of Hydrogen Energy, 2007.32(12): 1616-1624.
[6] L Valverde-Isorna, D.Ali, D Hogg.Modelling the performance of wind-hydrogen energy systems:Case study the Hydrogen Office in Scotland/UK [J] . Renewable and Sustainable Energy Reviews, 2016, 53:1313-1332
[7] JI Liqiang, ZHAO Yingpeng, WANG Fan. Status of hydrogen energy technology and its application in the field of energy storage and power generation [J].Functional Metal Materials, 2019. 26 (6): 23-31.
[8] Shaw S, Peteves E . Exploiting synergies in European wind and hydrogen sectors: a cost-benefit assessment [J]. International Journal of Hydrogen Energy, 2008, 33(13):3249-3263
[9] FANG R. Life cycle cost assessment of wind power –hydrogen coupled integrated energy system[J]. International Journal of Hydrogen Energy, 2019.44(56): 29399-29408.
[10] CAI Guowei, CHEN Chong, KONG Lingguo1, PENG Long ZHANG Hao2Modeling and Control of Grid-Connected System of Wind/PV/Electrolyzer and SC. Power System Technology, 2016, 40(10): 2982-2990
[11] Wu Yu, Zhang Peter, et Optimal configuration of micro-grid considering uncertainties of electric vehicles and PV/wind sources [J]. Electric Measuring & Instrumentation, 2016, 53 (16):39-44.
[12] D. Bertsimas D, E. Litvinov E, X. A. Sun, J. Zhao and T. Zheng. Adaptive Robust Optimization for the Security Constrained Unit Commitment Problem[J]. IEEE Transactions on Power Systems, 2013, 28(1): 52-63.
[13] Xu Qiushi, Deng Changhong, Zhao Weixing, et al. A Multi-Scenario Robust Dispatch Method for Power Grid Integrated With Wind Farms[J]. Power System Technology, 2014, 38 (3) :653-661.
[14] Y. Li et al., "Flexible scheduling of microgrid with uncertainties considering expectation and robustness", IEEE Trans. Ind. Appl, vol.54 no.4, pp. 3009-3018, Jul./Aug. 2018.
[15] Xiong P ,Jirutitijaroen P,Singh C . A distributionally robust optimization model for unit commitment considering uncertain wind power generation[J]. IEEE Transactions on Power Systems, 2016, 32(1): 39-49
[16] YUAN Yiping, WANG Jianxue, ZHOU Kun, et al. Monthly unit commitment model coordinated short-term scheduling and efficient solving method for renewable energy power system [J]. Proceedings of the CSEE, 2019, 39 (18): 5336-5345.
[17] B. Zeng, L. Zhao. Solving Two-Stage Robust Optimization Problems Using A Column-and-Constraint Generation Method[J]. Operations Research Letters, 2013, 41(5): 457-461
[18] Nader A. El-Taweel; Hadi Khani; Hany E. Z. Farag.Hydrogen Storage Optimal Scheduling for Fuel Supply and Capacity-Based Demand Response Program Under Dynamic Hydrogen Pricing[J]. IEEE Transactions on Smart Grid 2019,10, 4.
[19] Bei Lia, Robin Rochea, Damien Pairea,Sizing of a stand-alone microgrid considering electric power, cooling/heating and hydrogen. IEEE Manchester PowerTech.
[20] Shan Fuzhou, Li Xiaolu,Song Yanmin,Zhou Haiming.Economic dispatching of regional integrated energy system based on improvement two-stage robust optimization. Electrical Measurement & Instrumentation,2018,10.55.

Robust control-based power quality optimization strategy for inverters

Changbin Hu[1,a], Linshu Cheng[1*], Shanna Luo[1,b], Heming Yan[2,c]

[1]North China University of Technology, College of Electrical and Control, Beijing, China

[2]State Grid Corporation of China, Global Energy Interconnection Research Institute Co.,Ltd, Beijing, China

[a]changbinlove@163.com

[*] Corresponding author: 961094886@qq.com

[b]14815390@qq.com

[c]hemingxnuo@126.com

Abstract—A dynamic compensation control strategy based on a residual observer combined with a gradient descent optimization algorithm is proposed to address the power quality problem of distributed AC networks. The robust control structure based on the residual observer is designed according to the theory of Plug and Play control framework. The proposed structure is effective in improving the dynamic performance of the system without changing the structure and parameters. A gradient descent optimization algorithm is used to optimize the parameters of the rectified control structure online. Based on the need for fast response capability and stable performance of distributed grid systems, the gradient descent algorithm is optimized using the model matching principle, which provides a theoretical basis for the selection of initial points in the optimization algorithm. A low-order derivation form of the robust control structure is given through a step-down process to further enhance the speed of optimization under this strategy. The analysis shows that the strategy effectively enhances the robustness of the distributed grid system in comparison with the conventional double closed-loop control. Modelling simulation results verify the effectiveness of the proposed control strategy.

1. INTRODUCTION

In a distributed grid system, the switching devices of power electronic converters will generate harmonics to pollute the bus voltage, the temporary rise and fall of the output bus voltage when the load and distributed power supply cut in and out, and the unbalanced load and single-phase power supply access will cause the bus voltage to be unbalanced in three phases. The above power quality problems will affect the stable operation of the whole system. Therefore, it is very important to enhance the robustness of the distributed grid system and to improve the dynamic and steady-state performance of the system.

The paper [1] discusses the coordinated control of reactive, unbalanced and harmonic power in distributed grids, using unbalanced sag controllers and harmonic sag controllers in order to share harmonic and unbalanced power proportionally between distributed generation units and to effectively improve the bus voltage power quality. The paper [2] addresses the power quality problems caused by non-linearity and unbalanced load access in distributed grid systems, and introduces a 5th and 7th harmonic compensation control loop load negative sequence control loop on top of the traditional double closed loop to solve the three-phase unbalance and harmonic problems. In the paper [3], a control strategy of using the load current as a disturbance current and using the Disturbance Observer (DOB) as a feed-forward compensation of the voltage outer loop is proposed to solve the problem of output bus voltage fluctuation when the load is cut in and out, and a universal parameter design method for the Disturbance Observer is given. The design method of the disturbance observer with universal applicability is given. In [4], an observer-based voltage optimization control strategy is proposed for UPS systems, using feedback control and feedforward compensation to resolve the uncertainties caused by system errors and disturbances, effectively solving three typical power quality problems. In the paper [5], a voltage and current control strategy with internal mode control is proposed, and using small signal analysis, eigenvalue and sensitivity analysis is performed for islanded distributed grid systems, which further illustrates that internal mode control has good robustness and can effectively solve the power quality problems arising in distributed grid systems. In the paper [6], a robust voltage controller is transformed into a convex optimization problem solved for a linear objective function subject to a linear matrix inequality, which improves the dynamic response of the system and reduces the steady-state error of the system, using an islanded DC distributed grid system as the object.

In order to improve the robustness of distributed power grids against three typical power quality problems, a dynamic compensation control strategy based on a residual observer is proposed in this paper. Finally, the effectiveness of the strategy is verified by simulation.

2. Robust control structure

2.1. Robust control framework design based on residual observer

For a true rational transfer function G(s) has a minimal state space realization of the form

$$\boldsymbol{G}(s)=\begin{bmatrix} \boldsymbol{A} & \boldsymbol{B} \\ \boldsymbol{C} & \boldsymbol{D} \end{bmatrix} \tag{1}$$

Its left-right mutual mass decomposition (LCF and RCF) is

$$\boldsymbol{G}(s)=\hat{\boldsymbol{M}}^{-1}(s)\hat{\boldsymbol{N}}(s)=\boldsymbol{N}(s)\boldsymbol{M}^{-1}(s) \tag{2}$$

In which $\hat{\boldsymbol{M}}(s)$ and $\hat{\boldsymbol{N}}(s)$ are left coprime over RH∞, an $\boldsymbol{M}(s)$ and $\boldsymbol{N}(s)$ are right coprime over RH∞. If $\hat{\boldsymbol{X}}(s)$ and $\hat{\boldsymbol{Y}}(s)$ are left coprime over RH∞, $\boldsymbol{X}(s)$ and $\boldsymbol{Y}(s)$ are right coprime over RH∞, they can be shown:

$$\begin{cases} \begin{bmatrix} \hat{\boldsymbol{N}}(s) & \hat{\boldsymbol{M}}(s) \end{bmatrix} \begin{bmatrix} \hat{\boldsymbol{Y}}(s) \\ \hat{\boldsymbol{X}}(s) \end{bmatrix} = \boldsymbol{I}_m \\ \begin{bmatrix} \boldsymbol{X}(s) & \boldsymbol{Y}(s) \end{bmatrix} \begin{bmatrix} M(s) \\ N(s) \end{bmatrix} = \boldsymbol{I}_l \end{cases} \tag{3}$$

$$\begin{bmatrix} \boldsymbol{M}(s) & -\hat{\boldsymbol{Y}}(s) \\ N(s) & \hat{\boldsymbol{X}}(s) \end{bmatrix} = \begin{bmatrix} \boldsymbol{A}+\boldsymbol{BF} & \boldsymbol{B} & \boldsymbol{L} \\ \boldsymbol{F} & \boldsymbol{I} & \boldsymbol{o} \\ \boldsymbol{C}+\boldsymbol{DF} & \boldsymbol{D} & \boldsymbol{I} \end{bmatrix} \tag{4}$$

$$\begin{bmatrix} \boldsymbol{X}(s) & \boldsymbol{Y}(s) \\ -\hat{\boldsymbol{N}}(s) & \hat{\boldsymbol{M}}(s) \end{bmatrix} = \begin{bmatrix} \boldsymbol{A}-\boldsymbol{LC} & -(\boldsymbol{B}-\boldsymbol{LD}) & -\boldsymbol{L} \\ \boldsymbol{F} & \boldsymbol{I} & \boldsymbol{o} \\ \boldsymbol{C} & -\boldsymbol{D} & \boldsymbol{I} \end{bmatrix} \tag{5}$$

where $\boldsymbol{F}$ and $\boldsymbol{L}$ are chosen such that $\boldsymbol{A}+\boldsymbol{BF}$ and $\boldsymbol{A}-\boldsymbol{LC}$ are both stable.

Consider a standard feedback control loop as shown in Figure 1.

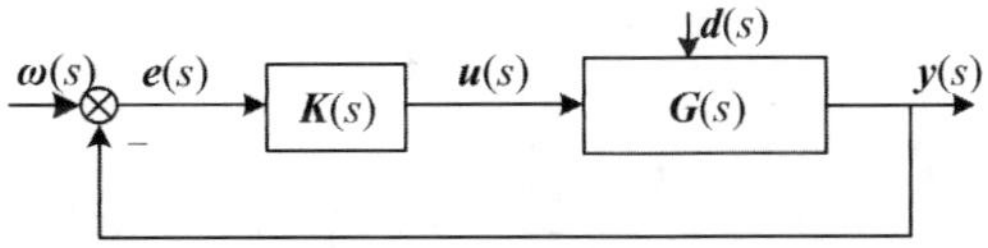

Fig. 1 Standard feedback control loop

Then the set of all proper controllers via Youla parameterization achieving internal stability can be parameterized by:

$$\begin{aligned} \boldsymbol{K}(s) &= -\left(\hat{\boldsymbol{Y}}(s)+\boldsymbol{M}(s)\boldsymbol{Q}(s)\right)\left(\hat{\boldsymbol{X}}(s)-\boldsymbol{N}(s)\boldsymbol{Q}(s)\right)^{-1} \\ &= -\left(\boldsymbol{X}(s)-\boldsymbol{Q}(s)\hat{\boldsymbol{N}}(s)\right)^{-1}\left(\boldsymbol{Y}(s)+\boldsymbol{Q}(s)\hat{\boldsymbol{M}}(s)\right) \end{aligned} \tag{6}$$

where, ***Q(s)*** is Youla parameterization matrix.

From the above theoretical derivation, it follows that for a true rational transfer function ***G(s)***, and an output control signal u_0, if this control system is internally stable, then all internally stable controllers can be parameterized as:

$$\boldsymbol{u}(s) = \boldsymbol{u}_0 + \boldsymbol{Q}(s)\boldsymbol{r} \tag{7}$$

where ***Q(s)*** is a stable parameterization matrix and ***r*** is the residual vector.

As shown in Figure 2, the essence of this robust control theory is that the actual and estimated values of the output are processed in the residual observer and the resulting residual signal is fed back into the controller to achieve compensated control of the disturbance.

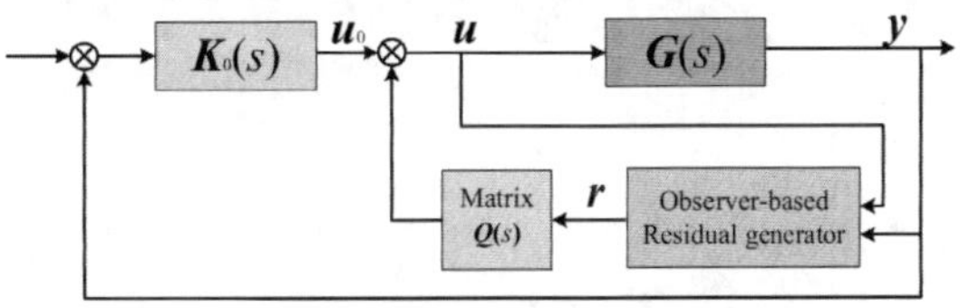

Fig. 2 Fault-tolerant control structure

2.2.Robust controller modelling

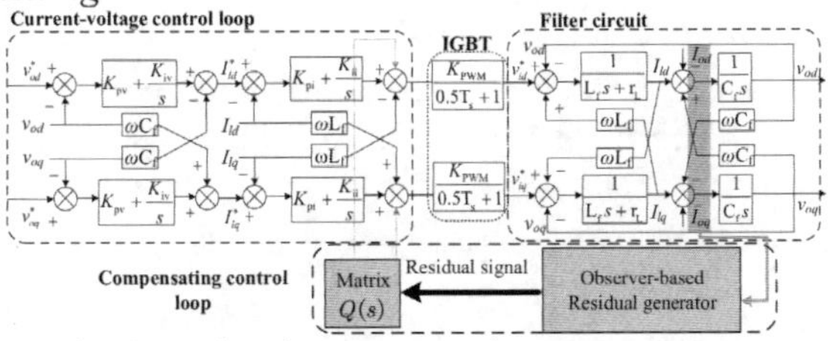

Fig. 3 Fault-tolerant control strategy structure

Based on the above theoretical proof, a robust control strategy for voltage inverters is proposed by combining the traditional inverter double closed-loop control strategy with a robust control architecture based on residual observers, whose architecture is shown in Figure 3. When the control performance of the controller fails to meet the standard conditions due to external disturbing effects, the original double closed-loop feedback controller can be redesigned without changing or redesigning, and only the parameter matrix ***Q(s)*** needs to be reconfigured.

The state space equations for the voltage and current control loop are as follows:

$$\begin{cases} \begin{bmatrix} \dot{\gamma}_{dq} \\ \dot{\varphi}_{dq} \end{bmatrix} = \boldsymbol{A}_{\mathrm{vc}} \begin{bmatrix} \gamma_{dq} \\ \varphi_{dq} \end{bmatrix} + \boldsymbol{B}_{\mathrm{vc}} \begin{bmatrix} e_{idq} \\ e_{vdq} \end{bmatrix} \\ \begin{bmatrix} v_{id}^{*} \\ v_{iq}^{*} \end{bmatrix} = \boldsymbol{C}_{\mathrm{vc}} \begin{bmatrix} \gamma_{dq} \\ \varphi_{dq} \end{bmatrix} + \boldsymbol{D}_{\mathrm{vc}} \begin{bmatrix} e_{idq} \\ e_{vdq} \end{bmatrix} \end{cases} \tag{8}$$

where $\boldsymbol{A}_{vc}$ is the 4X4 zero matrix and $\boldsymbol{B}_{vc}$ is the 4X4 unit matrix. $\boldsymbol{C}_{vc}$, $\boldsymbol{D}_{vc}$ are shown below.

$$\boldsymbol{C}_{vc} = \begin{bmatrix} K_{ii} & 0 & 0 & -L_f\omega K_{iv} \\ 0 & K_{ii} & L_f\omega K_{iv} & 0 \end{bmatrix},$$

$$\boldsymbol{D}_{vc} = \begin{bmatrix} K_{pi} & L_f\omega & L_fC_f\omega^2 & -L_f\omega K_{pv} \\ -L_f\omega & K_{pi} & L_f\omega K_{pv} & L_fC_f\omega^2 \end{bmatrix}$$

The original PI controller (20) can be reformulated into a discrete state-space representation as shown:

$$\begin{cases} \boldsymbol{x}_{c,k+1} = \boldsymbol{A}_{vcz}\boldsymbol{x}_{c,k} + \boldsymbol{B}_{vcz}\boldsymbol{e}_k \\ \boldsymbol{u}_{c,k} = \boldsymbol{C}_{vcz}\boldsymbol{x}_{c,k} + \boldsymbol{D}_{vcz}\boldsymbol{e}_k \end{cases} \tag{9}$$

The measured inverter output voltage differs from the estimated value when there is load cut-off, high harmonics from non-linear loads and load unbalance disturbances occur, resulting in a residual signal. Therefore, a Luenberger observer can be established. Then the state-space representation of the observer based residual generator can be expressed:

$$\begin{cases} \boldsymbol{z}_{k+1} = \boldsymbol{A}_z\boldsymbol{z}_k + \boldsymbol{B}_z\boldsymbol{u}_k + \boldsymbol{L}_z\boldsymbol{r}_k \\ \hat{\boldsymbol{y}}_k = \boldsymbol{C}_z\boldsymbol{z}_k \\ \boldsymbol{r}_k = \boldsymbol{y}_k - \hat{\boldsymbol{y}}_k \end{cases} \tag{10}$$

$$\boldsymbol{A}_z = \begin{bmatrix} -\frac{r_L}{L_f} & \omega & -\frac{1}{L_f} & 0 \\ -\omega & -\frac{r_L}{L_f} & 0 & -\frac{1}{L_f} \\ \frac{1}{C_f} & 0 & 0 & \omega \\ 0 & \frac{1}{C_f} & -\omega & 0 \end{bmatrix}, \boldsymbol{B}_z = \begin{bmatrix} \frac{1}{L_f} & 0 \\ 0 & \frac{1}{L_f} \\ 0 & 0 \\ 0 & 0 \end{bmatrix}$$

$$\boldsymbol{C}_z = \begin{bmatrix} 1 & 0 & 0 & 0 \\ 0 & 1 & 0 & 0 \\ 0 & 0 & 1 & 0 \\ 0 & 0 & 0 & 1 \end{bmatrix}$$

In order to design the parameter matrix $\boldsymbol{Q(s)}$ and reduce dynamic complexity, we can represent it in a controlled canonical form [7], which has fewer design parameters and has the following state space representation.

$$\begin{cases} \boldsymbol{x}_{r,k+1} = \boldsymbol{A}_r\boldsymbol{x}_{r,k} + \boldsymbol{B}_r\boldsymbol{r}_k \\ \boldsymbol{u}_{r,k} = \boldsymbol{C}_r\boldsymbol{x}_{r,k} + \boldsymbol{D}_r\boldsymbol{r}_k \end{cases} \tag{11}$$

$$
\boldsymbol{A}_r = \begin{bmatrix}
0 & 1 & \cdots & 0 & & & & & \\
0 & 0 & \ddots & \vdots & & & & & \\
\vdots & \vdots & \ddots & 1 & & & & & \\
* & * & * & * & \cdots & * & * & * & * \\
\vdots & \vdots & \vdots & \vdots & \vdots & \vdots & \vdots & \vdots & \vdots \\
 & & & & & 0 & 1 & \cdots & 0 \\
 & & & & & 0 & 0 & \ddots & \vdots \\
 & & & & & \vdots & \vdots & \ddots & 1 \\
* & * & * & * & \cdots & * & * & * & *
\end{bmatrix}, \quad \boldsymbol{B}_r = \begin{bmatrix}
0 & & \\
0 & & \\
\vdots & & \\
1 & & \\
\vdots & \vdots & \vdots \\
 & & 0 \\
 & & 0 \\
 & & \vdots \\
 & & 1
\end{bmatrix}
$$

$$
\boldsymbol{C}_r = \begin{bmatrix} c_{11} & \cdots & c_{1q} \\ \vdots & \ddots & \vdots \\ c_{l1} & \cdots & c_{lq} \end{bmatrix}, \boldsymbol{D}_r = \begin{bmatrix} d_{11} & \cdots & d_{1m} \\ \vdots & \ddots & \vdots \\ d_{l1} & \cdots & d_{lm} \end{bmatrix}
$$

Where * is a non-zero element of the matrix $\boldsymbol{A}_r$ and the state vector $\boldsymbol{x}_{r,k}$ is selectable by the designer.

Combining the above discrete state space expressions (9), (10) and (11) and Figure 2 the state space expression with the reference signal $\boldsymbol{w}_k$ and the residual signal $\boldsymbol{r}_k$ as inputs and the error tracking signal $\boldsymbol{e}_k$ and the control signal $\boldsymbol{u}_k$ as outputs can be established as:

$$
\begin{cases}
\begin{bmatrix} z_{k+1} \\ \boldsymbol{x}_{c,k+1} \\ \boldsymbol{x}_{r,k+1} \end{bmatrix} = \underbrace{\begin{bmatrix} \boldsymbol{A}_z - \boldsymbol{B}_z \boldsymbol{D}_{\text{vcz}} \boldsymbol{C}_z & \boldsymbol{B}_z \boldsymbol{C}_{\text{vcz}} & \boldsymbol{B}_z \boldsymbol{C}_r \\ -\boldsymbol{B}_{\text{vcz}} \boldsymbol{C}_z & \boldsymbol{A}_{\text{vcz}} & 0 \\ 0 & 0 & \boldsymbol{A}_r \end{bmatrix}}_{A_G} \begin{bmatrix} z_k \\ \boldsymbol{x}_{c,k} \\ \boldsymbol{x}_{r,k} \end{bmatrix} \\
\quad + \underbrace{\begin{bmatrix} \boldsymbol{B}_z \boldsymbol{D}_{\text{vcz}} & \boldsymbol{B}_z(\boldsymbol{D}_r - \boldsymbol{D}_{\text{vcz}}) + \boldsymbol{L}_z \\ \boldsymbol{B}_{\text{vcz}} & -\boldsymbol{B}_{\text{vcz}} \\ 0 & \boldsymbol{B}_r \end{bmatrix}}_{B_G} \begin{bmatrix} \boldsymbol{w}_k \\ \boldsymbol{r}_k \end{bmatrix} \\
\begin{bmatrix} \boldsymbol{e}_k \\ \boldsymbol{u}_k \end{bmatrix} = \underbrace{\begin{bmatrix} -\boldsymbol{C}_z & 0 & 0 \\ -\boldsymbol{D}_{\text{vcz}} \boldsymbol{C}_z & \boldsymbol{C}_{\text{vcz}} & \boldsymbol{C}_r \end{bmatrix}}_{C_G} \begin{bmatrix} z_k \\ \boldsymbol{x}_{c,k} \\ \boldsymbol{x}_{r,k} \end{bmatrix} \\
\quad + \underbrace{\begin{bmatrix} \boldsymbol{I} & -\boldsymbol{I} \\ \boldsymbol{D}_{\text{vcz}} & \boldsymbol{D}_r - \boldsymbol{D}_{\text{vcz}} \end{bmatrix}}_{D_G} \begin{bmatrix} \boldsymbol{w}_k \\ \boldsymbol{r}_k \end{bmatrix}
\end{cases} \tag{12}
$$

2.3. Gradient descent optimization s

This paper introduces a gradient descent online optimization algorithm, which can optimize the system parameters online in real time with greater flexibility and adaptability, and iteratively optimize the parameters in $\boldsymbol{Q(s)}$ to enhance the robustness of the inverter system. In order to minimize the interference of unknown disturbances and noise on the inverter output performance, the system tracking error and the control amount need to be as small as possible, so the error tracking signal and the control signal are chosen as the parameters of the objective function. The objective function of the optimization problem is

$$
\boldsymbol{J} = \frac{1}{2N} \sum_{k=k_0}^{N+k_0-1} \left[\boldsymbol{e}_k^T W_{e,k} \boldsymbol{e}_k + \boldsymbol{u}_k^T W_{u,k} \boldsymbol{u}_k \right] \tag{13}
$$

k_0 and N represent the starting point of the gradient descent optimization process and the time window of the whole optimization process respectively. $\boldsymbol{W}_{e,k} \geq 0$ and $\boldsymbol{W}_{u,k} \geq 0$ are weighting matrices of the tracking error and control signal, respectively.

The value of the parameter $\boldsymbol{\theta}$ is optimized by the gradient descent algorithm. In order to successfully find the optimal value of $\boldsymbol{\theta}$, the gradient algorithm is chosen to be designed in the steepest direction of descent, where the gradient function is:

$$-\frac{\partial \boldsymbol{J}}{\partial \boldsymbol{\theta}_j} = -\frac{1}{N}\sum_{k=k_0}^{N+k_0-1}\begin{bmatrix} \boldsymbol{e}_k^T \boldsymbol{W}_{e,k} & \boldsymbol{u}_k^T \boldsymbol{W}_{u,k} \end{bmatrix}\begin{bmatrix} \dfrac{\partial \boldsymbol{e}_k}{\partial \boldsymbol{\theta}_j} \\ \dfrac{\partial \boldsymbol{u}_k}{\partial \boldsymbol{\theta}_j} \end{bmatrix} \tag{14}$$

The update rule for the value of parameter $\boldsymbol{\theta}$ is

$$\boldsymbol{\theta}_{j+1} = \boldsymbol{\theta}_j - \boldsymbol{\Gamma}_j \frac{\partial \boldsymbol{J}}{\partial \boldsymbol{\theta}_j}, \quad \boldsymbol{\theta}_0 = 0 \tag{15}$$

where j is the number of iterations and $\boldsymbol{\Gamma}_j$ is the iteration step.

The size of the iteration step can determine the convergence speed of the objective function, so a suitable iteration step can be designed according to the optimization theory to speed up the convergence speed. The Taylor expansion of the objective function yields.

$$\boldsymbol{\Gamma}_j = \left[\frac{\partial^2 \boldsymbol{J}}{\partial \boldsymbol{\theta}_j \partial\left(\boldsymbol{\theta}_j\right)^T}\right]^{-1} \tag{16}$$

Where $\boldsymbol{\Gamma}_j$ is the Hessian matrix derived from the objective function $\boldsymbol{J}$.

According to equation (11), the $\boldsymbol{C}_r$ and $\boldsymbol{D}_r$ matrices contain unknown parameters c and d , and the derivatives of the error tracking signal $\boldsymbol{e}_k$ and the control signal $\boldsymbol{u}_k$ with respect to $\boldsymbol{\theta}$ for the $\boldsymbol{C}_r$ and $\boldsymbol{D}_r$ matrix row vectors (column vectors), respectively, can be found as:

$$\begin{cases} \begin{bmatrix} \dfrac{\partial \boldsymbol{z}_{k+1}}{\partial \boldsymbol{\theta}_{c_i}} \\ \dfrac{\partial \boldsymbol{x}_{c,k+1}}{\partial \boldsymbol{\theta}_{c_i}} \end{bmatrix} = \underbrace{\begin{bmatrix} \boldsymbol{A}_z - \boldsymbol{B}_z \boldsymbol{D}_{\mathrm{vcz}} \boldsymbol{C}_z & \boldsymbol{B}_z \boldsymbol{C}_{\mathrm{vcz}} \\ -\boldsymbol{B}_{\mathrm{vcz}} \boldsymbol{C}_z & \boldsymbol{A}_{\mathrm{vcz}} \end{bmatrix}}_{A_{\partial G}} \begin{bmatrix} \dfrac{\partial \boldsymbol{z}_k}{\partial \boldsymbol{\theta}_{c_i}} \\ \dfrac{\partial \boldsymbol{x}_{c,k}}{\partial \boldsymbol{\theta}_{c_i}} \end{bmatrix} + \begin{bmatrix} \boldsymbol{B}_z \boldsymbol{I}_{c_i} \\ 0 \end{bmatrix} \boldsymbol{x}_{r,k} \\ \begin{bmatrix} \dfrac{\partial \boldsymbol{e}_k}{\partial \boldsymbol{\theta}_{c_i}} \\ \dfrac{\partial \boldsymbol{u}_k}{\partial \boldsymbol{\theta}_{c_i}} \end{bmatrix} = \underbrace{\begin{bmatrix} -\boldsymbol{C}_z & 0 \\ -\boldsymbol{D}_{\mathrm{vcz}} \boldsymbol{C}_z & \boldsymbol{C}_{\mathrm{vcz}} \end{bmatrix}}_{C_{\partial G}} \begin{bmatrix} \dfrac{\partial \boldsymbol{z}_k}{\partial \boldsymbol{\theta}_{c_i}} \\ \dfrac{\partial \boldsymbol{x}_{c,k}}{\partial \boldsymbol{\theta}_{c_i}} \end{bmatrix} + \begin{bmatrix} 0 \\ \boldsymbol{I}_{c_i} \end{bmatrix} \boldsymbol{x}_{r,k} \end{cases} \tag{17}$$

Where, $\boldsymbol{I}_{c_i} = \dfrac{\partial \boldsymbol{C}_r}{\partial \boldsymbol{\theta}_{c_i}}$

$$\begin{cases} \begin{bmatrix} \dfrac{\partial z_{k+1}}{\partial \boldsymbol{\theta}_{d_i}} \\ \dfrac{\partial \boldsymbol{x}_{c,k+1}}{\partial \boldsymbol{\theta}_{d_i}} \end{bmatrix} = \underbrace{\begin{bmatrix} \boldsymbol{A}_z - \boldsymbol{B}_z \boldsymbol{D}_{\text{vcz}} \boldsymbol{C}_z & \boldsymbol{B}_z \boldsymbol{C}_{\text{vcz}} \\ -\boldsymbol{B}_{\text{vcz}} \boldsymbol{C}_z & \boldsymbol{A}_{\text{vcz}} \end{bmatrix}}_{A_{\partial G}} \begin{bmatrix} \dfrac{\partial z_k}{\partial \boldsymbol{\theta}_{d_i}} \\ \dfrac{\partial \boldsymbol{x}_{c,k}}{\partial \boldsymbol{\theta}_{d_i}} \end{bmatrix} \\ \qquad + \begin{bmatrix} B_z I_{d_i} \\ 0 \end{bmatrix} \boldsymbol{r}_k \\ \begin{bmatrix} \dfrac{\partial \boldsymbol{e}_k}{\partial \boldsymbol{\theta}_{d_i}} \\ \dfrac{\partial \boldsymbol{u}_k}{\partial \boldsymbol{\theta}_{d_i}} \end{bmatrix} = \underbrace{\begin{bmatrix} -\boldsymbol{C}_z & 0 \\ -\boldsymbol{D}_{\text{vcz}} \boldsymbol{C}_z & \boldsymbol{C}_{\text{vcz}} \end{bmatrix}}_{C_{\partial G}} \begin{bmatrix} \dfrac{\partial z_k}{\partial \boldsymbol{\theta}_{d_i}} \\ \dfrac{\partial \boldsymbol{x}_{c,k}}{\partial \boldsymbol{\theta}_{d_i}} \end{bmatrix} \\ \qquad + \begin{bmatrix} 0 \\ I_{d_i} \end{bmatrix} \boldsymbol{r}_k \end{cases} \tag{18}$$

Where, $\boldsymbol{I}_{d_i} = \dfrac{\partial \boldsymbol{D}_r}{\partial \boldsymbol{\theta}_{d_i}}$

By bringing equations (17) and (18) into the gradient expression (15), a suitable parameterisation matrix ***Q(s)*** can be obtained by the gradient descent algorithm. In the gradient descent algorithm, the choice of initial points also has an impact on the optimization time and optimization speed, so the choice of initial points can be optimized according to robust model matching theory to obtain the appropriate parameterization matrix ***Q(s)*** more quickly.

2.4Model matching problem solving

According to the dynamic compensation theory based on the residual generator in Fig. 3, the dynamic compensation structure is designed as follows:

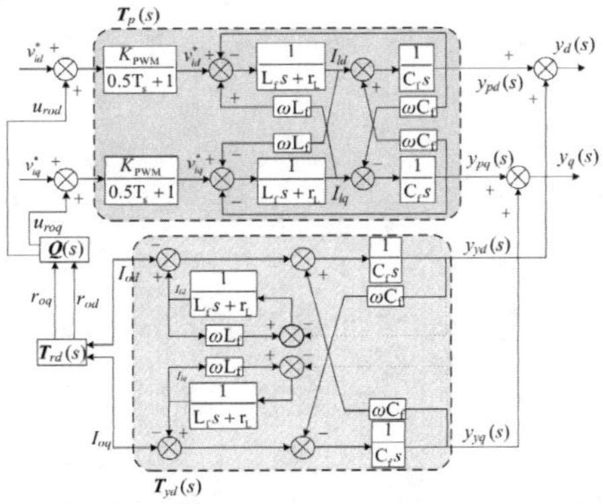

Fig. 4 Fault-tolerant control structure topology

where ***Q(s)*** is the desired voltage loop robust controller, $\boldsymbol{u}_0$ and $\boldsymbol{u}_r$ are the inputs to the PI controller and ***Q(s)*** respectively, so that $\boldsymbol{T}_p$, $\boldsymbol{T}_{yd}$ and $\boldsymbol{T}_{rd}$ are the transfer functions of the perturbation acting on the LC filter circuit of the controlled object, the transfer function of the LC filter circuit and the transfer function of the residual observer respectively. $\boldsymbol{y}_0, \boldsymbol{y}_r, \boldsymbol{y}_w$ are the output of $\boldsymbol{u}_0$ acting on $\boldsymbol{T}_p$, $\boldsymbol{u}_r$ acting on $\boldsymbol{T}_p$ and the output of perturbation d acting on $\boldsymbol{T}_{rd}$ respectively.

After the dynamic compensation structure in Fig. 4 is simplified and refined, and its compensation process is shown in Fig. 5.

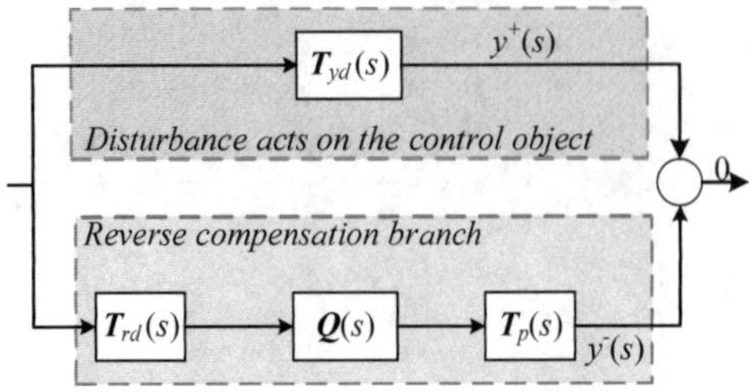

Fig. 5 Model matching structure

The controller ***Q(s)*** is shown in:

$$\min \| \boldsymbol{T}_{yd}(s) + \boldsymbol{G}_p(s)\boldsymbol{Q}(s)\boldsymbol{T}_{rd}(s) \|_\infty \tag{19}$$

Through model-matching theory, the parameterization matrix ***Q(s)*** is:

$$\begin{cases} \boldsymbol{Q}(1.1) = -3(s+1001)(s+12.5) \\ \boldsymbol{Q}(1.2) = -700(s+1001) \\ \boldsymbol{Q}(2.1) = 700(s+1001) \\ \boldsymbol{Q}(2.2) = -3(s+1001)(s+12.5) \end{cases} \tag{20}$$

It can be seen that the numerator order of the parameterized matrix controller ***Q(s)*** derived by the model matching method is higher than the denominator order. By introducing two non-dominated poles s=-1e^3 farther away from the imaginary axis, the denominator order can be made equal to the numerator order without affecting the performance of the original system, thus ensuring the physical realizability of ***Q(s)***.

$$\begin{cases} \boldsymbol{Q}(1.1) = \dfrac{-3(s+1001)(s+12.5)}{s^2 + 2000s + 1e^6} \\ \boldsymbol{Q}(1.2) = \dfrac{-700(s+1001)}{s^2 + 2000s + 1e^6} \\ \boldsymbol{Q}(2.1) = \dfrac{700(s+1001)}{s^2 + 2000s + 1e^6} \\ \boldsymbol{Q}(2.2) = \dfrac{-3(s+1001)(s+12.5)}{s^2 + 2000s + 1e^6} \end{cases} \tag{21}$$

Using the reduced-order theory of controllers proposed in the paper [8] ,***Q(s)*** can be simplified as:

$$\begin{cases} \boldsymbol{Q}(1.1) = -1.5 + \dfrac{580}{s+500} \\ \boldsymbol{Q}(1.2) = -0.35 - \dfrac{175}{s+500} \\ \boldsymbol{Q}(2.1) = 0.35 + \dfrac{175}{s+500} \\ \boldsymbol{Q}(2.2) = -1.5 + \dfrac{580}{s+500} \end{cases} \tag{22}$$

The parameter matrix ***Q(s)***, obtained according to the model matching theory, is processed by order reduction, causing the problem of inaccurate controller model, but simplifying the amount of system operations. Equation (22) can also be used as the initial point of the gradient descent algorithm for the purpose of optimizing the gradient descent algorithm within the margin of error allowed.

3. EXPERIMENTAL VERIFICATION AND ANALYSIS

In order to verify the correctness and effectiveness of the control strategy mentioned in this paper. Taking a single inverter as an example, the inverter system is built in Matlab/Simulink to analyze the suppression of current disturbances caused by switching of different types of loads under the isolated operation state of distributed power supply.

TABLE I. INVERTER PARAMETERS

Parameters	*Inverter value*
V_{dc}	700V
Ts	5e-6s
L_f	2mH
C_f	1500μF
r_f	0.01
K_{pv}	10
k_{iv}	100
k_{pi}	10
k_{ii}	100

3.1. Power Load Switching Experiment

In the simulation environment of Matlab/Simulink, when the inverter performs load shedding, there are transient fluctuations in the output voltage. In order to simulate the effect of the robust control structure on the perturbation suppression of load throwing, the inverter is connected to a 20kW active load and a 20kVar reactive load for stable operation, and the output d-axis voltage of the inverter with robust control is observed by throwing a 20kW active load and a 20kVar reactive load once per second for comparison with different initial points selected. The results of the simulation experiments are shown in Figure 5.

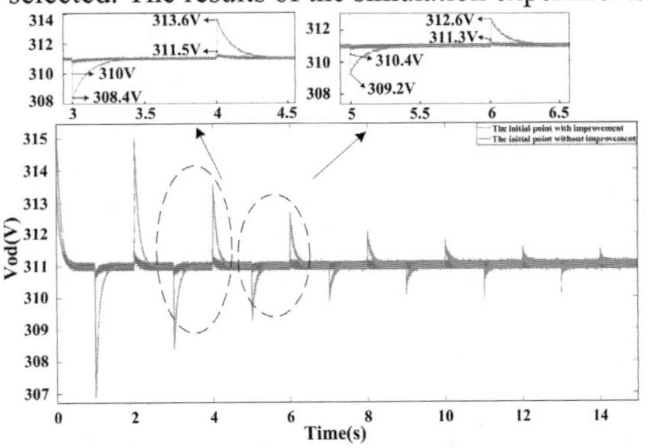

Fig. 6 d-axis output voltage comparison

From the comparison waveforms, it can be concluded that when the default starting point is 0, the voltage fluctuation of the final output d-axis voltage is ±0.6V when the simulation time is set to 15 seconds; when the starting point is obtained by the model matching method, the output voltage fluctuation is ±0.6V in 2 seconds when the simulation time is set to 15 seconds, and the final output voltage fluctuation is ±0.2V. Therefore, the simulation shows that the proposed optimization strategy not only effectively suppresses the voltage fluctuation of the load switching, but also speeds up the response time to return to the steady state value. Therefore, the simulation shows that the proposed optimization strategy not only effectively suppresses the voltage fluctuations arising from load switching, but also speeds up the response time of voltage recovery to the steady-state value and improves the robustness of the inverter.

Journal of Physics: Conference Series **2215** (2022) 012002 doi:10.1088/1742-6596/2215/1/012002

3.2.Nonlinear load switching Experiment

In the Matlab/Simulink simulation environment, harmonics are injected into both the three-phase output voltage and the d-axis output voltage when the inverter is connected to a non-linear load. To simulate the disturbance suppression of the injected harmonic sources by the robust control structure, the output d-axis voltages and THD with and without robust control are observed and compared by connecting the inverter to the 5th, 7th and 11th harmonic sources.

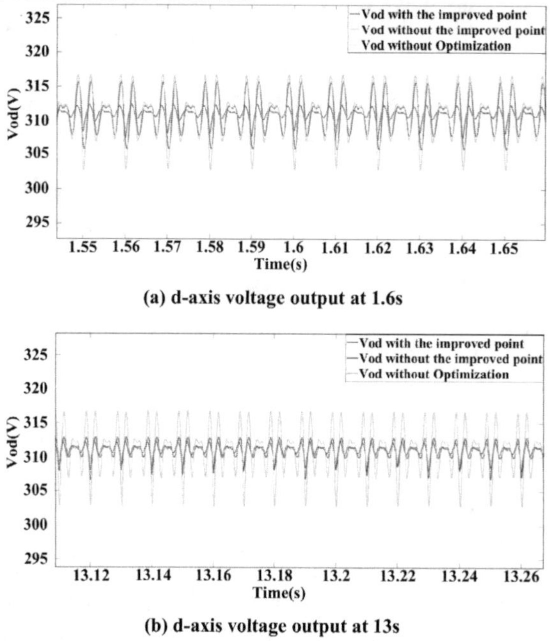

Fig. 7 Comparison of d-axis output voltage under three control strategies

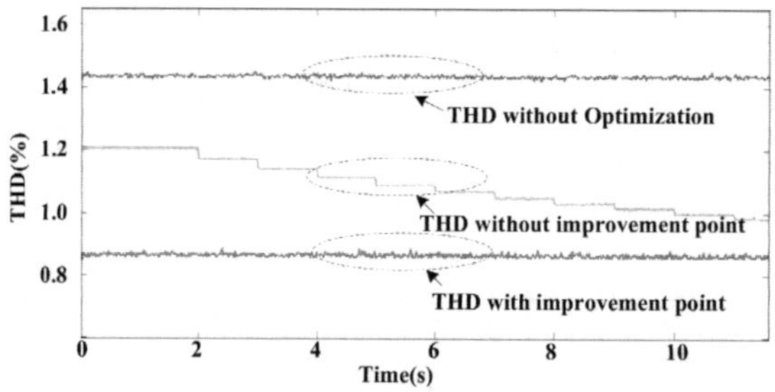

Fig. 8 Comparison of THD under three control strategies

A comparison of the above three-phase output voltage waveforms and THD shows that when the 5th, 7th and 11th harmonic sources are connected, the THD is 1.44% without the use of optimal control theory. With robust control, the THD drops to 0.94% when the initial point is selected as the default initial point and to 0.86% when the initial point is obtained by the model matching method with the simulation time set at 15 seconds; the comparison of the d-axis output voltage at 1.6s and 13s shows that the voltage disturbance suppression effect has improved. Therefore, it can be concluded from the simulated experimental waveforms that when the 5th, 7th and 11th harmonic sources are connected, the theory proposed in this paper effectively improves the system's immunity to the injected harmonics, enhances the robustness of the inverter and suppresses the impact of harmonic disturbances on the distributed grid.

4. Conclusion

The method proposed in this paper can effectively suppress voltage disturbances caused by load throwing and cutting, three-phase voltage unbalance caused by unbalanced loads and harmonic disturbances caused by non-linear loads, and improve the stability of inverter operation.

In the robust control structure based on the residual observer, a data-driven iterative optimization is carried out for the robust controller ***Q(s)*** parameter, on the basis of which the robust model matching principle is introduced, basically solving the problem that this iterative optimization algorithm is limited by the initial point.

One of the future research works in this paper is how to improve the proposed optimization strategy to obtain better optimization results that can be achieved by optimizing the iterative step size.

References

[1] E. Kim, F. Mwasilu, H. H. Choi and J. Jung, "An Observer-Based Optimal Voltage Control Scheme for Three-Phase UPS Systems," in *IEEE Transactions on Industrial Electronics*, vol. 62, no. 4, pp. 2073-2081, April 2015, doi: 10.1109/TIE.2014.2351777.J. Clerk Maxwell, A Treatise on Electricity and Magnetism, 3rd ed., vol. 2. Oxford: Clarendon, 1892, pp.68–73.

[2] Huo Qunhai,Kong Li,Tang Xisheng.Control of unbalanced non-linear mixed load of micro-source inverter[J].Proceedings of the Chinese Society of Electrical Engineering,2010,30(15):10-15.

[3] Cao Wenyuan, Han Minxiao, Xie Wenqiang, Li Rui, Yuan Dong. Load current feedforward control and parameter design method for voltage source inverter based on disturbance observer[J].Transactions of the Chinese Society of Electrical Engineering,2020,35(04):862-873.

[4] E. Kim, F. Mwasilu, H. H. Choi and J. Jung, "An Observer-Based Optimal Voltage Control Scheme for Three-Phase UPS Systems," in *IEEE Transactions on Industrial Electronics*, vol. 62, no. 4, pp. 2073-2081, April 2015, doi: 10.1109/TIE.2014.2351777.

[5] S. Leitner, M. Yazdanian, A. Mehrizi-Sani and A. Muetze, "Small-Signal Stability Analysis of an Inverter-Based Microgrid With Internal Model-Based Controllers," in *IEEE Transactions on Smart Grid*, vol. 9, no. 5, pp. 5393-5402, Sept. 2018, doi: 10.1109/TSG.2017.2688481.

[6] M. S. Sadabadi, Q. Shafiee and A. Karimi, "Plug-and-Play Robust Voltage Control of DC Microgrids," in *IEEE Transactions on Smart Grid*, vol. 9, no. 6, pp. 6886-6896, Nov. 2018, doi: 10.1109/TSG.2017.2728319.

[7] D. Luenberger, "Canonical forms for linear multivariable systems," in *IEEE Transactions on Automatic Control*, vol. 12, no. 3, pp. 290-293, June 1967, doi: 10.1109/TAC.1967.1098584.

[8] Y. Ohta, H. Maeda and S. Kodama, "Rational approximation of L/sub 1/ optimal controllers for SISO systems," in IEEE Transactions on Automatic Control, vol. 37, no. 11, pp. 1683-1691, Nov. 1992, doi: 10.1109/9.173139.

Joint Control Strategy of Microgrid Inverter Based on Optimal Control and Residual Dynamic Compensation

Changbin Hu[1,a], Chao Liu[1]*, Shanna Luo[1,b], Heming Yan[2,c]

[1]School of Electrical and Control Engineering, North China University of Technology, Beijing, China

[2]State Grid Corporation of China, Global Energy Interconnection Research Institute Co., Ltd., Beijing, China

[a]changbinlove@163.com

*Corresponding author: 965455306@qq.com

[b]14815390@qq.com

[c]hemingxnuo@126.com

Abstract—With the continuous development of microgrid, the inverter under traditional PID control is difficult to meet the power quality requirements of microgrid. In order to improve the anti-interference ability and transient response speed of the output voltage of Isolated AC microgrid inverter, a dynamic compensation control structure combining linear quadratic optimal control and disturbance residual generator is proposed in this paper. Firstly, the quadratic optimal controller is established based on the inverter model. Secondly, the residual generator is established to observe the residual value of the disturbed system. Then, the dynamic compensation controller is designed according to the superposition principle, and a control signal opposite to the influence caused by the disturbance is output at the output of the controller. The disturbance signals can cancel each other without knowing the specific disturbance form. Finally, the experiments of load switching, three-phase imbalance and nonlinear load are designed on MATLAB/Simulink simulation software to verify the effectiveness of the control strategy proposed in this paper.

1. INTRODUCTION

Since entering the 21st century, the development of power resources in the national market has made a qualitative leap, from the traditional power to the power market with modern power electronic technology. Distributed generation, energy storage devices, loads and various control systems are combined into micro grids. These micro grids can not only be connected with the large grid, but also operate independently. This not only puts forward the requirements of flexibility, controllability and strong elasticity for microgrid itself, but also brings great challenges to its safety and reliability.

Since the 1990s, the International Electrotechnical Commission (IEC) has established a number of national standards for bus voltage deviation, fluctuation and flicker, three-phase voltage imbalance and harmonic disturbance, which provides a development opportunity for power quality control. When the micro grid is connected to the grid, the large grid can stabilize its voltage and frequency, and has strong anti-interference ability. When the microgrid operates in island mode, its AC bus voltage is supported by many micro sources, which are mainly controlled by two-way AC/DC and DC/DC converters. At this time, due to the uncertainty of distributed micro source output, load switching and the characteristics of low inertia of microgrid system, its bus voltage is easy to be affected. Therefore, maintaining the voltage stability of isolated bus through reasonable control of AC/DC converter has gradually become one of the key problems [1].

In industrial practice, PID control is most widely used, and there are many design methods, but its closed-loop stability is still a very important problem. In addition, the stability region of PID control object is difficult to determine online, which limits the control performance [2]. Especially in the case of various disturbances, it is difficult to obtain ideal dynamic characteristics. In serious cases, the system will collapse. Reference [3] proposes a new voltage control strategy based on internal model robustness and compound control for hybrid microgrid, which improves the robustness and anti-interference ability of the whole microgrid. On the basis of droop control, references [4] and [5] adopt frequency division droop control method to mainly suppress the power quality problem of PCC terminal voltage after the inverter output voltage passes through a section of line. In reference [6], a multi resonant controller is used in the original primary control, and a compensator composed of multiple resonant voltage regulators is added on the basis of the original secondary control. However, the above methods can only eliminate fixed frequency harmonics and double frequency negative sequence voltage. Reference [7] proposed a voltage control method based on nonlinear disturbance observer, which can effectively suppress the fluctuation of bus voltage, but the inverter has the relationship of dq axis coupling, which will make the system very complex.

In order to solve the transient response speed and anti-interference ability of output voltage of AC microgrid inverter and improve the robustness of output voltage, the main research contents of this paper are as follows.

- After modeling the microgrid inverter, combined with state feedback, integral control and linear quadratic optimal control algorithm, a controller based on LQR is designed to improve the transient response speed of the inverter.
- After analyzing the output voltage fluctuation of the inverter, a dynamic compensation structure based on the disturbance residual generator is proposed. The structure is based on the robust double coprime decomposition and Euler parameterization theory. It can quickly extract and compensate the disturbance signal without additional voltage and current sensors, and enhance the anti-interference performance of the output voltage of the inverter.

Load switching, three-phase unbalanced load and harmonic load experiments are designed in MATLAB / Simulink to verify the correctness and effectiveness of the proposed control.

2. INVERTER CONTROL STRATEGY BASED ON LQR

2.1. Mathematical model of three-phase inverter

The state space equation of three-phase inverter circuit can be obtained by selecting load voltage u_{od}, u_{oq}, inductance current i_{Ld} and i_{Lq} as state variables, inverter bridge output voltage u_{id} and u_{iq} as input and load current i_{od} and i_{oq} as disturbance input d:

$$\begin{cases} \dot{\boldsymbol{x}} = \boldsymbol{A}\boldsymbol{x} + \boldsymbol{B}\boldsymbol{u} + \boldsymbol{E}\boldsymbol{d} \\ \boldsymbol{y} = \boldsymbol{C}\boldsymbol{x} \end{cases} \tag{1}$$

$$A=\begin{bmatrix} -\frac{R_f}{L_f} & \omega & -\frac{1}{L_f} & 0 \\ -\omega & -\frac{R_f}{L_f} & 0 & -\frac{1}{L_f} \\ \frac{1}{C_f} & 0 & 0 & \omega \\ 0 & \frac{1}{C_f} & -\omega & 0 \end{bmatrix},\ B=\begin{bmatrix} \frac{1}{L_f} & 0 \\ 0 & \frac{1}{L_f} \\ 0 & 0 \\ 0 & 0 \end{bmatrix}\ E=\begin{bmatrix} 0 & 0 \\ 0 & 0 \\ -\frac{1}{C_f} & 0 \\ 0 & -\frac{1}{C_f} \end{bmatrix},\ C=\begin{bmatrix} 0 & 0 & 1 & 0 \\ 0 & 0 & 0 & 1 \end{bmatrix}$$

Where, ω is the fundamental angular frequency.

2.2. inverter Establishment of optimal controller

In modern control theory, state feedback can comprehensively reflect the internal characteristics of the system, and good dynamic performance can be obtained through pole assignment. An integral controller can be added to eliminate the steady-state error. In addition, it is not necessary to use proportional integral controller instead of integral controller, because it will introduce a new zero point into the system and affect the dynamic response of the system [8]. Thus, the structure of the system can be obtained, as shown in Figure 1 (a).

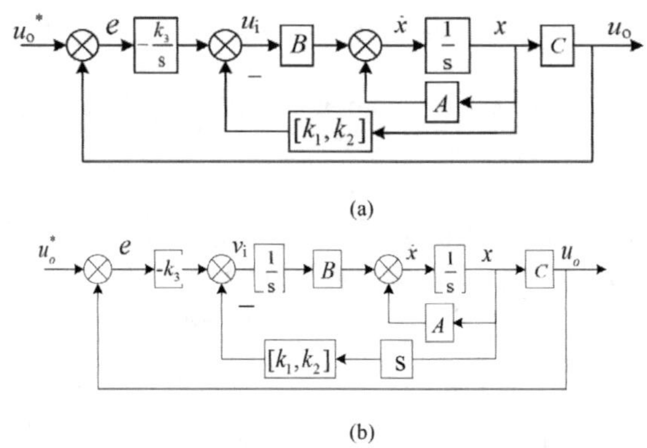

Figure 1. Three-phase voltage inverter control structure

In Figure 1, k_1 and k_2 are the state current and voltage feedback parameters, k_3 is the integral parameter, e is the difference between the given voltage and the actual output voltage, and u_o^* is the given input of the inverter circuit.

The optimal controller is usually solved by Riccati equation to obtain the optimal control input u_i^*. By selecting appropriate values of the error weighting matrix $\boldsymbol{Q}$ and the weighting matrix $\boldsymbol{R}$ of the control signal, the values of the optimal control parameters k_1, k_2 and k_3 can be solved. The solution of $\boldsymbol{u}_i^*$ is shown in equation (2).

$$\boldsymbol{u}_i^* = -\boldsymbol{R}^{-1}\boldsymbol{B}^{T}\boldsymbol{P} = -k_1x_1 - k_2x_2 - \int k_3x_3dt \tag{2}$$

Where,

$\boldsymbol{x}_1=\begin{bmatrix} i_{Ld} & i_{Lq}\end{bmatrix}^T$, $\boldsymbol{x}_2=\begin{bmatrix} u_{od} & u_{oq}\end{bmatrix}^T$, $\boldsymbol{x}_3=\begin{bmatrix} e_d & e_q\end{bmatrix}^T$, e_d and e_q are the difference between the given voltage and the actual voltage of axis d and q respectively. It is difficult to directly solve $\boldsymbol{u}_i^*$ through equation (2), and it is easier to obtain the optimal control input by redefining the state variables.Let,

$$\boldsymbol{A}^*=\begin{bmatrix} \boldsymbol{A} & 0 \\ \boldsymbol{-C} & 0 \end{bmatrix}, \boldsymbol{B}^*=\begin{bmatrix} \boldsymbol{B} \\ 0 \end{bmatrix}$$

$$\boldsymbol{x}^*=\begin{bmatrix} \dot{x}_1 & \dot{x}_2 & x_3 \end{bmatrix}^T,\ \boldsymbol{v}_i=\dot{\boldsymbol{u}}_i,\ \boldsymbol{K}=\begin{bmatrix} k_1 & k_2 & k_3 \end{bmatrix}$$

Without considering the disturbance, equation (1) can be transformed into the following new equation of state:

$$\begin{cases} \dot{\boldsymbol{x}}^* = \boldsymbol{A}^*\boldsymbol{x}^* + \boldsymbol{B}^*\boldsymbol{v}_i \\ \boldsymbol{v}_i = -\boldsymbol{K}\boldsymbol{x}^* \\ \boldsymbol{y} = \boldsymbol{C}\boldsymbol{x} \end{cases} \tag{3}$$

Where

$$\boldsymbol{k}_1 = \begin{bmatrix} k_{11} & k_{12} \\ k_{21} & k_{22} \end{bmatrix}, \boldsymbol{k}_2 = \begin{bmatrix} k_{13} & k_{14} \\ k_{23} & k_{24} \end{bmatrix}, \boldsymbol{k}_3 = \begin{bmatrix} k_{15} & k_{16} \\ k_{25} & k_{26} \end{bmatrix}$$

The system control block diagram after redefining the state variable is shown in Figure 1 (b), and the performance index of the new system can be expressed as:

$$\boldsymbol{J}(\boldsymbol{u}_i) = \int_0^\infty \boldsymbol{x}^*\boldsymbol{Q}\boldsymbol{x}^{*\mathrm{T}} + \boldsymbol{v}_i\boldsymbol{R}\boldsymbol{v}_i^{\mathrm{T}} dt \tag{4}$$

The resulting system control structure is shown in Figure 2.

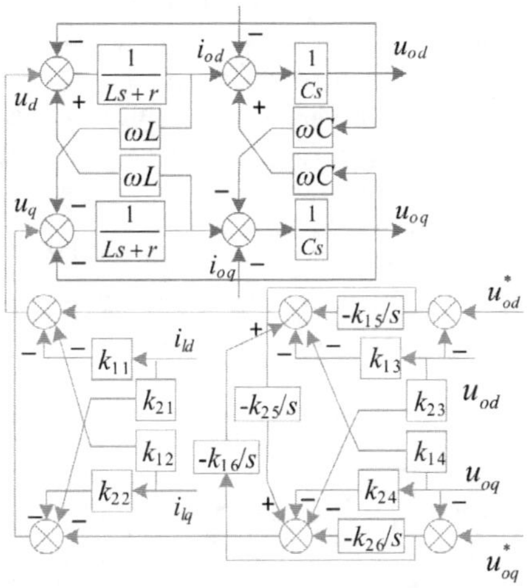

Figure 2. Three-phase inverter LQR control block diagram

The LQR controller can effectively improve the transient response speed and steady-state accuracy of the system. However, considering that there are various disturbances in the operation of the system, the dynamic compensation structure is added on the basis of the original controller, so that the system can reduce the output voltage fluctuation of the inverter without identifying the specific disturbance form, Improve the anti-interference ability and reliability of inverter.

3. Dynamic Compensation Control Strategy of Inverter Based on Optimal Controller

3.1. Disturbance analysis

In general, the ideal single-phase output voltage of the inverter can be expressed as:

$$u(t) = U_1 \sin(\omega t + \theta_1) \tag{5}$$

Where u_1 is the amplitude of the fundamental voltage, θ_1 is the initial phase angle. Disturbances are divided into two categories [9]:

- The amplitude U_1 of fundamental voltage changes, including voltage sag, overvoltage, undervoltage, fluctuation and flicker;
- Additive disturbance, including harmonics, three-phase imbalance, impulse voltage, etc.

The bridge arm current of three-phase inverter is transformed into the expression in dq coordinate system, as shown in equation (6).

$$\begin{cases} C_f \dfrac{du_{od}}{dt} = i_{Ld} + \omega C_f u_{oq} - i_{od} \\ C_f \dfrac{du_{oq}}{dt} = i_{Lq} + \omega C_f u_{od} - i_{oq} \end{cases} \tag{6}$$

According to equation (6), when the output voltage fluctuates, the output currents i_{od} and i_{oq} of the inverter are equivalent to the disturbance input. When the circuit performs load switching, the expression of output current of dq axis is:

$$\begin{cases} i_{od} = i_{od1} + I_{pd} 1(t - t_p) \sin\theta_1 \\ i_{oq} = -i_{oq1} - I_{pq} 1(t - t_p) \cos\theta_1 \end{cases} \tag{7}$$

Where i_{od1} and i_{oq1} are the amplitude of the fundamental current, I_{pd} and I_{pq} are the amplitude of fluctuating current, 1 $(t$-$t_p)$ is the step function, and t_p is the load switching time. According to equation (7), the switching load of inverter shows step disturbance in dq coordinate system.

When the load connected to the inverter is three-phase unbalanced load, the phase voltage can be decomposed into positive, negative and zero sequence voltages according to the symmetrical component method. After ignoring the harmonic component, the output expression of three-phase current in dq coordinate system is:

$$\begin{bmatrix} i_{od} \\ i_{oq} \\ i_{oo} \end{bmatrix} = I_m^+ \begin{bmatrix} \cos\theta^+ \\ -\sin\theta^+ \\ 0 \end{bmatrix} + I_m^- \begin{bmatrix} \cos(2\omega t + \theta^-) \\ -\sin(2\omega t + \theta^-) \\ 0 \end{bmatrix} + I_m^o \begin{bmatrix} 0 \\ 0 \\ \cos(\omega t + \theta^o) \end{bmatrix} \tag{8}$$

Where, i_{oo} is zero sequence current, and I_m^+ 、I_m^- 、I_m^0 are output positive, negative and zero sequence current amplitudes respectively. θ^+、θ^-、θ^0 are the initial phase angles of positive, negative and zero sequence current respectively.

It can be seen from equation (8) that the unbalanced voltage on dq axis is the superposition of DC component and 2 times of power frequency AC negative sequence component.

When the inverter has nonlinear load, the output voltage of the whole inverter can be expressed as the sum of the fundamental voltage and harmonic voltage. Among them, the positive sequence harmonic voltage will appear in the dq axis in the form of $6k$+1 (k is a positive integer) harmonic, the negative sequence harmonic voltage will appear in the dq axis in the form of $6k$-1 harmonic, and the zero sequence harmonic voltage is zero.

3.2. Dynamic compensation theory

If for a true rational transfer function ***G***(s), the controllability is observable, and the external controller ***K***(s) remains stable, as shown in Figure 3.

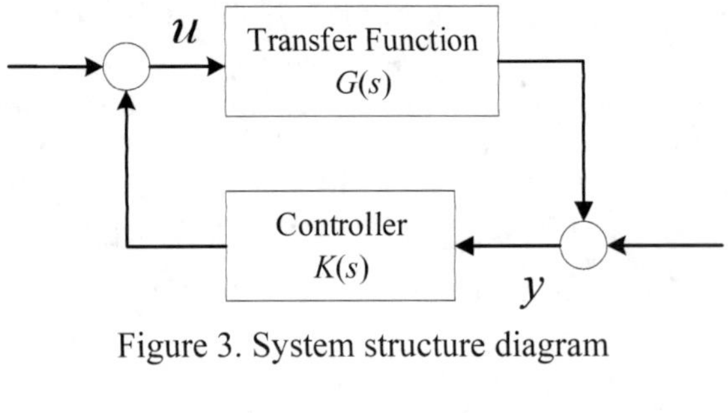

Figure 3. System structure diagram

Theorem 1 [10]: The control loop of the controlled object $\boldsymbol{G}(s)$ and a control signal $\boldsymbol{u}_1$ provided by the existing controller $\boldsymbol{K}$(s) are given. If the control loop is internally stable, all internally stable controllers can be parameterized as follows:

$$\boldsymbol{u}(s) = \boldsymbol{u}_1(s) + \boldsymbol{Q}_c(s)\boldsymbol{r}(s) \tag{9}$$

Where $\boldsymbol{u}_1$ is the output of the original controller, $\boldsymbol{u}$ is the output of the actual controller, $\boldsymbol{Q}_c$ is the transfer function matrix of the compensation signal, and $\boldsymbol{r}$ is the residual signal.

Therefore, when the system composed of the control object $\boldsymbol{G}_p(s)$ and the controller $\boldsymbol{K}(s)$ is stable, the system can still remain stable when the compensation controller $\boldsymbol{Q}_c(s)$ is added. The residual generator models are the original control object, observes the local original information and makes a difference with the actual output information to obtain the corresponding residual signal $\boldsymbol{r}(s)$. Without additional load sensors, it can not only reduce the equipment investment and communication cost, but also respond to the current state of the system in real time. When the disturbance $\boldsymbol{d}(s)$ is zero, the model in the residual generator corresponds to the actual control object, and the output residual signal $\boldsymbol{r}(s)$ is zero. When the disturbance $\boldsymbol{d}(s)$ comes, the residual generator can quickly detect the residual signal and output the compensation signal to the control end through $\boldsymbol{Q}_c(s)$, which is superimposed with the output signal of the original controller to realize the dynamic suppression of the disturbance.

According to the above theoretical analysis, the residual is generated by the difference between the observer output and the actual value, and its structure can be expressed as equation (10).

$$\begin{cases} \dot{\hat{x}} = (\boldsymbol{A} - \boldsymbol{LC})\hat{x} + \boldsymbol{Bu} + \boldsymbol{Ly} \\ \hat{y} = \boldsymbol{C}\hat{x} \\ \boldsymbol{r} = \boldsymbol{y} - \hat{y} \end{cases} \tag{10}$$

Where $\boldsymbol{L}$ is the observer gain matrix, $\hat{x}$ is the reconstructed state vector and $\hat{y}$ is the reconstructed output. With the continuous increase of $\boldsymbol{L}$, the response will gradually accelerate, but unnecessary noise and other signal disturbances will also be introduced. Therefore, the poles of the observer are usually 2-5 times of the real part of the system poles.

The dynamic compensation structure based on the residual generator is shown in Figure 4,

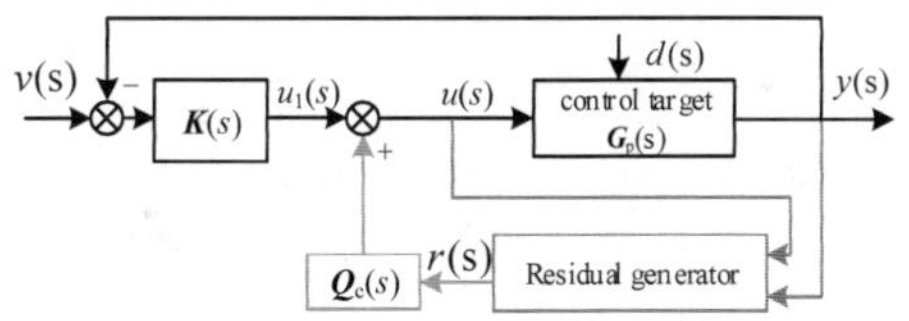

Figure 4. Dynamic compensation control framework based on robust residual generator

According to the structure of Figure 4, the disturbance caused by the outside can be extracted by the residual generator without knowing the specific disturbance type in advance. If you want the output not to be affected by the disturbance, you only need to meet the following requirements:

$$\boldsymbol{Q}_c(s)\boldsymbol{r}(s) - \boldsymbol{d}(s) = 0 \tag{11}$$

3.3. Dynamic compensation of inverter based on LQR controller

3.3.1. Voltage dynamic compensation

According to the inverter and the above compensation theory, let $e_x = x - \hat{x}$, by making a difference between equations (1) and (10), the calculation expression of inverter residual generator shown in equation (12) can be obtained:

$$\begin{cases} \dot{\boldsymbol{e}}_x = (\boldsymbol{A} - \boldsymbol{LC})\boldsymbol{e}_x + \boldsymbol{Ed} \\ \boldsymbol{r} = \boldsymbol{Ce}_x \end{cases} \tag{12}$$

Where $\boldsymbol{d}$ is the disturbance input of the equivalent residual generator. $\boldsymbol{r}$ is the residual value output by the residual generator. The residual generator of equation (10) is equivalent to the calculated output of equation (12), so the residual value of inverter output can be obtained without actually measuring the size of $\boldsymbol{d}$. Since the dq axis of the inverter is structurally symmetrical, the dynamic compensation structure taking the d axis as an example is shown in Figure 5. In Figure 5, u_{od}^* is the input reference voltage of axis d, u_{od} is the output voltage of axis d, i_{Ld} is the inductance current, I_o is the disturbance input, r_d is the output residual signal of axis d residual generator, $\boldsymbol{Q}_c$ is the transfer function of voltage compensation signal, u_{rd} is the voltage compensation output, u_{id} is the controller output.

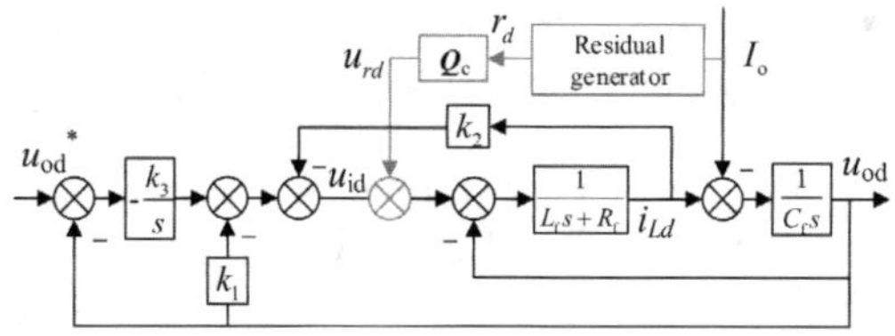

Figure 5. Decoupling d-axis dynamic compensation structure

As can be seen from Figure 5, k_1 and k_2 are the state feedback part of the optimal controller, $-k_3/s$ is the integral control part of the controller, u_{id} is the output control signal of the controller, when there is no disturbance Current I_o, the output of the residual generator is 0, and the output compensation voltage signal of $\boldsymbol{Q}_c$ is 0. When there is disturbance Current I_o, the residual generator can immediately detect the disturbance and output the residual signal and compensate through $\boldsymbol{Q}_c$. However, due to the feedback branch of inductive current between the inverter and the controller, some control objects form an inner loop feedback with k_2, so that the compensation signal of $\boldsymbol{Q}_c$ is fed back to the input of the control signal through k_2 to form a secondary disturbance. At this time, the disturbance should be fully compensated, must meet:

$$\frac{\frac{1}{L_f s+R_f}}{1+\frac{k_2}{L_f s+R_f}} Q_c(s) r_d(s) - I_o(s) = 0 \tag{13}$$

According to equation (12), it can be solved that the disturbance action is the input and the residual is the output transfer function $\boldsymbol{G}_{rd}$. The solution of $\boldsymbol{Q}_c$ in equation (13) can be simplified as:

$$Q_c(s) = \frac{L_f s + R_f + k_2}{G_{rd}} \tag{14}$$

3.3.2. Secondary compensation

It can be seen from equation (14) that the unpacking of $\boldsymbol{Q}_c$ includes a part of LQR controller. In order to make the compensation controller operate independently and not affected by the optimal controller, a compensation controller H is designed to offset the secondary disturbance signal before it is affected by the optimal controller. The compensation structure is shown in Figure 6.

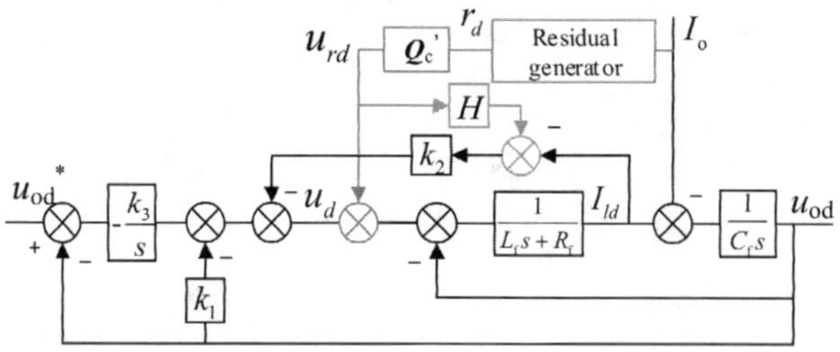

Figure 6. System dynamic compensation structure after secondary compensation

Based on the above theoretical analysis and derivation, a complete inverter dynamic compensation architecture based on linear quadratic optimal control and the parameters to be solved are obtained, which is extended to the dual axis control structure, as shown in Figure 7.

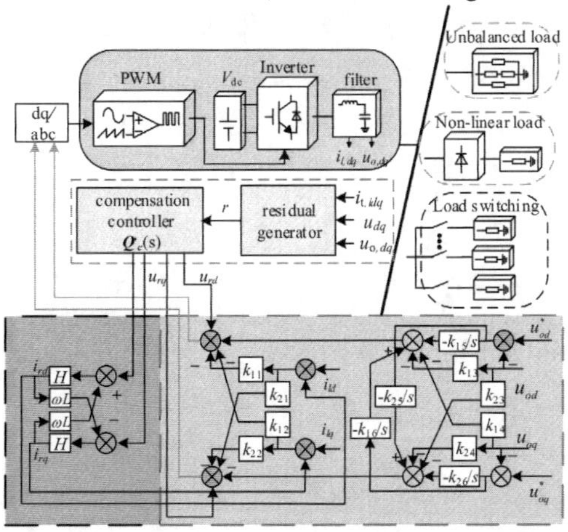

Figure 7. Three-phase inverter dynamic compensation structure

4. SOLUTION OF DYNAMIC COMPENSATION CONTROLLER

4.1. Q'c solution of voltage dynamic compensation

According to the above theoretical derivation and model matching theory [11], the model matching structure of inverter dynamic compensation controller is obtained, as shown in Figure 8.

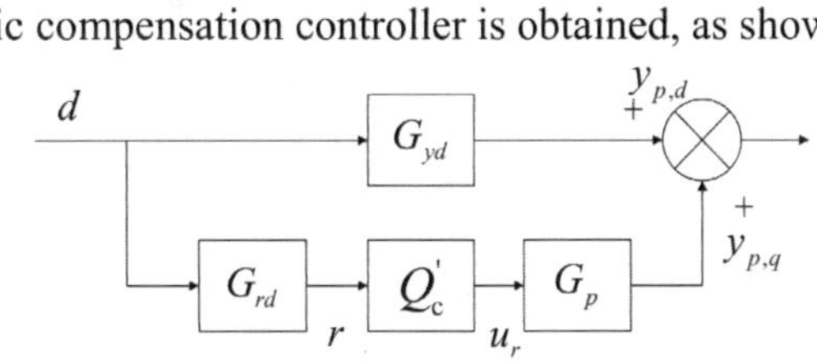

Figure 8. Model matching structure

Where $\boldsymbol{G}_{\mathrm{yd}}$ is the transfer function matrix of the disturbance directly acting on the system, $\boldsymbol{G}_{\mathrm{rd}}$ is the transfer function matrix of the residual generator, and $\boldsymbol{G}_{\mathrm{p}}$ is the compensation signal ur output by $\boldsymbol{Q}'_{\mathrm{c}}$ as the transfer function matrix of the system input.

According to the superposition principle of linear system, when the disturbance action is used for the control object, it will produce a disturbance output $y_{\mathrm{p,d}}$, and the compensation system will also produce a compensation output $y_{\mathrm{p,q}}$. when $y_{\mathrm{p,d}}+y_{\mathrm{p,q}}=0$, the effect of any disturbance on the system can be offset theoretically. The solution process of $\boldsymbol{Q}'_{\mathrm{c}}$ is shown in equation (15).

$$\boldsymbol{G}_{\mathrm{yd}} + \boldsymbol{G}_{\mathrm{rd}}\boldsymbol{Q}'_{\mathrm{c}}\boldsymbol{G}_{\mathrm{p}} = 0 \tag{15}$$

$\boldsymbol{G}_{\mathrm{yd}}$ input is $\boldsymbol{d}$, output is $y_{\mathrm{p,d}}$, and state space expression is:

$$\begin{cases} \dot{\boldsymbol{x}}_{\mathrm{d}} = \boldsymbol{A}_{\mathrm{d}}\boldsymbol{x}_{\mathrm{d}} + \boldsymbol{E}_{\mathrm{d}}\boldsymbol{d} \\ \boldsymbol{y}_{\mathrm{p,d}} = \boldsymbol{C}_{\mathrm{d}}\boldsymbol{x}_{\mathrm{d}} \end{cases} \tag{16}$$

Where,

$$\boldsymbol{A}_{\mathrm{d}}=\begin{bmatrix} -\frac{R_{\mathrm{f}}}{L_{\mathrm{f}}} & \omega & -\frac{1}{L_{\mathrm{f}}} & 0 \\ -\omega & -\frac{R_{\mathrm{f}}}{L_{\mathrm{f}}} & 0 & -\frac{1}{L_{\mathrm{f}}} \\ \frac{1}{C_{\mathrm{f}}} & 0 & 0 & \omega \\ 0 & \frac{1}{C_{\mathrm{f}}} & -\omega & 0 \end{bmatrix}, \boldsymbol{E}_{\mathrm{d}}=\begin{bmatrix} 0 & 0 \\ 0 & 0 \\ -\frac{1}{C_{\mathrm{f}}} & 0 \\ 0 & -\frac{1}{C_{\mathrm{f}}} \end{bmatrix} \boldsymbol{C}_{\mathrm{d}}=\begin{bmatrix} 0 & 0 & 1 & 0 \\ 0 & 0 & 0 & 1 \end{bmatrix}$$

$\boldsymbol{x}_d=[i_{\mathrm{Ld}} \quad i_{\mathrm{Lq}} \quad u_{\mathrm{od}} \quad u_{\mathrm{oq}}]^{\mathrm{T}}$ is the state quantity and $\boldsymbol{d}=[i_{\mathrm{od}} \quad i_{\mathrm{oq}}]^{\mathrm{T}}$ is the disturbance current input.

$\boldsymbol{G}_{\mathrm{rd}}$ input is $\boldsymbol{d}$ and output is $\boldsymbol{r}$. The state space expression is shown in equation (15). According to the system dimension, it can be obtained that the gain matrix of the observer is a 4 × 2, so let the observer gain matrix be:

$$\boldsymbol{L}=\begin{bmatrix} l_{11} & l_{12} & l_{13} & l_{14} \\ l_{21} & l_{22} & l_{23} & l_{24} \end{bmatrix}^{\mathrm{T}}$$

Where,

$\boldsymbol{e}_x=\begin{bmatrix} i_{\mathrm{Ld}}-\hat{i}_{\mathrm{Ld}} & i_{\mathrm{Lq}}-\hat{i}_{\mathrm{Lq}} & u_{\mathrm{od}}-\hat{u}_{\mathrm{od}} & u_{\mathrm{oq}}-\hat{u}_{\mathrm{oq}} \end{bmatrix}^{\mathrm{T}}$ is the state quantity and $\boldsymbol{d}=[i_{\mathrm{od}} \quad i_{\mathrm{oq}}]^{\mathrm{T}}$ is the disturbance current input.

$\boldsymbol{G}_{\mathrm{p}}$ input is u_r, output is $\boldsymbol{y}_{\mathrm{p,q}}$, and state space expression is:

$$\begin{cases} \dot{\boldsymbol{x}}_{\mathrm{p}}=\boldsymbol{A}_{\mathrm{p}}\boldsymbol{x}_{\mathrm{p}}+\boldsymbol{B}_{\mathrm{p}}\boldsymbol{u}_r \\ \boldsymbol{y}_{\mathrm{p,q}}=\boldsymbol{C}_{\mathrm{p}}\boldsymbol{x}_{\mathrm{p}} \end{cases} \tag{17}$$

Where,

$$\boldsymbol{A}_{\mathrm{p}}=\begin{bmatrix} -\frac{R_{\mathrm{f}}}{L_{\mathrm{f}}} & \omega & -\frac{1}{L_{\mathrm{f}}} & 0 \\ -\omega & -\frac{R_{\mathrm{f}}}{L_{\mathrm{f}}} & 0 & -\frac{1}{L_{\mathrm{f}}} \\ \frac{1}{C_{\mathrm{f}}} & 0 & 0 & \omega \\ 0 & \frac{1}{C_{\mathrm{f}}} & -\omega & 0 \end{bmatrix}, \boldsymbol{B}_{\mathrm{p}}=\begin{bmatrix} \frac{1}{L_{\mathrm{f}}} & 0 \\ 0 & \frac{1}{L_{\mathrm{f}}} \\ 0 & 0 \\ 0 & 0 \end{bmatrix}$$

$$\boldsymbol{C}_{\mathrm{p}}=\begin{bmatrix} 0 & 0 & 1 & 0 \\ 0 & 0 & 0 & 1 \end{bmatrix}$$

$x_{\mathrm{p}}=[i_{\mathrm{Ld}} \quad i_{\mathrm{Lq}} \quad u_{\mathrm{od}} \quad u_{\mathrm{oq}}]^{\mathrm{T}}$ is the state quantity and $u_r=[u_{r\mathrm{d}} \quad u_{r\mathrm{q}}]^{\mathrm{T}}$ is the compensation voltage input.

According to equation (15), the transfer function matrix of the compensation controller $\boldsymbol{Q}'_{\mathrm{c}}$ can be solved through matrix operation, and the simplest expression matrix element expression can be obtained after simplification

$$\begin{cases} \boldsymbol{Q}'_c(1,1) = -C_f L_f s^2 + (-C_f R_f - C_f L_f l_{13})s - \\ \quad L_f l_{11} - C_f R_f l_{13} - 1 + C_f L_f \omega^2 + C_f L_f l_{14}\omega \\ \boldsymbol{Q}'_c(1,2) = (2C_f L_f \omega - C_f L_f l_{23})s + C_f R_f \omega - \\ \quad C_f R_f l_{23} - L_f l_{21} + C_f L_f l_{24}\omega \\ \boldsymbol{Q}'_c(2,1) = (-C_f L_f l_{14} - 2C_f L_f \omega)s - L_f l_{12} - \\ \quad C_f R_f l_{14} - C_f R_f \omega - C_f L_f l_{13}\omega \\ \boldsymbol{Q}'_c(2,2) = -C_f L_f s^2 + (-C_f R_f - C_f L_f l_{24})s - \\ \quad L_f l_{22} - C_f R_f l_{24} - 1 + C_f L_f \omega^2 - C_f L_f l_{23}\omega \end{cases} \tag{18}$$

According to the results of the above formula, the numerator order is greater than the denominator order. According to the linear system control principle, the denominator must be compensated to make the denominator order greater than or equal to the numerator, so as to obtain the simplest matrix expression of $\boldsymbol{Q}'_c$.

$$\boldsymbol{Q}'_c(s) = \frac{1}{(\eta s+1)^2}\begin{bmatrix} \boldsymbol{Q}'_c(1,1) & \boldsymbol{Q}'_c(1,2) \\ \boldsymbol{Q}'_c(2,1) & \boldsymbol{Q}'_c(2,2) \end{bmatrix} \tag{19}$$

Where, η Is the compensation factor.

4.2. Quadratic compensation solution

When only the single axis of the inverter is considered and the outer loop controller is not considered, the following can be obtained by solving the transfer function:

$$\begin{cases} G_{yd} = -\dfrac{L_f s + R_f}{L_f C_f s^2 + C_f R_f s + 1} \\ G_p = \dfrac{1}{L_f C_f s^2 + C_f R_f s + 1} \end{cases} \tag{20}$$

According to equation (16), it can be solved as follows:

$$Q'_c = \frac{-G_{yd}}{G_p G_{rd}} = \frac{L_f s + R_f}{G_{rd}} \tag{21}$$

When considering the outer loop controller and feedback, the feedback of inductive current will cause secondary disturbance to the system. In order to eliminate the influence of secondary disturbance on the system, secondary compensation must be carried out. According to the structure of Figure 6, the transfer function can be expressed as:

$$\begin{cases} G_{yd1} = -\dfrac{s(L_f s + R_f + k_2)}{L_f C_f s^3 + C_f(R_f + k_2)s^2 + k_1 s + 1 - k_3} \\ G_{p1} = \dfrac{s}{L_f C_f s^3 + C_f(R_f + k_2)s^2 + k_1 s + 1 - k_3} \end{cases} \tag{22}$$

$$Q_c = \frac{-G_{yd1}}{G_{p1} G_{rd}} = \frac{L_f s + R_f + k_2}{G_{rd}} \tag{23}$$

The solution expression of H can be obtained by simultaneous equations (21) and (23).

$$Q_c = Q'_c + Q'_c k_2 H \tag{24}$$

Solution:

$$H = \frac{1}{L_f s + R_f} \tag{25}$$

Where G_{yd1} is the transfer function of the disturbance directly acting on the whole system, and G_{p1} is the transfer function of the output compensation signal acting on the whole system.

The expression of the final secondary compensation can be obtained by extending to the two axis decoupling model of three-phase inverter.

$$\begin{cases} \dot{i}_{rd} = -\frac{R_f}{L_f} i_{rd} + \omega i_{rq} + \frac{1}{L_f} u_{rd} \\ \dot{i}_{rq} = -\frac{R_f}{L_f} i_{rq} - \omega i_{rd} + \frac{1}{L_f} u_{rq} \end{cases} \tag{26}$$

Where, i_{rd}、 i_{rq} are d-axis and q-axis secondary compensation outputs respectively, and u_{rd}、 u_{rq} are d-axis and q-axis voltage compensation outputs respectively.

5. EXPERIMENTAL VERIFICATION

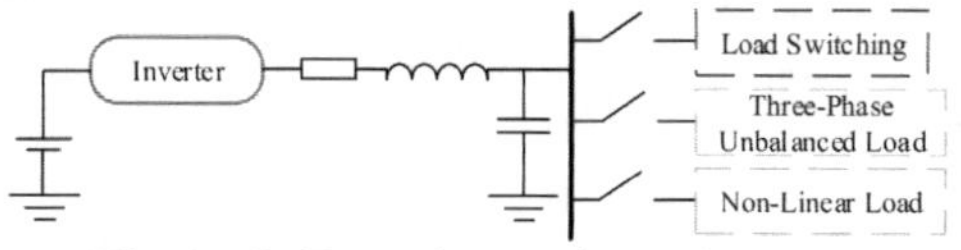

Figure 9. Experimental topology

In order to verify the compensation effect of the inverter dynamic compensation architecture based on linear quadratic control designed in this paper, different kinds of experiments are designed, and MATLAB / Simulink simulation is used to verify the rationality of the compensation architecture. Parameter settings are shown in Table 1.

TABLE I. SIMULATION PARAMETERS

Parameter Name And Unit	Value
Given voltage at DC side/V	700
AC side line voltage/V	380
System frequency/Hz	50
IGBT switching frequency/kHz	50
LC filter inductance/H	2×10^{-3}
LC filter capacitor/C	15×10^{-4}
Filter inductance parasitic resistance/Ω	0.01
Linear quadratic Q matrix	Diag [10^2 10^2 10^2 10^2 10^7 10^7]
Linear quadratic R matrix	[1]

5.1. Step disturbance experiment

Switch the load with active power of 15kW and reactive power of 8kVar to ensure that other parameters of the line remain unchanged. Through MATLAB / Simulink simulation, the d-axis voltage waveform is obtained, as shown in Figure 10.

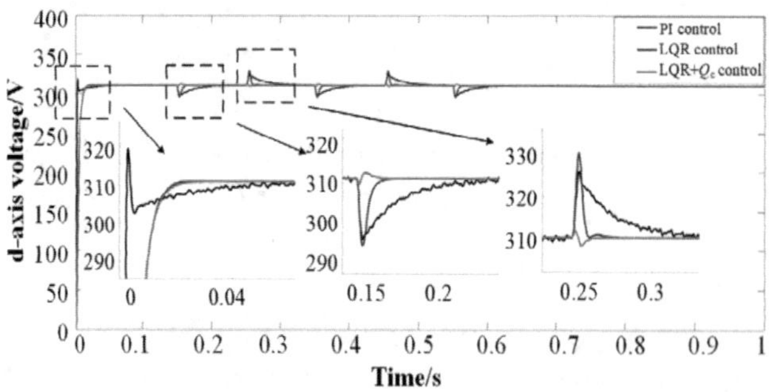

Figure 10. Comparison of step disturbance d-axis voltage waveform

It can be seen from Figure 10 that under the double closed-loop PI control, the inverter is stable within 50ms after startup, and the overshoot is obvious. Under the LQR control, the system can be stable without overshoot within 20ms, which significantly improves the transient response speed of the system. After 0.15s of load input, due to the influence of disturbance current, the d-axis voltage is temporarily reduced to 295V, The system voltage under double closed-loop PI control is stabilized again after 500ms, and the system voltage under LQR control is stabilized after 20ms, which greatly improves the transient response speed of the system, but the sag amplitude does not change significantly. After adding the dynamic compensation structure of disturbance residual generator, the voltage sag amplitude of the system is greatly reduced and restored to stability after 20ms. After the load is cut off in 0.25s, the response effect is similar to the input load.

The comparison of three-phase voltage waveforms is shown in Figure 11. When the load is put into operation in 0.15s, when there is only a double closed-loop PI controller, the three-phase voltage amplitude temporarily decreases to 300V and slowly recovers to stability after 500ms. After the load is put into operation, the three-phase voltage under LQR control temporarily decreases to 302V and recovers to stability after 20ms. After adding the dynamic compensation structure of disturbance residual generator, the fluctuation amplitude of three-phase voltage is greatly reduced and can be restored to stability in 20ms. When the load is cut off for 0.25s, the three-phase voltage response effect is similar to that when the load is input. The experiment shows that LQR control can effectively improve the transient response speed of the system, and LQR+Q_c control strategy can greatly reduce the amplitude of voltage fluctuation and improve the rapidity and immunity of the inverter system without changing the rapidity of the system.

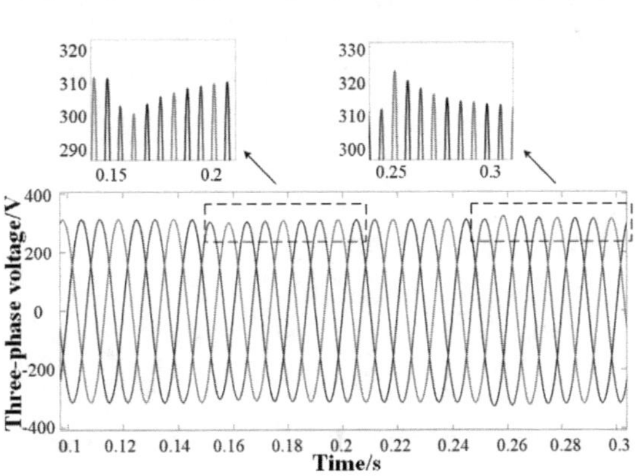

(a) Three phase voltage under PI control

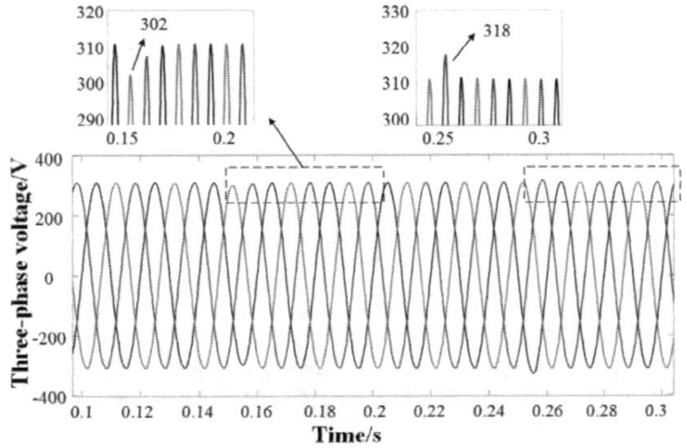

(b) Three phase voltage under LQR control

Journal of Physics: Conference Series **2215** (2022) 012003 doi:10.1088/1742-6596/2215/1/012003

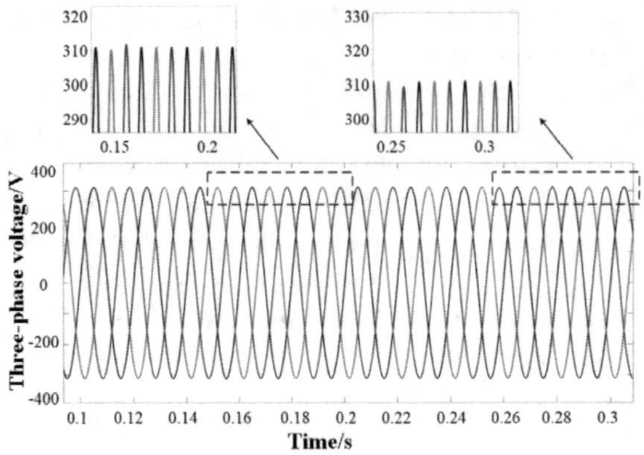

(c) Three phase voltage under LQR + Q_c control

Figure 11. Step disturbance three-phase voltage waveform

5.2. Three phase unbalance disturbance

Input unbalanced three-phase load at 0.2s to ensure that other parameters remain unchanged. Through MATLAB / Simulink simulation, the d-axis voltage waveform is obtained, as shown in Figure 12.

It can be seen from Figure 12 that after the unbalanced load is input, the d-axis voltage under LQR control has an obvious double power frequency sinusoidal fluctuation, and the fluctuation amplitude is 323V. After adding the LQR + Q_c control structure proposed in this paper, the fluctuation amplitude is reduced to 312V.

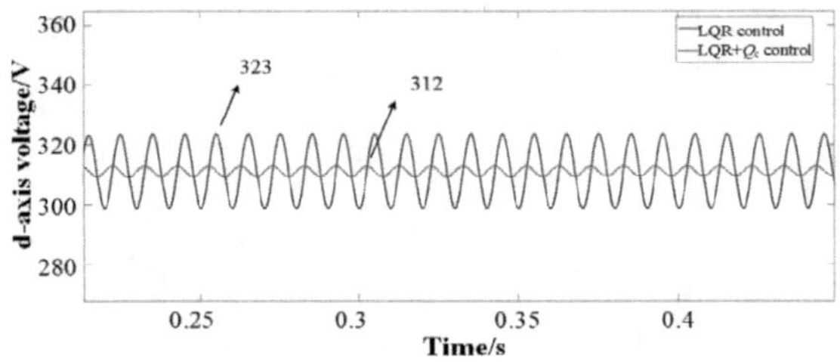

Figure 12. Three-phase unbalanced d-axis voltage waveform comparison

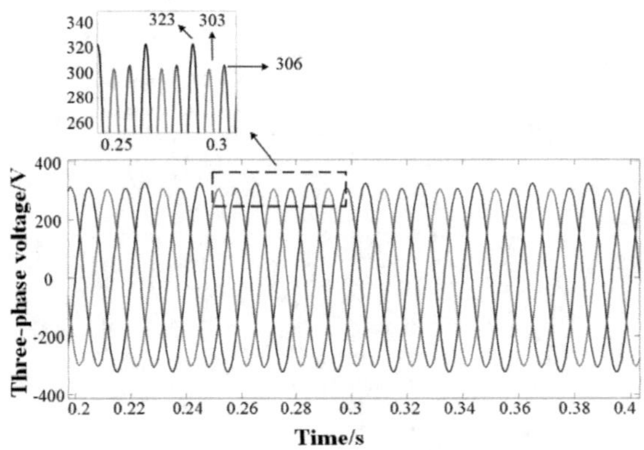

(a) Three phase voltage under LQR control

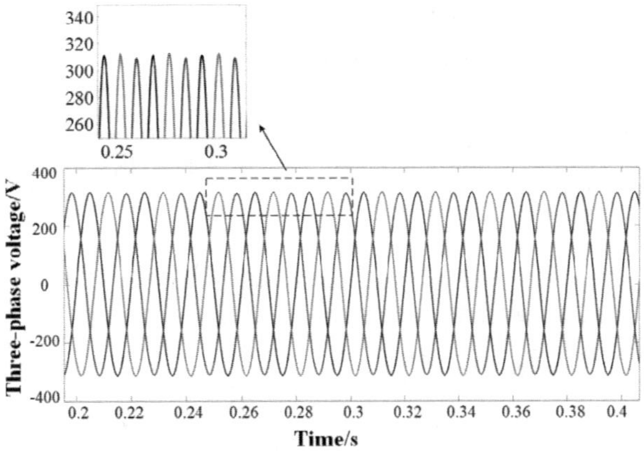

(b) Three phase voltage under LQR + Q_c control

Figure 13. Three-phase voltage waveform under unbalanced load

The three-phase voltage waveform under unbalanced load is shown in Figure 13. Under the influence of unbalanced load, the amplitude of three-phase sinusoidal voltage changes to varying degrees with different three-phase loads. After adding the compensation structure proposed in this paper, the three-phase voltage amplitude is basically the same. Figure 14 shows the three-phase voltage unbalance. The three-phase voltage unbalance under LQR control is 4.03%. After adding the compensation structure proposed in this paper, the voltage unbalance is reduced to 0.58%.

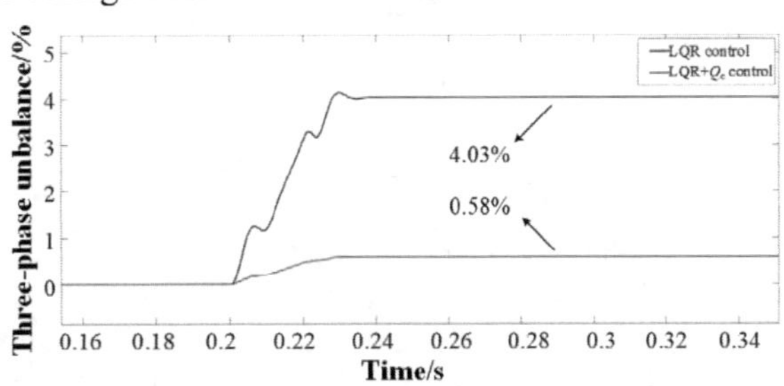

Figure 14. Waveform comparison of voltage unbalance

The simulation results show that the LQR+Q_c control structure can effectively suppress the influence of three-phase unbalanced load on the output voltage waveform and improve the reliability and robustness of the output voltage.

5.3. Harmonic disturbance

Harmonic disturbance is mainly caused by nonlinear load. Through MATLAB simulation, input nonlinear load at 0.2s, and the d-axis voltage waveform can be obtained, as shown in Figure 15.

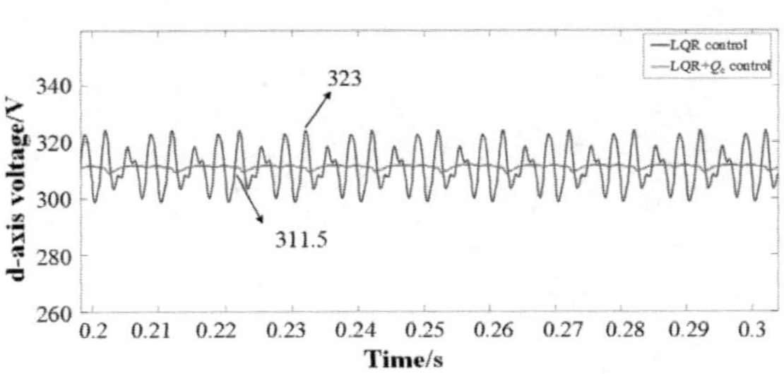

Figure 15. Comparison of d-axis voltage waveforms of nonlinear loads

It can be seen from Figure 15 that after adding nonlinear load, the d-axis voltage fluctuates obviously under the control of LQR, and the fluctuation amplitude is 323V. After adding the control structure proposed in this paper, the d-axis voltage fluctuation is significantly improved, the amplitude is reduced to 311.5V, and the anti-interference ability is greatly improved.

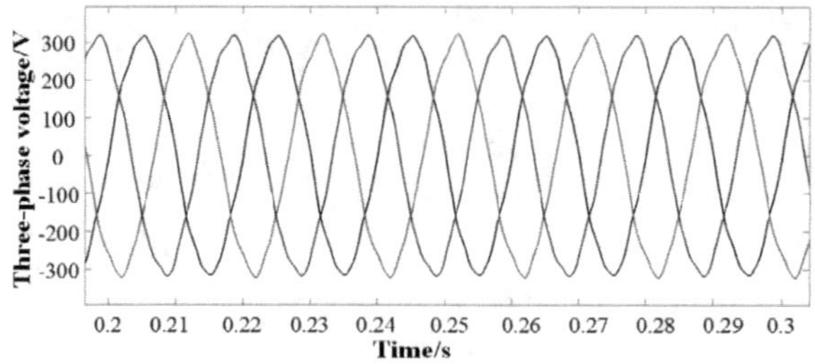

(a) Three phase voltage under LQR control

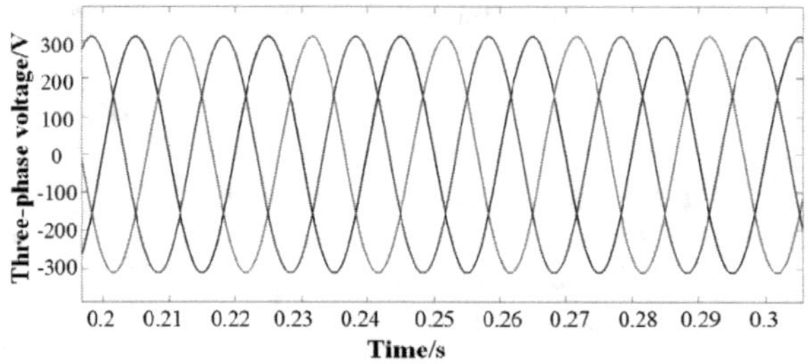

(b) Three phase voltage under LQR + Q_c control

Figure 16. Three-phase voltage waveform comparison under non-linear load

Figure 16 (a) shows the three-phase voltage output waveform under LQR control after adding nonlinear load. It can be seen from the simulation waveform that the three-phase voltage has obvious distortion. The waveform after adding dynamic compensation structure is shown in Figure 16 (b). The voltage waveform has been significantly improved and is close to the standard three-phase sinusoidal waveform. After that, the voltage is analyzed by using the FFT analysis provided by Matlab / Simulink, and the harmonic analysis diagram shown in Figure 178 is obtained. It can be seen from Figure 17 (a) that under the control of LQR, the distortion rate of the inverter output voltage is 3.25%. After adding the dynamic compensation structure, the output voltage is as shown in Figure 17 (b), and the distortion rate of the output voltage becomes 0.21%. Experiments show that the dynamic compensation structure can effectively suppress the influence of nonlinear load on the output voltage and improve the robustness and reliability of the output voltage.

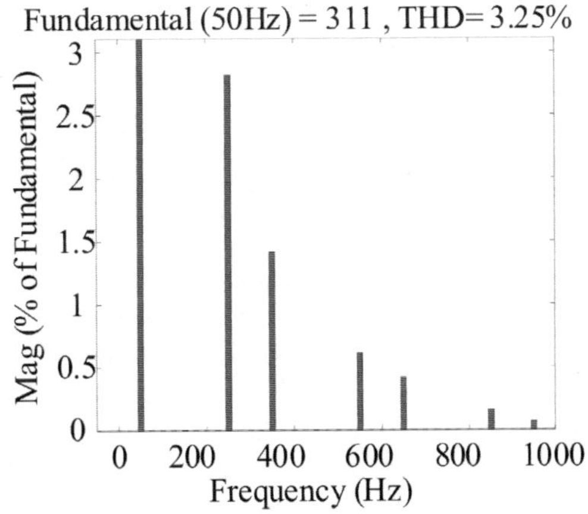

(a) FFT analysis of three-phase voltage under LQR control

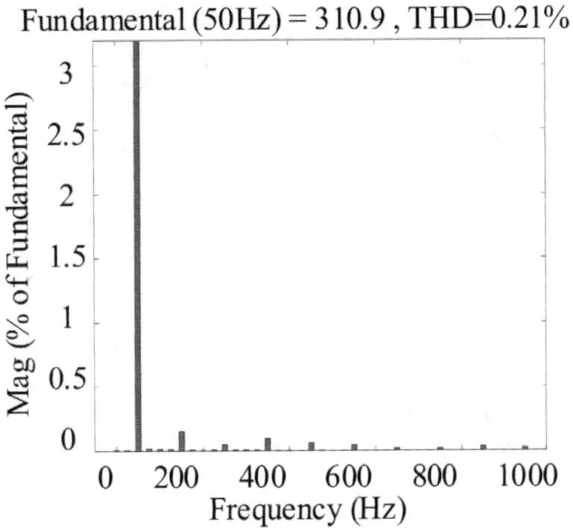

(b) FFT analysis of three-phase voltage under LQR control

Figure 17. Experimental waveform of nonlinear load

6. Conclusions

Under the background of microgrid, an inverter output voltage control strategy based on optimal control and disturbance residual generator is proposed, which realizes the rapidity of output voltage response and immunity. The specific conclusions are as follows.

- Through the state feedback of output voltage and inductance current, the inverter can obtain good dynamic performance. After adding integral control, the steady-state error of the system can be eliminated. The two are combined and solved by LQR, so that the voltage can follow quickly without overshoot when the inverter is started, and the transient response speed of the system is improved.
- When the disturbance occurs, the load current as a disturbance causes the fluctuation of the inverter output voltage. The compensation control strategy proposed in this paper can quickly extract the disturbance and realize compensation without additional voltage and current sensors, which can improve the anti-interference ability of the inverter and enhance the robustness of the inverter system without changing the transient response speed.

ACKNOWLEDGMENT
PR China, the National Key R&D Program of China, Key Special Projects for International Cooperation in Science and Technology Innovation between Governments (No.2021YFE0103800).

REFERENCES
[1] Li Xiaoye，Li Yongli，Zhang Weiya，et al. A power quality control strategy based on multi-functional grid-connected inverter[J]. Power System Technology，2015，39(2)：556-562
[2] Luo H, Krueger M, Koenings T, et al. Real-Time Optimization of Automatic Control Systems With Application to BLDC Motor Test Rig[J]. IEEE Transactions on Industrial Electronics, 2017, 64(5):4306-4314.Luo H, Krueger M,Koenings T ,et al. Real-Time Optimization of Automatic Control Systems With Application to BLDC Motor Test Rig[J].IEEE Transactions on Industrial Electronics, 2017, 64(5):4306-4314.
[3] Wang Can，Yang Ping，Ye Chao，et al. Voltage Control Strategy for Three/Single Phase Hybrid Multimicrogrid [J]. IEEE Transactions on Energy Conversion，2016，31(4)：1498 - 1509.
[4] Zhong Qingchang . Harmonic droop controller to reduce the voltage harmonics of inverters[J] . IEEE Transactions on Industrial Electronics，2013，60(3)：936-945.
[5] Dong H，Yuan S，Han Z，et al. A comprehensive strategy for power quality improvement of multi-inverter-based microgrid with mixed loads[J]. IEEE Access，2018(6)：30903-30916.
[6] Feng W，Sun K，Guan Y，et al. Active power quality improvement strategy for grid-connected microgrid based on hierarchical control[J]. IEEE Transactions on Smart Grid，2018，9(4)：3486-3495.
[7] WANG C, LI X L, GUO L, et al. Anonlinear-disturbance observer-based DC-bus voltage control for a hybrid AC/DC microgrid [J]. IEEE Transactions on Power Electronics,2014，29（11）：6162-6177
[8] Hasanzadeh A, Edrington C S, Liu Y, et al. An LQR based optimal tuning method for IMP-based VSI controller for electric vehicle traction drives. IEEE, 2011.
[9] Tang Yi，Liu Hao，Fang Yongli. A power quality disturbance classification method based on the analysis in time domain [J]. Automation of Electric Power Systems，2008，32(17)：50-54(in Chinese).
[10] Luo Hao，Yang Xu，Krueger Minjia，et al. A plug-and-play monitoring and control architecture for disturbance compensation in rolling mills[J]. IEEE/ASME Transactions on Mechatronics，2018，23(1)：200-210.
[11] B A Francis. A Course in H1 Control Theory[M]. Berlin：Springer， 19.

Dynamic Compensation Control Strategy for DC Converter Based on Linear Optimal Quadratic

Changbin Hu [a], **Huiru Wang***, **Shanna Luo** [b], **Yufei Huang** [c]

School of Electrical and Control Engineering, North China University of Technology, Beijing, China

[a]changbinlove@163.com

*Corresponding author: 1710430907@qq.com

[b]14815390@qq.com

[c] 476889002@qq.com

Abstract—Aiming at the problem of DC bus voltage fluctuation, a DC-DC converter control method combining LQR (Linear Optimal Quadratic) and dynamic compensation control is proposed. Firstly, the state space model of DC-DC converter is established. Secondly, according to the following characteristics of LQR controller, the error of the controlled output is introduced into the integral control as an extended state to eliminate the output tracking error. Then, in order to suppress the voltage fluctuation caused by load switching of DC-DC converter during power grid operation, a dynamic compensation structure based on disturbance observer is designed through robust coprime decomposition and Euler parameterization theory. The structure compensates directly at the input of the current loop, and the compensation controller $Q(s)$ is calculated by model matching. Finally, the simulation is carried out by Matlab / Simulink. The results show that without changing the structural parameters of the original system, the new control architecture can effectively suppress the DC bus voltage fluctuation caused by load theft and power fluctuation, and enhance the anti-interference performance and dynamic performance of the system.

1. INTRODUCTION

After entering the 21st century, with the progress of science and technology, human economy and industry have also entered a stage of rapid development. However, under this overall booming development, the problem of environmental degradation has gradually become non negligible, and the development of new energy has become the focus of scientific research all over the world. As an indispensable part of new energy power generation, new energy vehicles, information equipment and smart grid, power electronic devices, among which DC-DC converter is the core component of new energy management such as hybrid electric vehicle and photovoltaic inverter [1-3]. With the wide application of DC-DC converter in, academic and engineering circles at home and abroad are constantly exploring and studying the control methods of DC-DC converter, and are committed to maintaining constant output even in the event of sudden situations such as input and large-scale load change during system operation [4].

There are many control methods of DC-DC converter, traditional analog compensation network control, PID control, fuzzy control, adaptive control, synovial control, robust control, neural network control, hysteresis control and so on. PID control method is widely used in practical engineering because of its simplicity, high stability and high reliability. However, the tuning process of traditional PID control parameters largely depends on engineering experience, with poor adaptability and complex adjustment process. However, the actual system is often a model combining linear and non-linear. For a long time, many scholars have proposed new control algorithms to control DC-DC converters. In reference [5] and reference [6], yfoulise et al. Proposed a new state feedback control method for boost DC-DC converter, which solved the problem of model switching. The control method has strong robustness and commercial efficiency. But its closed-loop stability is still a very important problem. In general, a trade-off must be made between dynamic performance and steady-state performance.

In modern control theory, state feedback can comprehensively reflect the internal characteristics of the system, and good dynamic performance can be obtained through pole assignment. The design of LQR is based on state space technology. The main problem is to select the optimal control input under the constraints of linear system to minimize the quadratic objective function. Therefore, in order to solve the power quality problem of DC microgrid system, a controller based on state feedback and integral control is proposed for DC microgrid.

According to the characteristics of LQR control, the differential of output voltage, output current and error differential of the system are re-selected as new state variables, and the optimal controller parameters are obtained by using Riccati equation, so as to improve the dynamic performance and steady-state performance of the system.At the same time, a dynamic compensation architecture is proposed for the influence of switching load on the output voltage of the converter, which can further improve the anti-interference ability of DC microgrid without changing the structure of the existing controller. However, in previous studies, the proposed compensation architecture is a control structure for the output voltage, which has the problem of obvious response lag. Aiming at this problem, this paper proposes and designs a control structure for inductive current, which can quickly compensate and adjust after the disturbance, improve the response speed of the system and improve the dynamic performance of the whole system.

2. CONTROL STRATEGY OF DC CONVERTER BASED ON LQR

2.1. Mathematical Model of DC Converter

The state space equation of buck converter is:

$$\begin{cases} \dot{x} = Ax + Bu + Ew \\ y = Cx \end{cases} \tag{1}$$

Its input, output and state matrix respectively are: $u=[V_r]$、$y=[U_c]$、$x=[U_c \quad I_L]$, V_r is the output modulation wave of the current loop; w is disturbance.

$$A = \begin{bmatrix} 0 & \frac{1}{C} \\ -\frac{1}{L} & -\frac{r}{L} \end{bmatrix} ; \quad B = \begin{bmatrix} 0 & \frac{1}{L} \end{bmatrix}^T ;$$
$$C = \begin{bmatrix} 1 & 0 \end{bmatrix} ; \quad E = \begin{bmatrix} -\frac{1}{C} & 0 \end{bmatrix}^T ; \tag{2}$$

In the formula: r, L and C are the inductance parasitic resistance, inductance and capacitance of the converter respectively. U_t and U_c are the instantaneous voltage and output voltage after the switching transistor of the converter respectively; I_L and I_o are the inductance current and output current of the converter respectively.

Boost DC-DC converter is a non-minimum phase system. Select a static stable working point to linearize its state space, and select the difference between the actual value of inductance current and the steady-state value as the state quantity. Modify the parameter matrix in equation (2) according to the topology of boost converter, as shown in equation (3).

$$A=\begin{bmatrix} 0 & \frac{1}{C} \\ -\frac{1}{L} & -\frac{r}{L} \end{bmatrix} \quad ; \quad B=\begin{bmatrix} 0 & \frac{1}{L} \end{bmatrix}^T \quad ;$$

$$C=\begin{bmatrix} 1 & 0 \end{bmatrix} \quad ; \quad E=\begin{bmatrix} -\frac{1}{C} & 0 \end{bmatrix}^T \quad ; \tag{3}$$

Where the input, output, disturbance and state matrices are modified to $u=[d-d_0]$, $y=[U_c-U_{c0}]$, $w=[I_o-I_{o0}]$ and $x=[U_c-U_{c0} \quad I_L-I_{L0}]$. Where U_{c0}, I_{L0}, $I_{o0}$and d_0 are the corresponding output voltage, inductance current, load current and duty cycle in steady state respectively.

2.2. Linear Optimal Control Strategy

When a linear system model is known, if you want the system to achieve an optimal control effect, you can introduce the optimal quadratic control index of the linear system, and equation (4) is the optimal objective function.

$$J(u)=\frac{1}{2}\int_0^{\infty}\left[x^T(t)Qx(t)+u^T(t)Ru(t)\right]dt \tag{4}$$

In the formula: Q is a semi positive definite real symmetric matrix, which is the weighting matrix of the target requirement term; R is a positive definite real symmetric matrix, which is the weighting matrix of the control signal term.

In LQR theory, the necessary and sufficient condition for its optimal control is

$$u^*(t)=-R^{-1}(t)B^T(t)P(t)x(t)=-Kx(t) \tag{5}$$

In the formula: K is the optimal feedback gain matrix $K=[k_1\ k_2\ k_3]^T$, a is a constant valued positive definite matrix, and must satisfy the Riccati algebraic equation.

$$PA+A^TP-PBR^{-1}BP+Q=0 \tag{6}$$

Therefore, we can transform the problem of solving the optimal control into solving the Riccati equation, that is, we can find the feedback gain matrix K.

While the microgrid operates stably, the switched DC-DC converter will inevitably be disturbed by some known or unknown disturbances. In order to ensure the stable operation of a system, even when the unknown disturbance is a continuous and slowly changing disturbance, it is very necessary to apply a controller to the system.

In practical engineering, the types of systems are often distinguished and determined by different forms of open-loop transfer function of the system, which are generally divided into type 0, type I and type II. The buck and boost DC-DC converters discussed in this paper are 0-type systems, that is, when the input is a step signal, the system output will have steady-state error. In order to eliminate this steady-state error, it is necessary to include an "integrator" in the feedback path.

In order to realize the simultaneous control of state feedback parameters and integrator parameters by LQR, a new system model must be reconstructed. That is, first define a new state variable:

$$X^*=\begin{bmatrix}\dot{X} & e\end{bmatrix}=\begin{bmatrix}\dot{U}_c & \dot{I}_L & e\end{bmatrix}^T \tag{7}$$

Then the new system state equation can be written as:

$$\begin{cases} X^*=A^*X^*+B^*v \\ Y=C^*X^*+D^*v \end{cases} \tag{8}$$

In the formula:

$$v=\dot{u}=-KX^*=-\begin{bmatrix}k_1 & k_2 & k_3\end{bmatrix}\begin{bmatrix}\dot{U}_c & \dot{I}_L & e\end{bmatrix}^T \tag{9}$$

Then the new system is represented as equation (10):

$$A^* = \begin{bmatrix} A & 0 \\ -C & 0 \end{bmatrix} , \quad B^* = \begin{bmatrix} B \\ 0 \end{bmatrix}$$
$$C^* = \begin{bmatrix} C & 0 \end{bmatrix} , \quad D^* = D \tag{10}$$

Then the system control block diagram after defining the new state variable can be represented as Figure 1.

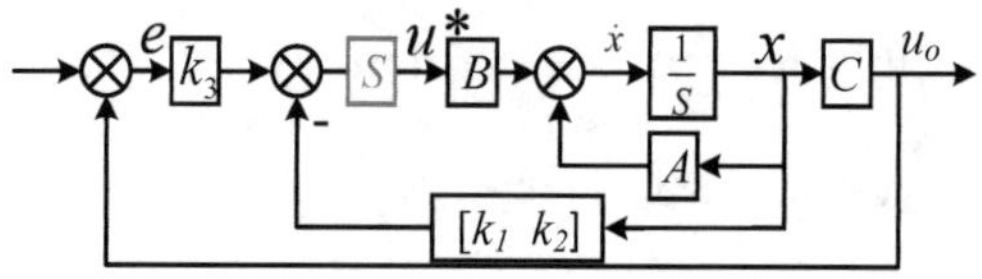

Figure 1. System control block diagram after defining new state variables

In the control block diagram of Figure 1, the control variable becomes by transformation, so the optimal control problem can be expressed by the following formula:

$$\begin{aligned} J(u) &= \int_0^\infty \left(X^{*T} Q^* X^* + \dot{u}^T R^* \dot{u} \right) dt \\ &= \int_0^\infty \left(he^2 + \dot{u}^2 \right) dt \\ &= \int_0^\infty \left(he^2 + v^2 \right) dt \end{aligned} \tag{11}$$

In the formula; $Q = \begin{bmatrix} 1 & 0 & 0 \\ 0 & 1 & 0 \\ 0 & 0 & h \end{bmatrix}$, h s the weighting coefficient of the required control target; $R^* = 1$.

Through the Q matrix, the tracking problem is transformed into a calibration problem that can be solved by LQR. In the LQR of the new system, three relevant parameter K lues can be solved by calling relevant statements through MATLAB.

3. DYNAMIC COMPENSATION CONTROL STRATEGY OF DC MICROGRID BASED ON LQR

3.1. Disturbance Suppression Theory Based on Residuals

Residual information is often used for fault diagnosis and location, as well as disturbance detection and suppression. Its expression is:

$$r(s) = y(s) - \hat{y}(s) \tag{12}$$

In the formula: $r(s)$ is the residual signal, $y(s)$ is the actual output of the system, and $\hat{y}(s)$ is the estimated output of the system.

The residual information can reflect the influence degree of disturbance and other faults on the system. Through the performance of the residual, the working state of the system can be judged. When $r(s) \neq 0$, it indicates that the actual output of the system deviates from the estimated output, which means that the system is disturbed or the system has faults. On the contrary, when $r(s)=0$, it means that the system is not affected by disturbance or fault.

Based on the Luenberger observer, the gain matrix L is calculated by pole assignment method, and the disturbance residual generator is established. The expression is equation (13).

$$\begin{cases} \dot{e}(s) = (A - LC)e(s) + Ed(s) \\ r(s) = Ce(s) \\ e(s) = x(s) - \hat{x}(s) \end{cases} \tag{13}$$

Where: A, C and E are the system matrix, output matrix and disturbance input matrix of the controlled object; Is the deviation between the actual output value and the estimated value of the system, that is, the state quantity of the disturbance observer; Is the differential of; L is the gain matrix of the observer, and its value is generally 3 ~ 4 times of the real part of the poles of the original system.

According to the robust double coprime decomposition and Euler parameter stabilization controller theory.

$$u(s) = u_o(s) + Q(s)r(s) \tag{14}$$

In the formula, $u(s)$ is the input value of the controlled object; $u_o(s)$ is the output of the original controller; $Q(s)$ is the compensation controller.

Without affecting the stability of the original system, the residual information $r(s)$ generated by the disturbance observer is applied to the original system through the compensation controller $Q(s)$ and superimposed with the output signal of the integral controller to quickly suppress the disturbance of the system. That is, on the basis of equation (14), the LQR dynamic disturbance suppression structure based on residual is designed. The control structure is shown in Figure 2.

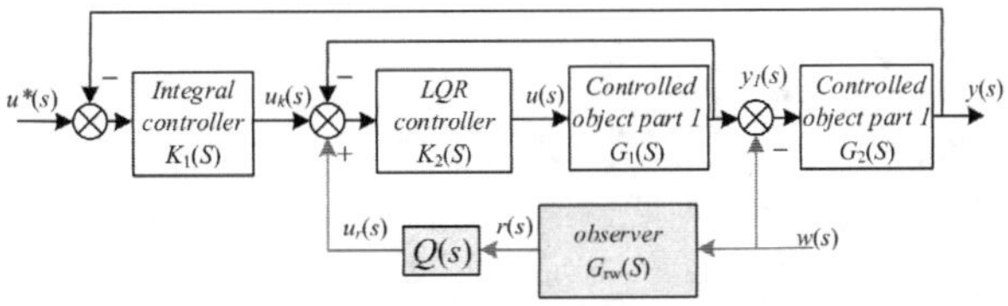

Figure 2. Dynamic disturbance suppression structure based on residual

When the disturbance $w(s)$ occurs, the compensation controller $Q(s)$ will quickly output the compensation signal $u_r(s)$ according to the disturbance and conduct disturbance suppression compensation through the LQR control loop. According to Figure 2, the overall transfer function expression of the disturbance suppression compensation structure based on the residual is obtained as formula (15).

$$\begin{cases} y_1(s) = \left[u_k(s) + G_{rw}(s)Q(s)w(s)\right]G_T(s) \\ G_T(s) = \dfrac{K_2(s)G_1(s)}{1 + K_2(s)G_1(s)} \\ y(s) = \left[y_1(s) - w(s)\right]G_2(S) \end{cases} \tag{15}$$

In the formula, $G_T(s)$ is the closed-loop transfer function of the combination of LQR controller and controlled object.

It can be seen from Figure 2 that the output of $Q(s)$ is not directly superimposed with the output of the converter double closed-loop controller, but the signal action point of $Q(s)$ is adjusted to the input side of the LQR controller and superimposed with the output signal of the integral controller of the original system to realize the suppression and compensation of system disturbance. Therefore, the compensation structure of Figure 2 can be equivalent to that shown in Figure 3.

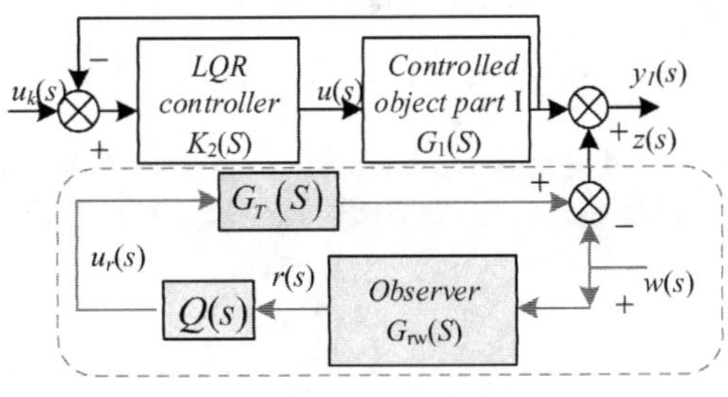

Figure 3. Equivalent block diagram of dynamic compensation structure

In Figure 3, the red dotted line is the equivalent block diagram of the compensation structure proposed in this paper. According to the superposition theorem, it is superimposed with the disturbance under the action of the compensation controller, so that $Z(s)=0$, so as to achieve the compensation effect of restraining the disturbance. That is, the compensation control structure can be expressed by formula (16).

$$z(s) = G_T(s)Q(s)G_{rw}(s)w(s) - w(s) = 0 \tag{16}$$

The solution expression (17) of the compensation controller can be written out from formula (16):

$$Q(s) = \frac{1}{G_T(s)G_{rw}(s)} \tag{17}$$

As can be seen from equation (17) and Figure 3, the compensation structure proposed in this paper only considers LQR controller and some control objects. When solving the transfer function of the compensation controller, only the transfer function of the inner loop of the controller and some compensation objects are required, which simplifies the solution of the compensation controller.

3.2. Dynamic Compensation Structure and Controller Solution of DC Microgrid Based on LQR

According to the above theory, the control block diagram of disturbance suppression compensation structure based on Buck and Boost DC-DC converters can be obtained, as shown in Figure 4.

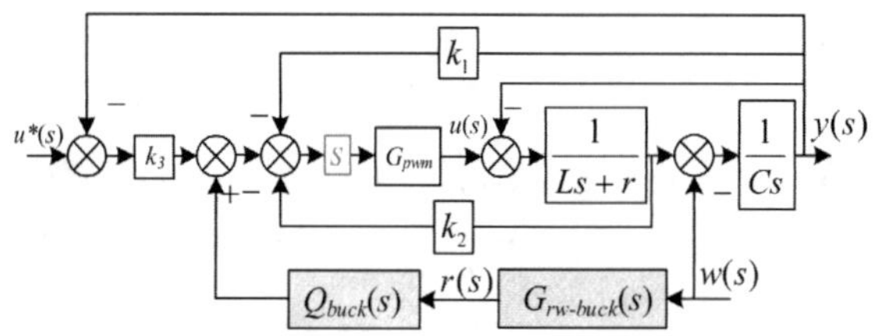

(a) Buck DC-DC converter

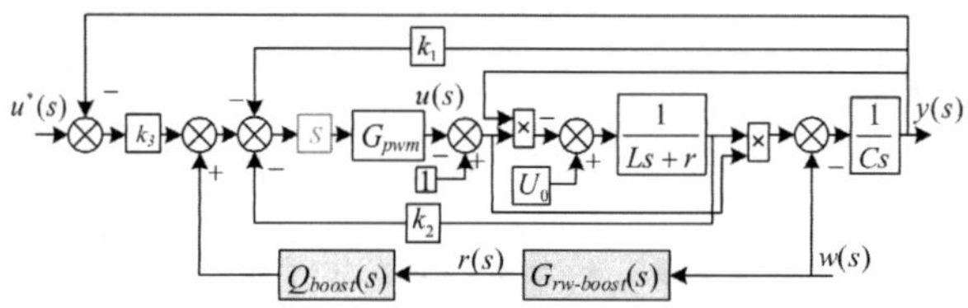

(b) Boost DC-DC converter

Figure 4. Control block diagram of disturbance suppression compensation structure of two DC-DC converters

G_{pwm} in Figure 4 is the equivalent module of PWM link, where K_{pwm} in formula $G_{pwm} = \frac{K_{pwm}}{0.5Ts+1}$ is the equivalent gain of PWM, and according to PWM principle, when the load amplitude value is equal to the input value of voltage source, K_{pwm} can be equivalent to 1. T is the sampling period of the system.

According to the control block diagram shown in Figure 4, the closed-loop transfer function $G_{T\text{-}Buck}(s)$ of the inner loop controller corresponding to the Buck DC-DC converter can be obtained from the Mason Gain Formula.

$$G_{T\text{-}buck}(s) = \frac{CG_{pwm}s}{LCs^2 + (rc + k_2 CG_{pwm})s + k_1 G_{pwm} + 1} \tag{18}$$

Because boost DC-DC converter is a non-minimum phase system, when model matching is used to solve compensator $Q(s)$, a positive pole will be generated in the transfer function of the model, which will affect the stability of the system. However, the compensation position of the compensation controller $Q(s)$ to the system compensation is adjusted to the input node of the LQR controller, which avoids the problem of converting the positive zero point into the positive point in the boost DC-DC converter and ensures the stability of the system.

$$G_{T-boost} = \frac{U_c^* d^* G_{pwm} Cs}{LCs^2 + (rC - U_c^* G_{pwm} Ck_1)s - U_c^* d^* G_{pwm} k_2} \tag{19}$$

Based on the linearized system state space model, a disturbance observer is established and transformed into a transfer function model. Obviously, the system is completely controllable, so the pole assignment method can be used to determine the state feedback matrix $L=[l_1 \quad l_2]^{T}$ of the disturbance observer. Therefore, the transfer function expression of the DC-DC converter disturbance observer is formula (20).

$$\begin{aligned} G_{rw-buck}(s) &= C\left(sI - A + LC\right)E \\ &= \frac{-(Ls+r)}{LCs^2 + C(Ll_1 + r)s + Cl_1 r + 1 + Ll_2} \end{aligned} \tag{20}$$

At the same time, considering the method of pole assignment, a state observer is established for boost DC-DC converter, and the transfer function of the state observer can be obtained as formula (21).

$$G_{rw-boost} = \frac{-(Ls+r)}{LCs^2 + C(Ll_1 + r)s + Cl_1 r + \frac{Ll_2}{2} + \frac{1}{4}} \tag{21}$$

From equation (17) and the red dashed box in the equivalent block diagram of dynamic compensation structure in Figure 4, two solution structures of DC converter disturbance suppression compensator can be obtained.

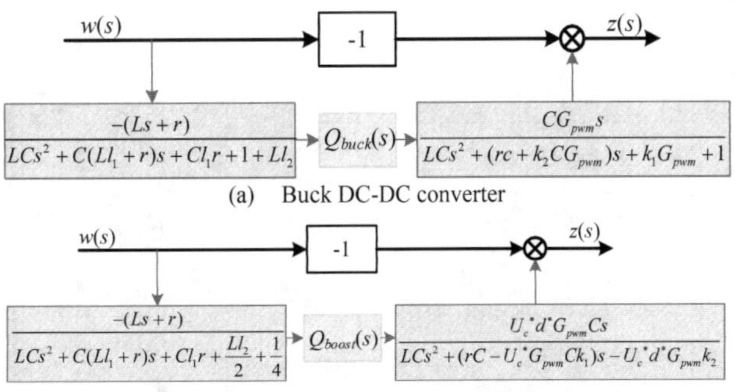

(a) Buck DC-DC converter

(b) Boost DC-DC converter

Figure 5. solution structure of disturbance suppression compensator

From (a) in the Figure 5, the compensation controller formula is equation (22).

$$Q_{buck}(s) = \frac{1}{G_{T-buck} G_{rw-buck}} = \frac{m_4 s^3 + m_3 s^2 + m_2 s + m_1}{n_2 s + n_1} \tag{22}$$

See equation (23) below for each matrix element in formula (22).

$$\begin{cases} m_1 = (r - G_{pwm} k_2)(rcl_1 + 1 + Ll_2) \\ m_2 = L(rCl_1 + 1 + Ll_2) + C(r - k_2 G_{pwm}) \\ m_3 = L^2 Cl_1 + 2LCr - LCG_{pwm} k_2 \\ m_4 = L^2 C \end{cases} \\ \begin{cases} n_1 = LG_{pwm} \\ n_2 = rG_{pwm} \end{cases} \tag{23}$$

From (b) in the Figure 5, the compensation controller formula is equation (24).

$$Q_{boost}(s) = \frac{1}{G_{T-boost}G_{rw-boost}} = \frac{m_4 s^3 + m_3 s^2 + m_2 s + m_1}{n_2 s + n_1} \tag{24}$$

The matrix elements in formula (24) are shown in the following formula.

$$\begin{cases} m_1 = (r + U_c^* G_{pwm} k_1)(rCl_1 + \frac{1}{2}Ll_2 + \frac{1}{4}) \\ m_2 = L(rCl_1 + \frac{1}{2}Ll_2 + \frac{1}{4}) + C(r + U_c^* G_{pwm} k_1)(Ll_1 + r) \\ m_3 = LC(Cl_1 + 2r + U_c^* G_{pwm} k_1) \\ m_4 = L^2 C \end{cases}$$
$$\begin{cases} n_1 = -d^* U_c^* G_{pwm} r \\ n_2 = -d^* U_c^* G_{pwm} L \end{cases} \tag{25}$$

At the same time, considering the physical realizability, the second-order link $G_\delta = \frac{1}{(\delta s + 1)^2}$ is introduced, which is a very small constant. That is, the new formulas of the compensation controllers of the two DC converters can be expressed in the following equation forms respectively.

4. EXPERIMENTAL SIMULATION VERIFICATION

4.1. Parameter Design

In the experimental verification, the parameter size of the converter has a non-negligible impact on the stable operation of the converter. According to the parameter design principle and calculation method, the relevant parameters of the main circuit of the simulation experiment are determined, as shown in TABLE Ⅰ.

TABLE I. MAIN CIRCUIT PARAMETERS OF SIMULATION EXPERIMENT

parameter	**Buck DC converter**	**Boost DC converter**
Inductance resistance /Ω	0.1	0.01
inductance /H	0.002	0.004
capacitance /F	0.001	0.03
Duty cycle	0.5	0.5
input voltage /V	100	100
DC voltage /V	200	50
load /Ω	5	5
Switching load /Ω	5	5

4.2. Simulation Verification

In order to verify the compensation effect of the DC converter dynamic compensation architecture based on linear optimal quadratic control designed in this paper, different kinds of experiments are designed, and MATLAB/Simulink simulation is used to verify the rationality of the compensation architecture.

4.2.1. Comparison Between PID Control and LQR Control

According to the main circuit parameters listed in Table Ⅰ, the control matrix K can be obtained by calling related statements in MATLAB. In order to highlight the advantages of LQR control, the comparison test between LQR control and classical PID control is designed to verify. In order to verify the effectiveness of the proposed dynamic compensation structure, the load is switched in a certain time to simulate the sudden disturbance of the system, and compare the operation state of the converter with or without the compensation structure. There are two different stages in this simulation: the first stage is when the load changes from 5 Ω to 2.5 Ω at 0.4s; In the second stage, the load changes from 2.5 Ω to 5 Ω at 0.7s. At the same time, the other parameters of the circuit remain unchanged during the whole simulation stage. The transient performance of output side voltage and current of the two DC converters during load switching in two stages and four control modes are compared respectively.

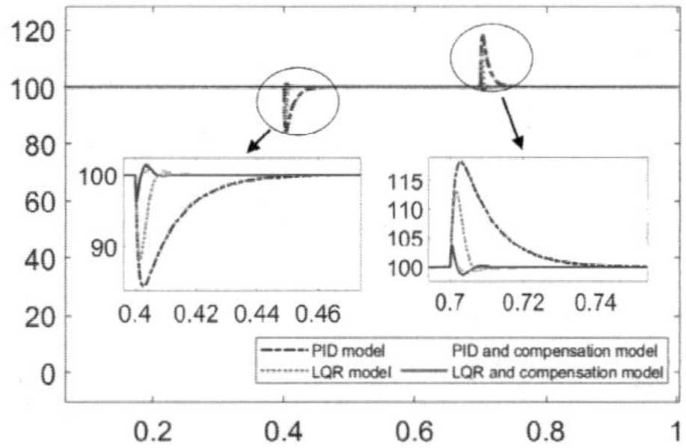

(a) Output voltage diagram

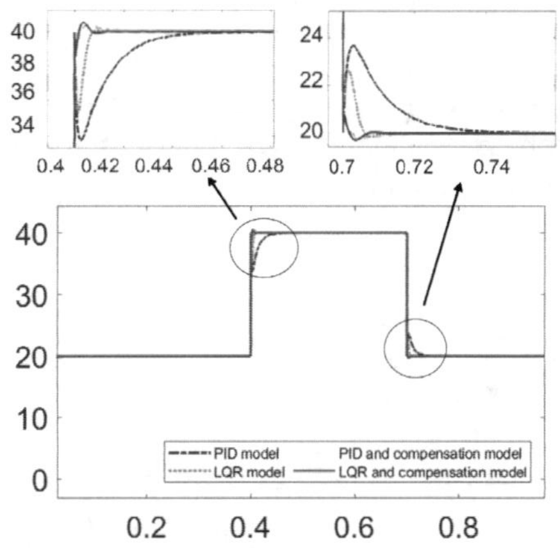

(b) Output current diagram

Figure 6. Comparison of four control modes of buck DC converter

As can be seen in Figure 6, when the 0.4s load changes from 5 Ω to 2.5 Ω, the load resistance decreases and the steady-state operation of the converter is disturbed. At the moment of switching the load, the output side voltage remains unchanged, the load becomes the original load value, and the output current rises from 20a to 40A. Affected by the disturbance current, the output voltage drops, and then the capacitor releases energy, and the output voltage gradually rises to 100V. It can be clearly seen in Figure 6 that there are obvious differences in the control effects under the four control modes. In order to make the simulation results more convincing, the simulation data under four control modes are listed.

TABLE II. SIMULATION DATA OF FOUR CONTROL MODES OF BUCK DC CONVERTER

		PID model	**LQR model**	**PID and compensation model**	**LQR and compensation model**
The load changes from 5 Ω to 2.5 Ω in 0.4 seconds	Maximum voltage drop /V	84.39	88.19	96.23	96.3
	Stability	0.42	0.41	0.4003	0.4003

	time /S				
	Maximum current drop /A	33.76	35.27	38.49	38.51
	stability time /S	0.4061	0.4015	0.4004	0.4004
The load changes from 2.5 Ω to 5 Ω in 0.7 seconds	Maximum voltage rise /V	118.3	113.2	103.8	103.7
	stability time /S	0.704	0.7016	0.7004	0.7004
	Maximum current rise /A	23.66	22.64	20.77	20.75
	stability time /S	0.7156	0.704	0.7004	0.7004

In the experimental simulation data in TABLE Ⅱ, it can be seen that after 0.4s load change, the output voltage and output current under PID control mode drop to 84.39v and 33.76a, while the voltage drop under LQR control, PID control compensation and LQR compensation is more than 90.

Therefore, both PID compensation and LQR compensation have a certain inhibitory effect on the voltage and current drop amplitude. Compared with the three control modes of PID control, LQR control and PID compensation, LQR compensation has a very obvious improvement on the compensation speed of voltage and current. The superiority and the advanced nature of LQR compensation control mode are proved.

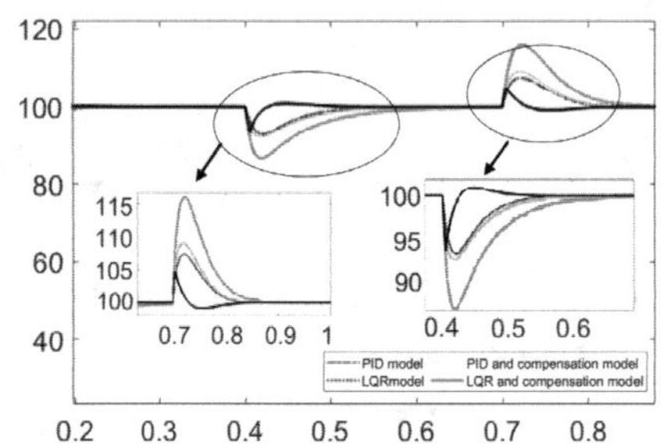

(a). Output voltage diagram

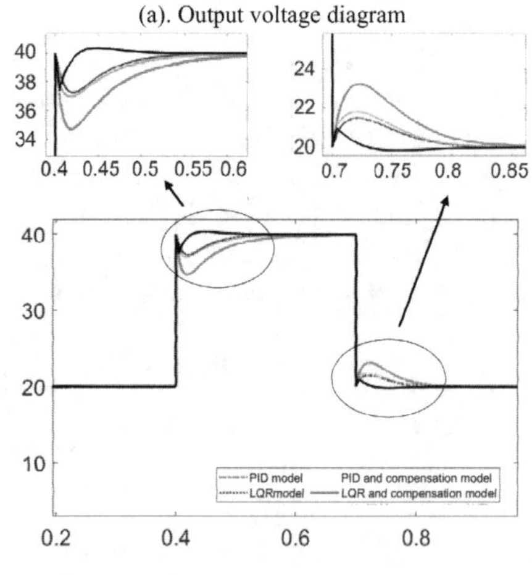

(b). Output current diagram

Figure 7. Comparison of four control modes of boost DC converter

As can be seen in Figure 7, when the 0.4s load changes from 5 Ω to 2.5 Ω, the load resistance decreases and the steady-state operation of the converter is disturbed. The output current rises from 20a to 40A. Affected by the output current, the output voltage drops. Correspondingly, the capacitor immediately releases energy, and the output voltage gradually rises to 100V. However, it is obvious that due to different control modes, there are also obvious differences in the voltage and current drop amplitude at the output side and the compensation speed between the two. In order to more intuitively reflect the differences between the four control modes, the experimental data are listed and compared to realize quantitative analysis.

TABLE III. SIMULATION DATA OF FOUR CONTROL MODES OF BOOST DC CONVERTER

		PID model	**LQR model**	**PID and compensation model**	**LQR and compensation model**
The load changes from 5 Ω to 2.5 Ω in 0.4 seconds	Maximum voltage drop /V	86.75	92.48	93.12	93.66
	stability time /S	0.48	0.45	0.44	0.40
	Maximum current drop /A	34.61	36.9	37.23	37.36
	stability time /S	0.48	0.45	0.44	0.41
The load changes from 2.5 Ω to 5 Ω in 0.7 seconds	Maximum voltage rise /V	116	109	107.4	104.9
	stability time /S	0.78	0.76	0.74	0.7
	Maximum current rise /A	23.23	21.4	21.84	20.96
	stability time /S	0.78	0.76	0.75	0.7

In the experimental simulation data in TABLE III, it can be seen that after 0.4s load change, the output voltage under PID control mode drops to 86.75v, while the voltage drop under LQR control, PID control compensation and LQR compensation is more than 90. Therefore, PID control, LQR control and compensation structure have a very good inhibitory effect on the voltage and current drop amplitude. Compared with the three control modes of PID control, LQR control and PID compensation, LQR compensation has a very obvious improvement on the compensation speed of voltage and current. The superiority and the advanced nature of LQR compensation control mode are proved.

4.2.2. Power Fluctuation Experiment

By harmonic injection on the right side of the converter, the power fluctuation of the DC microgrid is simulated, and the running state of the converter under different control modes is compared, and the superiority and the advanced nature of the LQR compensation control mode are verified.

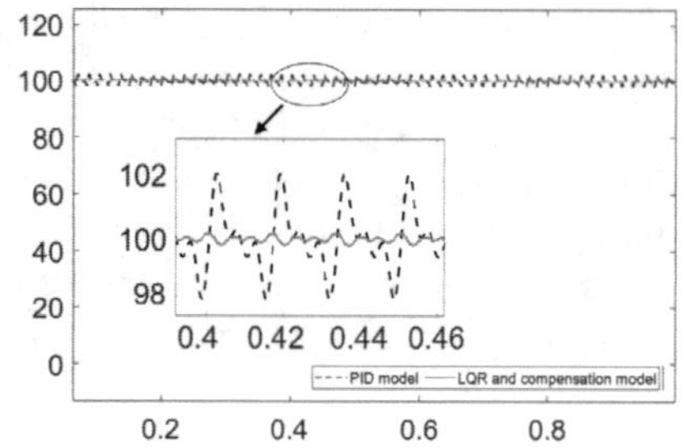

(a). Comparison diagram of output voltage of buck converter

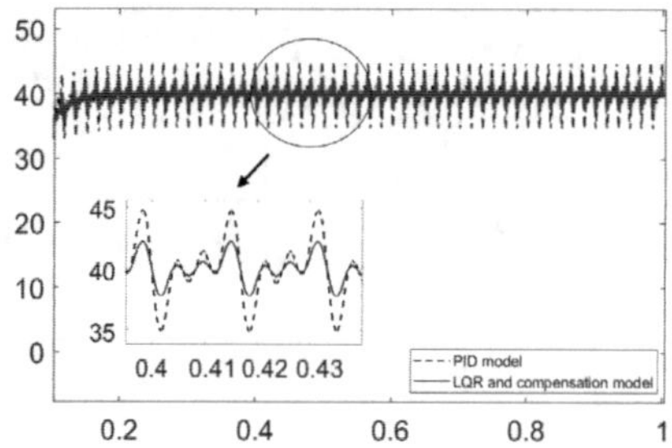

(b) Comparison diagram of output current of boost converter

Figure 8. Operation state of converter under two control modes

As shown in Figure8 (a) and (b), after injecting harmonics into the two DC converters, compared with the PID control mode, the LQR control compensation structure has a better suppression effect on the fluctuation amplitude of voltage and current. The superiority and the advanced nature of LQR compensation control mode are proved.

5. Conclusions

Under the background of microgrid, this paper selects boost and buck DC-DC converters in DC microgrid as the research object. The main purpose is to solve the power quality problem of microgrid, solve the optimal parameters of controller through LQR, so that the output voltage of DC converter can obtain the best steady-state performance and dynamic performance, and add dynamic compensation structure, The anti-interference performance of DC microgrid is further improved, the impact of load switching on output voltage is effectively solved, and the stability of microgrid itself is enhanced.

Acknowledgment

PR China, the National Key R&D Program of China, Key Special Projects for International Cooperation in Science and Technology Innovation between Governments (No.2021YFE0103800).

References

[1] Samanta S, Mishra J P, Roy B K. Virtual DC machine: an inertia emulation and control technique for a bidirectional DC–DC converter in a DC microgrid[J].

[2] Li Xialin, Guo Li, Zhang Shaohui, et al. Observer-based DC voltage droop and current feed-forward control of a DC microgrid[J]. IEEE Transactions on Smart Grid, 2018, 9(5): 5207-5216.

[3] H. Luo, X. Yang, M. Krueger, S. X. Ding and K. Peng, "A Plug-and-Play Monitoring and Control Architecture for Disturbance Compensation in Rolling Mills," in IEEE/ASME Transactions on Mechatronics, vol. 23, no. 1, pp. 200-210, Feb. 2018, doi: 10.1109/TMECH.2016.2636337.

[4] H. Luo, M. Krueger, T. Koenings, S. X. Ding, S. Dominic and X. Yang, "Real-Time Optimization of Automatic Control Systems With Application to BLDC Motor Test Rig," in IEEE

Transactions on Industrial Electronics, vol. 64, no. 5, pp. 4306-4314, May 2017, doi: 10.1109/TIE.2016.2577623.

[5] C. Yfoulis , D. Giaouris , F. Stergiopoulos , C. Ziogou , S. Voutetakis and S. Papadopoulou, "Optimal switching Lyapunov-based control of a Boost DC-DC converter," 2015 23rd Mediterranean Conference on Control and Automation (MED), 2015, pp. 304-309, doi: 10.1109/MED.2015.7158767.

[6] Frank H. F. Leung , Peter K. S. Tam, d C. K. Li. An Improved LQR-based Controller for Switching Dc-dc Converters [J]. IEEE Transactions on Industrial Electronics,1993,40(5):521-528.

Journal of Physics: Conference Series **2215** (2022) 012005 doi:10.1088/1742-6596/2215/1/012005

Energy Saving Control of Integrated Energy Multi-Energy Complementary Coupling Space Heating System Based on Fuzzy Theory

Changbin Hu[a], Xinyu Zhao*, Shanna Luo [b], Xiaojun Lu [c]

North China University of Technology, College of Electrical and Control Engineering, Beijing, China

[a]changbinlove@163.com

* 1621482395@qq.com

[b]14815390@qq.com

[c]863292186@qq.com

Abstract—Because the traditional solar-ground source heat pump heating system uses temperature difference controllers to control most, it causes the problem of large energy consumption of the equipment. In order to improve such problems, the complexity of the space heating control system is considered comprehensively, and the fuzzy strategy is used to improve the previous control system. TRNSYS was used to build a heating system simulation model, MATLAB was used to build a solar heating system fuzzy controller model, and the two models were coupled to jointly simulate the internal temperature change and energy consumption of the space under the same environmental conditions and different controllers. Experiments show that under the design conditions, the solar collector is greatly affected by the light, the temperature of the hot water storage tank is controlled between 58°C and 65°C. Using the traditional temperature difference controller, the system COP (Coefficient Of Performance) is low and the energy consumption is large, and the economic benefits are not high; the use of the fuzzy controller can increase the system COP by 6.1%, reduce the energy consumption by 22.85%, and save 8.71 yuan in the daily economic benefits of the system.

1. INTRODUCTION

At present, building energy consumption is increasing year by year, among which the energy consumption of building heating and cooling is the most, accounting for 55% of the total building energy consumption. Compared with the last century, the proportion of building energy consumption in the total energy consumption has increased from 10% to 46%. Therefore, reducing building energy consumption is a problem to be solved in recent years. The 14th Five-year Plan emphasizes the modern energy system of "clean, low-carbon, safe and efficient", speeds up the realization of energy transformation, strengthens the mutual combination and complement of "wind, light, water, fire and storage" to enhance the utilization rate of energy, solves the problem of the difficulty of absorbing renewable energy, and reduces energy consumption and energy cost. Therefore, strengthening the application of solar energy and ground source heat pump technology in buildings has become the main popular way of cooling and heating. Solar energy is an inexhaustible natural resource. It is clean, renewable and easy to collect. Sunlight can be irradiated to the absorber through the transparent cover plate and then converted into heat energy for heating. The ground source heat pump can transfer energy from low temperature heat source to high temperature heat source after the input of a small amount of high grade energy, to achieve cooling or heating, not only good stability, energy saving and efficient, long service life, but also a multi-purpose machine, environmental protection and renewable. But correspondingly it is very limited by the field, there is not enough space to realize the energy exchange is very difficult. Therefore, the combined application of solar energy and ground source heat pump for heating can solve each other's shortcomings, and give full play to each other's advantages [1].

In terms of single solar space heating system, literature [2] has analyzed the heating in cold regions, and the heat collection can be effectively improved from the Angle of installation angle. As for the heating system of a single ground source heat pump, literature [3] optimized the analysis from the depth of buried pipe, and concluded that the heat exchange effect of buried pipe increases with the increase of drilling depth, geothermal gradient and soil thermal conductivity. Coupling space in solar energy - ground source heat pump heating technology, the literature [4] studied the heating characteristics, indicates that the system, when combined with heating only high soil temperature can better ensure its effective operation, improve the COP of the system, the solar heating subsystem can make the soil temperature quick recovery after joining, the problem of soil temperature imbalance after long-term operation of the unit is avoided.

In recent years, in most space heating coupling systems, the traditional temperature closed-loop control method is basically used to achieve the system temperature variation flow control function, which is not high in intelligence, high energy consumption and uneconomic. Therefore, this paper takes a building in Beijing as the object to study the control strategy of comprehensive energy multi-energy complementary coupling space heating system. The TRNSYS model and MATLAB model are used for mutual coupling [5-7], and the traditional temperature closed-loop control method and fuzzy control algorithm [8] are proposed for joint control. The simulation effect of the system is analyzed. Compared with the traditional control results, the effectiveness and advantages of the proposed algorithm are verified.

2. SPACE HEATING SYSTEM

2.1. Composition of space heating system

In this paper, the system is mainly composed of solar collector, layered heat storage water tank, variable frequency water pump, fixed frequency water pump, flow divider, confluent three-way valve, heater, buried pipe and heat pump, which makes full use of natural resources to realize heating in winter and realize the heating demand of users all year round.

As showned in Figure 1, there are mainly two heating modes in this system. The first heating mode is night heating, when the solar heating system is closed, and the heater heating system assists the ground source heat pump heating system. If the heating quantity is sufficient, the heating system is closed, otherwise it is turned on. The second kind of daytime heating, at this time, the solar heating system is open, mainly by the heater heating system auxiliary ground source heat pump and solar heating system combined heating, if the solar energy and the ground source heat pump heating volume is sufficient, the heater will not start, otherwise open. The specific process is that the solar collector absorbs the heat energy released by the sun and heats the fluid internal heat source, while the fluid heat source is heated by the heater. The heat source of the two parts is gathered into the stratified water tank 1 through the confluence valve, and the cold working medium after heat exchange flows to the respective heating system for heating again through the shunt valve. In addition, the heating medium at the outlet of the stratified water tank 1 and the ground source heating medium simultaneously enter the heat pump for heat exchange, and the heating medium at the outlet of the heat pump enters the stratified heat storage tank 2 for heat supply to the room.

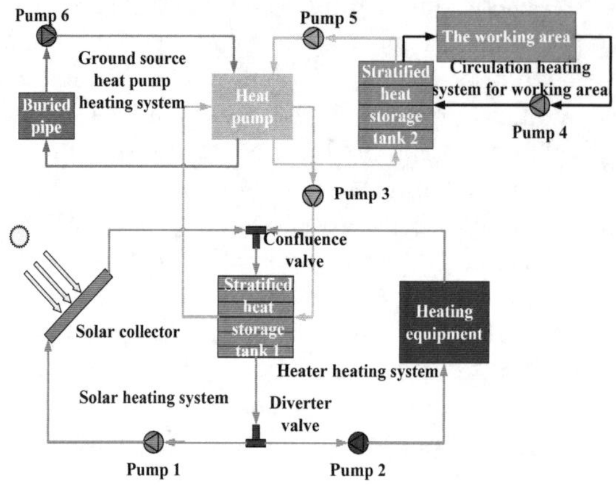

Figure 1. Topology of working district heating system

2.2. Control principle of space heating system

The operation of each subsystem of the space heating system is showed in Figure 2-5.

2.2.1. Solar heating system

Figure 2 is the logical topology of the solar heating system. The controller collects two kinds of data, namely, the outlet temperature of the heat transfer working medium of the solar collector and the outlet temperature of the heat transfer working medium of the heat storage tank, calculates the logical difference value, and controls the start and stop of the water pump, so that the heat transfer working medium can carry out circulation heating in the system.

2.2.2. Ground source heat pump heating system

Figure 3 is the logical topology of the ground source heat pump heating system. The input of the controller has two kinds of data, namely the temperature of the heat transfer working medium and the expected temperature. The 0/1 signal is calculated by the controller to control the start and stop of the water pump.

2.2.3. Heater heating system

Figure 4 is the logical topology of the heating system of the heater. The main controller receives two signals, namely the temperature of the heat exchange medium required by the heat storage tank and the expected temperature, and outputs the control signal through internal calculation to control the start and stop of the water pump and auxiliary heating equipment at the same time.

2.2.4. Circulation heating system for working area

Figure 5 logical topology for work area circulation heating system, heating area air temperature set point and the working area of the actual air temperature feedback quantity input to the controller at the same time, through the controller calculates the signal, the start-stop control pump.

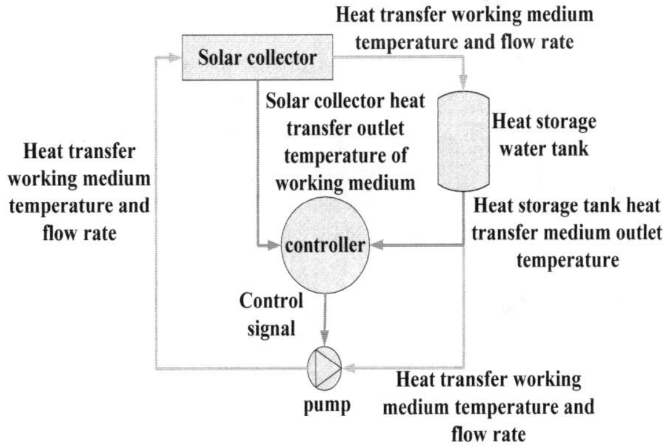

Figure 2. Logic diagram of solar heating system

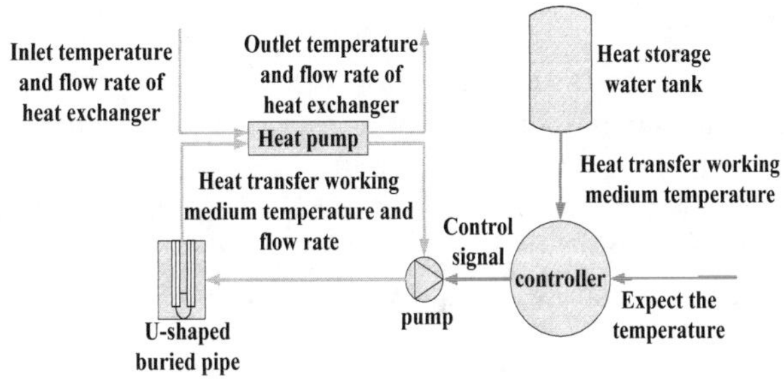

Figure 3. Logic diagram of ground source heat pump heating system

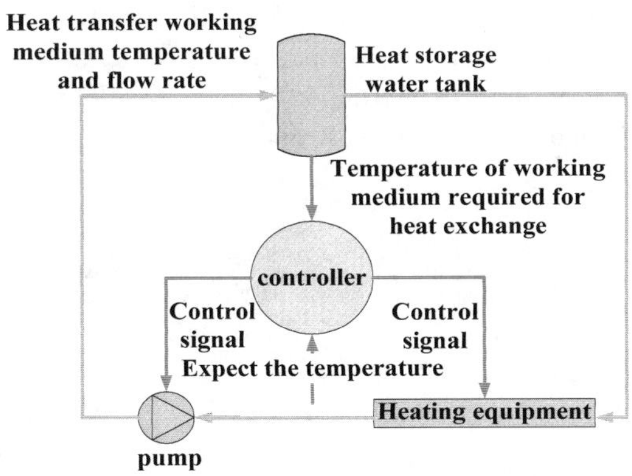

Figure 4. Logic diagram of heating system of heater

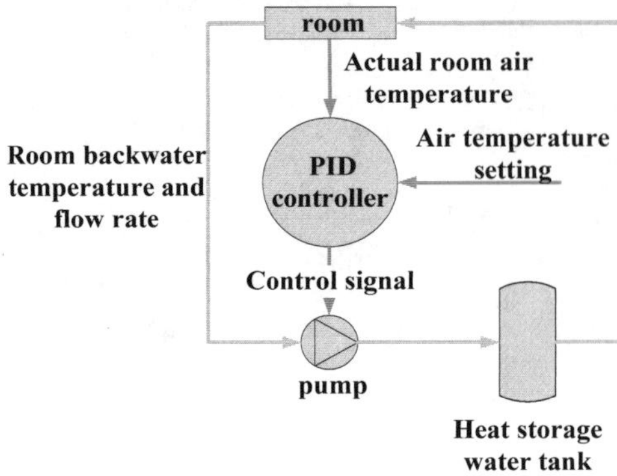

Figure 5. Logic diagram of circulation heating system in working area

2.3. Mathematical model of solar heating system

2.3.1. Mathematical model of solar collector

Solar collector is a collection of optical radiation energy and convert it into heat energy of a collection device, if according to the solar collector internal vacuum space form classification, mainly divided into two major categories of flat type vacuum tube collector and collector, it is on the basis of the first law of thermodynamics, the collection efficiency as an important index of performance evaluation of collectors, The collector efficiency is shown as follows:

$$\eta_c = F\left[\alpha\tau - U_L\left(\frac{T_p - T_a}{H_T}\right)\right] \tag{1}$$

Where, H_T solar radiation flux, W/m^2; T_p is the average temperature of solar collector, ℃;T_a is the ambient temperature,℃;U_L is heat dissipation loss coefficient, W/(m^2·℃);α is the absorption rate of solar radiation by the endothermic plate; τ is the transmittance of solar radiation from endothermic plate; F is heat collection efficiency factor; η_c is the thermal efficiency of solar collector.

The mathematical model of solar collector is expressed as [9] :

$$C_e K\frac{dT_{f,o}}{dt} = F_R\left[S - U_L\left(T_{f,i} - T_a\right)\right] + Gc_p\left(T_{f,i} - T_{f,o}\right) \tag{2}$$

Where, C_e is the effective heat capacity per unit area of the solar collector, *kJ/(m^2·℃)*; F_R is the heat transfer factor of solar collector; S is solar irradiance, *kJ/m^2*;T_a ambient temperature, ℃; $T_{f,i}$ is the inlet temperature of the solar collector, ℃; $T_{f,o}$ is the outlet temperature of the solar collector, ℃; C_p is the specific heat capacity of water at constant pressure, *kJ/(kg·℃)*;G is the flow rate of solar collector per unit area, *kg/h.*

2.3.2. Mathematical model of stratified heat storage tank

Stratified heat storage tank is a kind of heat storage device which can not only store excess heat but also make the working medium temperature of heating side and load side in a reasonable range. And on the basis of heat transfer working medium is different, the heat storage tank has a sensible heat storage, phase change thermal storage, chemical storage hot three thermal storage mode, the most common heat transfer working medium is water, due to poor thermal effect causes high temperature and low density of water to rise, high density in low temperature water drop, at this point in the heat storage water tank in the upper temperature is high, the lower the temperature low stratification, which is beneficial to improve the heating performance of the system, Improve the heat utilization rate. If the heat transfer working medium in the heat storage tank is evenly mixed, the temperature balance equation is [9] :

$$\left(mc_p\right)_s \frac{dT_s}{d\tau} = F_1\left(\dot{m}c_p\right)_G\left(T_{f,o} - T_s\right) - \left(UA\right)_s\left(T_s - T_a\right) - Q \quad (3)$$

Where, Q is the heating load provided by the heat storage tank, kW; F_1 is the collector control function; T_s is the temperature of heat storage tank, ℃; T_a is the ambient temperature, ℃; U is the average heat loss coefficient of water tank, $W/(m^2 \cdot K)$; A is the external area of the water tank, m^2;m is the water volume of the tank, kg; $(US)_s(T_s\text{-}T_a)$ refers to heat loss of heat storage tank.

2.3.3. Mathematical model of water pump

Water Pump in HVAC is an essential type of circulation equipment, which is mainly by receiving the control signal of the controller, judging and adjusting its speed, and then regulating the output flow, so as to meet the change of the load side demand. Pumps can be divided into fixed frequency pumps and frequency conversion pumps according to frequency, fixed frequency pump speed, energy consumption, has a greater impact on energy; Frequency conversion pump speed can be regulated according to the load, variable speed, low energy consumption, is the main trend of the industrial and agricultural field for circulation system equipment in today's era.

Variable frequency pump characteristic curve equation :

$$h = k_1 q^3 + k_2 q^2 + k_3 q + k_4 \quad (4)$$

The relationship between power, flow, head and speed of variable frequency water pump in operation is as follows:

$$\frac{P_1}{P_2} = \left(\frac{n}{n'}\right)^3 \quad (5)$$

$$\frac{h_1}{h_2} = \left(\frac{n}{n'}\right)^2 \quad (6)$$

$$\frac{q_1}{q_2} = \frac{n}{n'} \quad (7)$$

Where, k is the fitting coefficient of the pump characteristic curve equation; P is water pump power, kW; N is the pump motor speed,(r/min);H is the pump head; Q is the flow rate, kg/h.

Variable frequency pump effective power equation:

$$P_e = \frac{\rho g q h_1}{3.6 \times 10^6} \quad (8)$$

Where, P_e is the effective power of the variable frequency water pump, kW; ρ is the fluid density of heat transfer working medium, kg/m^3; G is the acceleration of gravity, m/s^2.

Energy dissipation equation of variable frequency water pump:

$$P = \frac{P_e}{\eta}\left(\frac{f}{f_0}\right)^3 + P_n \quad (9)$$

Where, f_0 is the power supply frequency, Hz; P_n is the frequency converter power, kW.

3. CONTROLLER DESIGN

3.1. Principle of fuzzy control

When using fuzzy controller, it is important to determine the input and output signals of the controller. The heating system is in a variable flow conditions, the fuzzy controller of the input signal is the temperature setting of the system and solar collector export working medium temperature, the output signal is of variable frequency pump control signal, when the fuzzy controller input signal is detected, the blur processing first, then on the basis of the designed fuzzy rules of fuzzy reasoning, Then the specific output control signal of the variable frequency pump is obtained by anti-fuzzy processing, and then the speed of the variable frequency pump is adjusted to obtain the different outlet flow rate of working medium. The fuzzy control block diagram is shown in Figure 6.

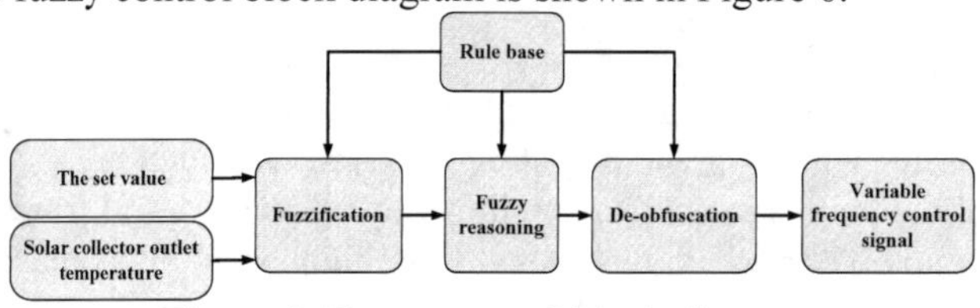

Figure 6. Fuzzy control block diagram

3.2. Fuzzy model building

3.2.1. Define the membership function

The triangular membership function is used to fuzzily transform the input variable set value (trimf) and divides the set value temperature into three fuzzy sets: SD (low temperature), MD (moderate temperature) and LD (high temperature). The value range is [65,80]. The membership function is as following:

$$\mu_{SD}(x)=\frac{73-x}{73} \tag{10}$$

$$\mu_{MD}(x)=\begin{cases}\dfrac{x}{73}\\ \dfrac{80-x}{73}\end{cases} \tag{11}$$

$$\mu_{LD}(x)=\frac{x-73}{73} \tag{12}$$

The input variable solar collector outlet temperature (trimf) is blurred by triangular membership function. The outlet working medium temperature of the solar collector is divided into three fuzzy sets: NG (low temperature), MG (medium temperature), and LG (high temperature). The value range is [-5, 95], and the membership function is as follows:

$$\mu_{NG}(y)=\frac{50-y}{50} \tag{13}$$

$$\mu_{MG}(y)=\begin{cases}\dfrac{y}{50}\\ \dfrac{95-y}{50}\end{cases} \tag{14}$$

$$\mu_{LG}(y)=\frac{y-50}{50} \tag{15}$$

The triangle membership function is used to fuzzier the output variable control signal (trimf), and the control signal is divided into five fuzzy sets: VS (very small), S (small), M (medium), L (large), VL (large), and the value range is [0, 1]. The membership function is as follows:

$$\mu_{VS}(z)=\frac{0.1-z}{0.2} \tag{16}$$

$$\mu_{S}(z)=\begin{cases}\frac{z}{0.2}\\ \frac{0.4-z}{0.25}\end{cases} \tag{17}$$

$$\mu_{M}(z)=\begin{cases}\frac{z-0.2}{0.25}\\ \frac{0.7-z}{0.25}\end{cases} \tag{18}$$

$$\mu_{L}(z)=\begin{cases}\frac{z-0.4}{0.25}\\ \frac{1-z}{0.3}\end{cases} \tag{19}$$

$$\mu_{VL}(z)=\frac{z-0.7}{0.3} \tag{20}$$

The fuzzy controller rules are shown in Table 1 below:

TABLE I. FUZZY CONTROL RULES TABLE

Control signal z		**Set temperature x/℃**		
		SD	***MD***	***LD***
Solar collector temperature y/℃	***NG***	VS	M	L
	MG	S	M	L
	LG	M	L	VL

3.2.2. Fuzzy control reasoning

- Rule matching

Assume that the sampling is carried out on the SDHW system at a certain time, the set temperature is 33℃, and the working medium temperature at the outlet of the solar collector is 30℃. After calculation, the membership degree can be obtained as following:

$$\mu_{SD}(33)=\frac{40}{73}\ \mu_{MD}(33)=\frac{33}{73}\ \mu_{NG}(30)=\frac{2}{5}\ \mu_{MG}(30)=\frac{3}{5}$$

TABLE II. SDHW SYSTEM RULE MATCHING TABLE

Control signal z		**Set temperature x/℃**		
		SD	***MD***	***LD***
Solar collector temperature y/℃	***NG***	$\mu_{VS}(z)$	$\mu_M(z)$	0
	MG	$\mu_S(z)$	$\mu_M(z)$	0
	LG	0	0	0

- Determination of rules

According to the fuzzy controller rule table, the following four rules can be triggered:

If (x is SD) and (y is NG) then (z is VS)

If (x is SD) and (y is MG) then (z is M)

If (x is MD) and (y is NG) then (z is S)

If (x is MD) and (y is MG) then (z is M)

- Credibility of rules

In the same rule, the rule conclusion can be obtained through the relationship between the premises and. The credibility of the general premise of each rule can be obtained by taking small operation between the premises: the credibility of the premise of rule 1 is $min\left(\frac{40}{73},\frac{2}{5}\right) = \frac{2}{5}$.The credibility of the premise of rule 2 is$min\left(\frac{40}{73},\frac{3}{5}\right) = \frac{40}{73}$......Thus, the prerequisite credibility table of SDHW system rules is shown in Table 3:

TABLE III. RELIABILITY TABLE

Control signal z		**Set temperature x/℃**		
		SD	***MD***	***LD***
Solar collector temperature y/℃	***NG***	$\frac{2}{5}$	$\frac{2}{5}$	0
	MG	$\frac{40}{73}$	$\frac{33}{73}$	0
	LG	0	0	0

- The reasoning of each rule.

Combine the above two tables to get the output of each rule (the confidence of the rule)

TABLE IV. RULES CREDIBILITY TABLE

Control signal z		**Set temperature x/℃**		
		SD	***MD***	***LD***
Solar collector temperature y/℃	***NG***	$\min(\frac{2}{5}, \mu_{VS}(z))$	$\min(\frac{2}{5}, \mu_M(z))$	0
	MG	$\min(\frac{40}{73}, \mu_S(z))$	$\min(\frac{33}{73}, \mu_M(z))$	0
	LG	0	0	0

- The total output of the fuzzy system (aggregation), take the union of the inference results of each rule, namely:

$$\mu_{all} = \max\{\min(\frac{2}{5}, \mu_{VS}(z)), \min(\frac{2}{5}, \mu_M(z)), \min(\frac{40}{73}, \mu_S(z)), \min(\frac{33}{73}, \mu_M(z))\} \quad (21)$$

- Deblurring.

The total output of the fuzzy system is actually the union of the inference results of the three rules. Use the maximum average method.

4. Simulation system analysis based on TRNSYS

4.1. Analysis of solar heating system

The figure below shows the temperature change of the heat transfer working fluid at the outlet of the solar collector and the hot water storage tank during a certain whole day during the heating season. It can be seen from Figure 7 that at night from 0:00 to 6:00, because there is no sun, the temperature of the working fluid at the outlet of the heat collector is continuously decreasing. At this time, the solar heat collection system is not circulating, and the control signal of the variable frequency pump is 0 as shown in Figure 8.After 6 o'clock, the temperature of the outlet of the heat collector gradually rises, and the control signal gradually when the temperature reaches about 50 ℃ at 11 o'clock, the control signal tends to be between 0.9 and 1. The temperature reaches about 90℃ at 15 o'clock. At this time, because the temperature is too high, the control signal of the variable frequency pump is reduced to reduce the flow output; After 15:00, the temperature of the collector outlet gradually decreases, and the control signal increases to meet the heating demand; Until the variable-frequency pump stops running at night, the collector temperature approaches 0°C. The hot water storage tank has been kept between 58°C and 65°C because the solar heat collection system is turned off at night, and the ground source heat pump alone may cause insufficient heat supply. At this time, the heater is turned on to keep the temperature of the water tank stable and supply Enough heat.

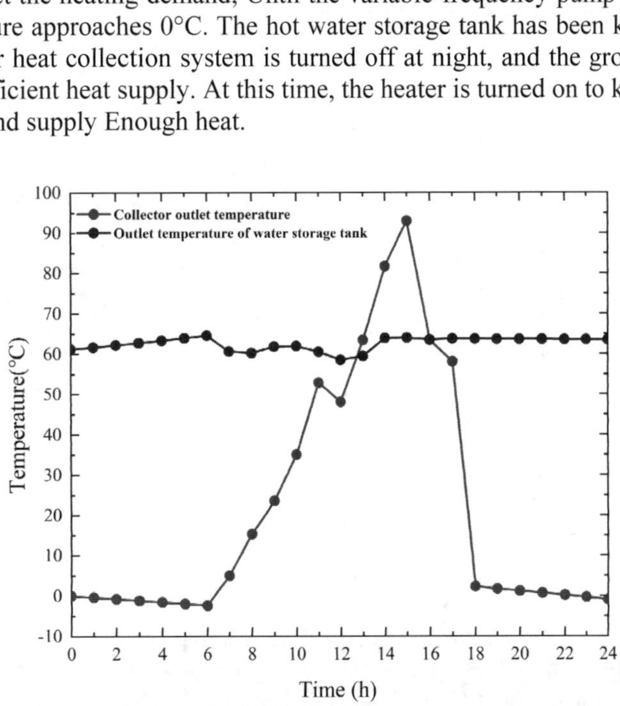

Figure 7. Temperature changes heating collecting system

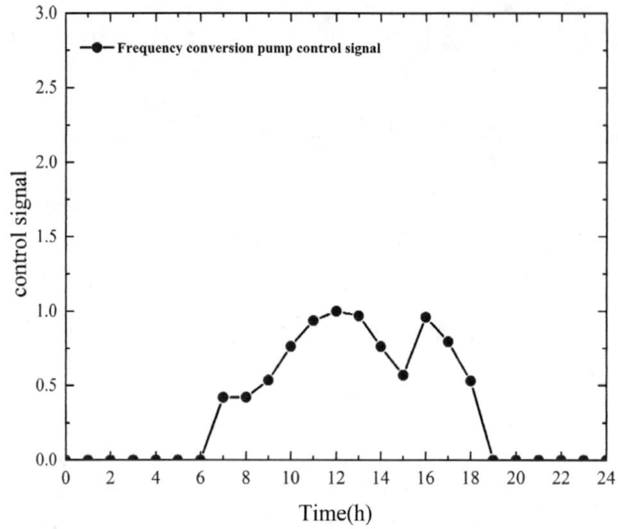

Figure 8. Frequency conversion control signals

4.2. System performance analysis

4.2.1. System COP comparison

COP of a heat pump system is a value used to evaluate whether a system is efficient and energy saving. The value is the ratio of the total heating power of the heat pump to the power of the compressor. The larger COP of the system is, more energy saving the system will be.

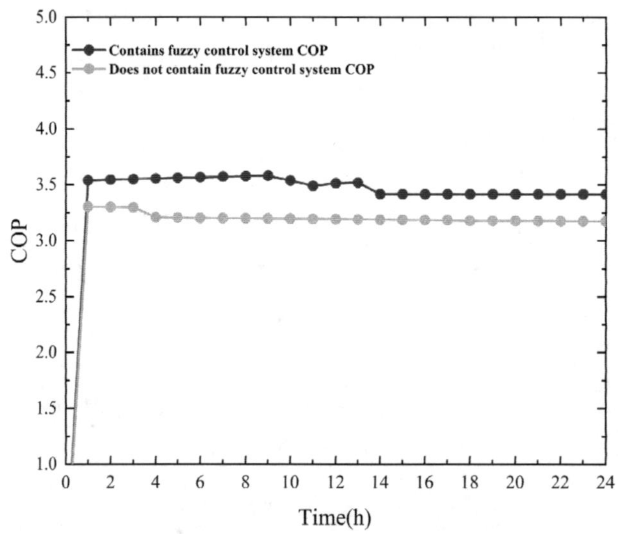

Figure 9. COP comparisons of the system

Figure 9 shows the comparison of COP changes of heating units in a whole day with and without the fuzzy control system. The fuzzy controller controls the frequency conversion system, while the system without the fuzzy control is controlled by temperature difference control, which controls the fixed-frequency system. When the controller detects that the water supply temperature is higher than the set value of water supply temperature by 1℃, the heating side of the system stops working, and when the water supply temperature is lower than the set value of water supply temperature, the system is in working state. In the figure, the green line is the change of COP without fuzzy control system, and the blue line is the change of COP with fuzzy control system. Can be clearly in the figure than the fixed frequency inverter system COP to an average about 0.2, this is because solar collector stop working at night, the ground source heat pump and auxiliary heater to work, if the amount of ground source heat pump heating is enough, the auxiliary heater does not need to open, or the auxiliary heating auxiliary heater. During the day, when the solar heat collection system is on, solar heating is preferred, and auxiliary heating is carried out on the ground source side. Making full use of solar energy to reduce the use of the ground source side can promote the recovery of underground temperature and improve the COP of the system. On the other hand, when operating under partial load, the flow rate of heat transfer working medium decreases and the flow resistance of heat transfer working medium in the heat exchanger decreases, making the decrease rate of heat supply lower than that of compressor power, so the COP of unit is improved.

4.2.2. Total energy consumption analysis of the system

Figure 10 heat pump for heating during the day, energy consumption, red curve with fuzzy control system part total energy consumption, all pumps blue curve for fuzzy control system does not contain all the pump total energy consumption, the figure can be seen that the energy dissipations of the red curve in the day is about 38.5 kWh, and blue curve energy consumed in a day is about 47.3 kWh. Pump energy consumption per day can reduce energy consumption by 22.85%. Figure 11 shows the total price in different time periods in Beijing and the total price of the whole system under different controls within a day. The blue curve is the case without fuzzy control of the total electricity price, the total electricity price in a day is 112.78 RMB, the yellow curve is the case with fuzzy control of the total electricity price, the total electricity price in a day is 104.07 RMB, which can reduce 8.71 RMB a day. In the same working condition, when the frequency system is given the control signal, because the unit water temperature is fixed, when the temperature changes, the pump speed is still constant, often running at full speed, resulting in energy waste; The frequency conversion system can change the flow according to the required temperature, so as to control the pump speed can change freely between zero and rated speed according to the demand, and reduce the energy consumption of the system.

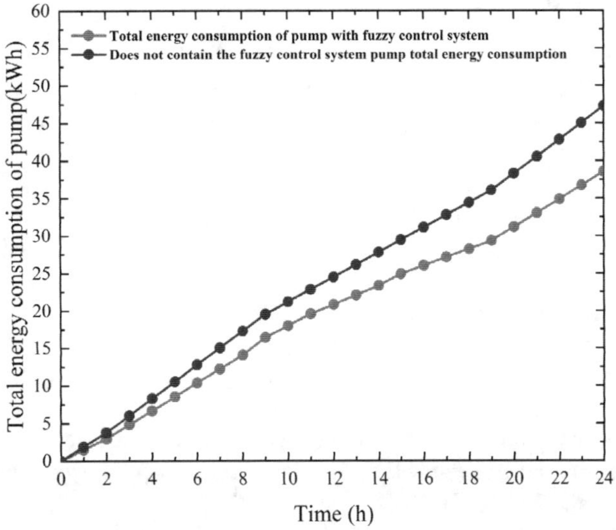

Figure 10. Comparison of total energy consumption of pumps in the system

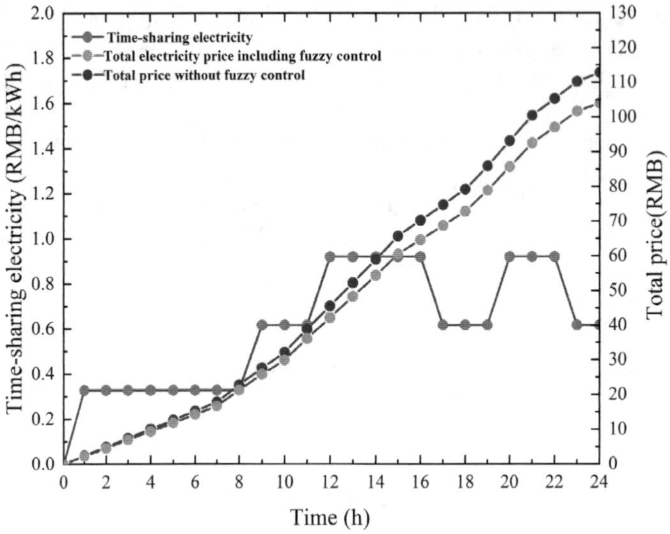

Figure 11. System total electricity price comparisons

5. Conclusions

In view of the operation characteristics of multi-energy complementary coupling space heating system of comprehensive energy, this paper proposes the control combining traditional control and fuzzy algorithm, and uses TRNSYS and MATLAB platform to carry out simulation verification, and draws the following conclusions:

(1) For the solar heating subsystem, the fuzzy algorithm can effectively control the operation of the water pump, realize the variable flow operation of the system, and effectively reduce the energy consumption of the system.

(2) For the overall multi-energy complementary heating system, the combined control of fuzzy and traditional control, compared with the single traditional control, better reflects the advantages of collaborative control, energy saving and emission reduction, system COP increased by 6.1%, overall energy consumption reduced by 22.85%, saving 8.71 yuan per day.

Acknowledgment

PR China, the National Key R&D Program of China, Key Special Projects for International Cooperation in Science and Technology Innovation between Governments (No.2021YFE0103800).

References

[1] Aleksandar Georgiev,Rumen Popov,Emil Toshkov. Investigation of a hybrid system with ground source heat pump and solar collectors: Charging of thermal storage and space heating[J]. Renewable Energy,2020,147(Pt 2).
[2] Liu Shaoliang. Study on rural residential solar heating System in Hebei [D]. Hebei University of Engineering,2012.
[3] Zhang Shan. Analysis and Optimization of Heat Supply System for Middle-deep Buried Pipe Heat Pump [D]. Shandong Jianzhu University,2020.
[4] Xu Hui. Study on characteristics of Solar energy and Ground source Heat Pump combined Heating for residential Buildings [D]. Shanghai University of Engineering Science,2016.
[5] KLEIN S, BECKMAN W, MITCHELLJ, et al.TRNSYS 16—of transient system simulation program, user manual[M]. Madison: University of Wisconsin-Madison, 2004.
[6] Farzin M. Rad,Alan S. Fung,Wey H. Leong. Feasibility of combined solar thermal and ground source heat pump systems in cold climate, Canada[J]. Energy & Buildings,2013,61.

[7] Lu Shiying, AI Wei, Chen Liding. Research on iterative learning control of central air conditioning based on Trnsys platform [J]. Computer measurement and control,2018,26(03):78-81+111.
[8] Zhang Weiguo, Yang Xiangzhong. Fuzzy Control Theory and Application [M]. Xi 'an: Northwestern Polytechnical University Press,1999.
[9] Li Shanke. Research on Solar-water Source Heat Pump Auxiliary Heating System [D]. Shenyang Jianzhu University,2011.

Endowing BIM Model with Mechanical Properties -- Finite Element Simulation Analysis of Long-Span Corrugated Steel Web Continuous Beam Bridge

X Y Zhang [1], Q L Qu[2], D Liang[3], T C Liu[2]

[1] Shanghai Interlink Road & Bridge Engineering Co.,Ltd., Shanghai 201315, China [2] CCCC Highway Bridges National Engineering Research Centre, Ltd. Beijing 100088, China

[3] Inner Mongolia Senior Highway Construction and Development Co., Ltd. Huhhot 010051, China

corresponding author's email: quqianglong@bnerc.com

Abstract: Taking the main bridge of Zhaojun Yellow River Super Large Bridge as the engineering background, this paper explores the application of BIM technology in the finite element simulation analysis of long-span corrugated steel web continuous beam bridge, and proposes to establish the model of corrugated steel web continuous beam bridge by using the parametric modeling method of Revit software, Dynamo visual programming software is used to extract the geometric parameter information of BIM model, and the method of data conversion between BIM model and finite element model is realized through Python language programming, so as to endow BIM model with mechanical properties, avoid repeated modeling, and improve modeling efficiency and accuracy. The finite element simulation of the converted model in Midas/Civil provides and support for bridge construction monitoring, and provides new ideas for the application of BIM technology and structural calculation in similar projects.

1. Introduction

BIM is the information technology revolution for the next time led by the engineering construction industry after CAD. It has the characteristics of visualization of operation, completeness of information, scalability, collaboration, simulation and drawing of information. Its application scope and depth in the entire life cycle of building structures have been deepened.

Structural finite element simulation is one of the key works of engineering design and construction monitoring. How to combine with structural finite element simulation is one of the key research directions of the in-depth application of BIM technology. Since the introduction of BIM technology into China, with the wide application of BIM technology, designers have more and more strong demand for data conversion between BIM model and finite element model. Many scholars have studied BIM Technology in structural finite element simulation[1][2][3][4]. Such as He Xiangping, Wang Hao and others[5] proposed a fast conversion method from BIM model to structural finite element model by using Revit API and C# language under the development environment of Visual Studio, and designed a model conversion program from Revit to MIDAS / Civil; taking ANSYS and ABAQUS software as the finite element analysis platform, Zhang Zheyuan[6] studied the transformation and analysis framework between BIM model and finite element model; Wang Xiongjue[7] tried to convert the BIM model established in rhino software into MCT command flow through programming, so as to realize the conversion between BIM model and Midas/Civil finite element model; Chen Zhiwei, Wu Kun and others[8] realized the conversion between Revit model and ANSYS finite element model in the form of APDL command flow; Chen Hao[9]studied the application of BIM in the finite element simulation calculation of Chaobaihe bridge. With the deepening of research and application, the difference of data format is still the biggest obstacle to the transformation between BIM model and finite element calculation model. How to transfer the three-dimensional model to finite element calculation software completely and efficiently through BIM modeling software, so as to realize the integration of design and calculation, is still the most concerned problem of designers.

2. Difficulties in construction monitoring

Corrugated steel web slab bridge arose in the 1980s. Since then, scholars at home and abroad have carried out a lot of research and practice on this new bridge form. In 2005, China independently built the first corrugated steel web plate bridge Liangguangshanpo River Highway Bridge. Since then, Inner Mongolia Changchonggou Bridge, Juancheng Yellow River Bridge, Xingtai main canal bridge of South-to-North Water Transfer and other bridges have been built, and the bridge span is also increasing. The practice shows that the corrugated steel web slab bridge has the advantages of good economic benefit, good durability, more reasonable structural stress, simple construction and short construction period.

Zhaojun Yellow River Super Large Bridge is a control project under construction for Baotou West connecting line of Baotou Dongsheng section reconstruction and expansion project of Baotou Maoming expressway. It is located at the junction of Baotou City and Ordos. The whole bridge has a total length of 4.3km and 20 connections in total. It is divided into main bridge, North Cross embankment bridge and South Cross embankment bridge, and beach approach bridge on the north and south banks. The length of the first connection of the main bridge is 1520m, and 85m + 9 × 150m + 85m corrugated steel web continuous beam, and the superstructure is an 11 span corrugated steel web prestressed concrete continuous beam bridge with a main span of 150m. The single girder adopts the form of single box and single chamber straight web, and the left and right box girders are symmetrical about the bridge centerline. The top plate width of single width box girder is 12.75m and the bottom plate width is 6.75m. After completion, it will become the longest beam bridge in China.

The PC composite box girder with corrugated steel web of the superstructure of the main bridge of Zhaojun Yellow River Super Large Bridge adopts the top and bottom plate dislocation construction method as shown in figure 1. During the construction process, the concrete top plate pouring of section n-1, the concrete bottom plate pouring of section n and the installation of corrugated steel web of section n + 1 are carried out. The difficulty of construction monitoring of Zhaojun Yellow River Super Large Bridge is the monitoring of internal force and geometric shape during the cantilever pouring of the main beam.

(a) (b)

Figure 1. Schematic diagram of the synchronous ectopic construction method.

3. Bridge finite element simulation based on BIM

With the help of Revit and its plug-in Dynamo, the data conversion between BIM model and Midas/Civil finite element model is realized by using MCT command flow as the intermediate medium.

3.1. BIM modeling software selection

The choice of BIM core modeling software is the first problem to realize the application of BIM technology. There are many kinds of BIM application software on the market, such as Revit software of Autodesk company, CATIA software of Dassault company, Bentley series software, etc. Revit software was first applied in the construction engineering industry. It is a modeling software widely used in the engineering construction industry. It has powerful functions of three-dimensional parametric modeling, data management and sharing. In recent years, its application in the bridge construction industry is also increasing. Dynamo, a powerful visual programming plug-in, can carry out special-shaped structure modeling through visual programming combined with Python language, and assist Revit management model engineering data management, making Revit more flexible in use, so as to meet the needs of practical engineering applications.

For the continuous beam bridge with corrugated steel webs of Zhaojun Yellow River Super Large Bridge, it is necessary to establish BIM models of concrete structure of roof and bottom plate and corrugated steel webs. In order to facilitate subsequent application, this paper establishes a parameterized BIM model of the main bridge of Zhaojun Yellow River Super Large Bridge by using the BIM modeling software Revit of Autodesk company as shown in figure 2, figure 3, figure 4, and gives relevant parameters to the model.

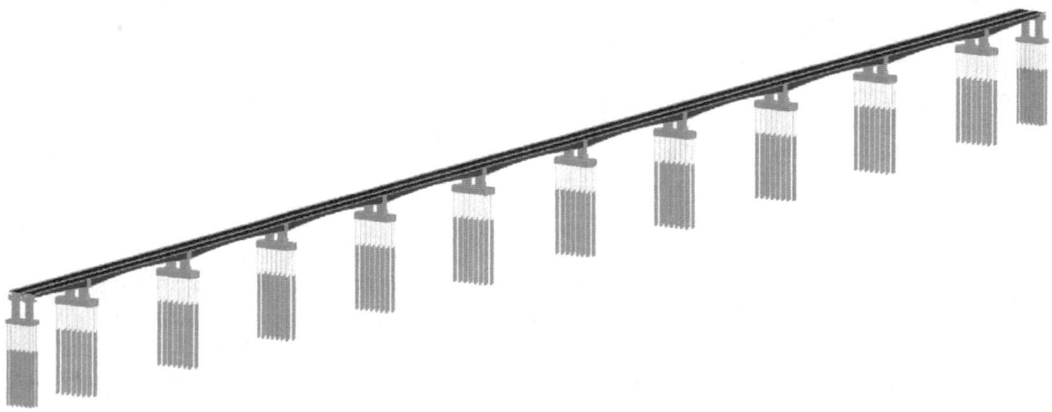

Figure 2. Whole BIM model of the main bridge of Zhaojun Yellow River Super Large Bridge.

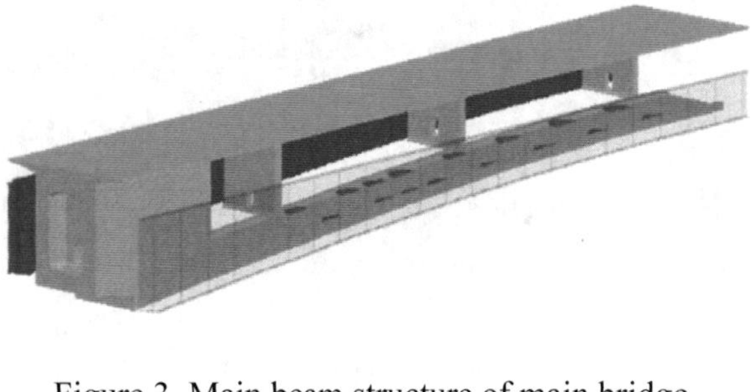

Figure 3. Main beam structure of main bridge.

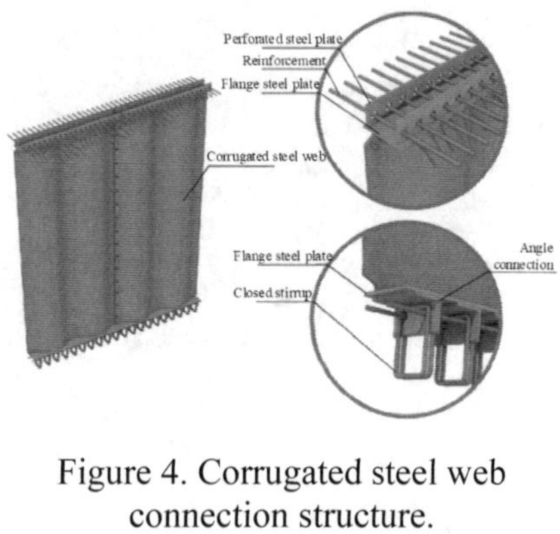

Figure 4. Corrugated steel web connection structure.

3.2. Data conversion between BIM model and finite element model

The traditional finite element simulation modeling is directly carried out in the finite element simulation analysis software (such as Midas/Civil). The modeling process needs more time and energy, and it is not easy to modify, resulting in problems such as heavy workload and poor accuracy of finite element analysis. This paper explores the conversion of BIM model data into MCT command flow that Midas/Civil can recognize, so as to realize the data conversion between BIM model and Midas/Civil model, reduce the workload of finite element modeling and analysis, and improve the accuracy. The parametric BIM model is established by Revit, and the parametric driving of BIM model is realized. Each beam section of the main beam has the geometric parameter information of the box girder section. In addition, each beam section adds the component code parameter information as the unique identification. The idea of data conversion between BIM model and finite element model is shown in figure 5.

Taking Zhaojun Yellow River Super Large Bridge as an example, the specific process of finite element simulation calculation based on BIM is as follows: firstly, through Dynamo visual programming, the component code is used as the identification code to extract the component geometric parameter attribute information in Revit model as shown in figure 6; secondly, the python program is written through the python script node in Dynamo and converted to the MCT command stream recognized by Midas/Civil as shown in figure 7; finally, open the generated MCT command stream with Midas/Civil and correct it to complete the data conversion from BIM model to finite element model.

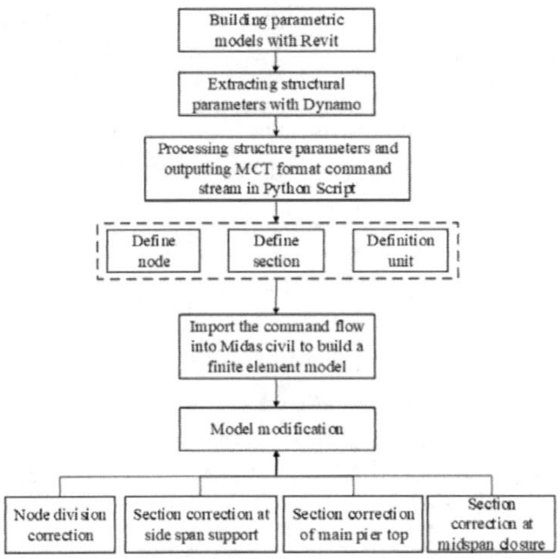

Figure 5. The idea of data conversion between BIM model and finite element model.

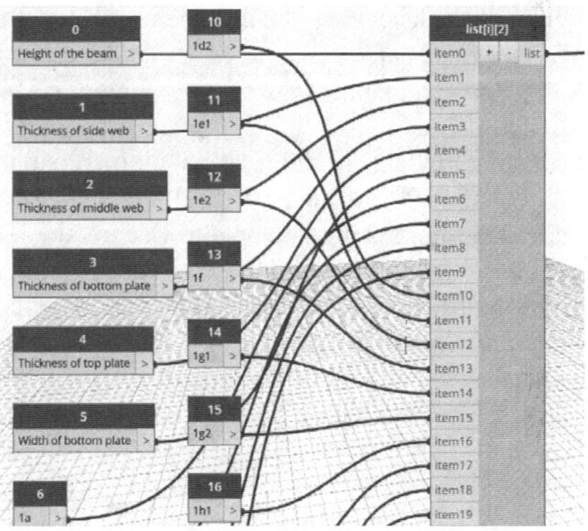

Figure 6. Extraction of parameter information.

Journal of Physics: Conference Series **2215** (2022) 012006 doi:10.1088/1742-6596/2215/1/012006

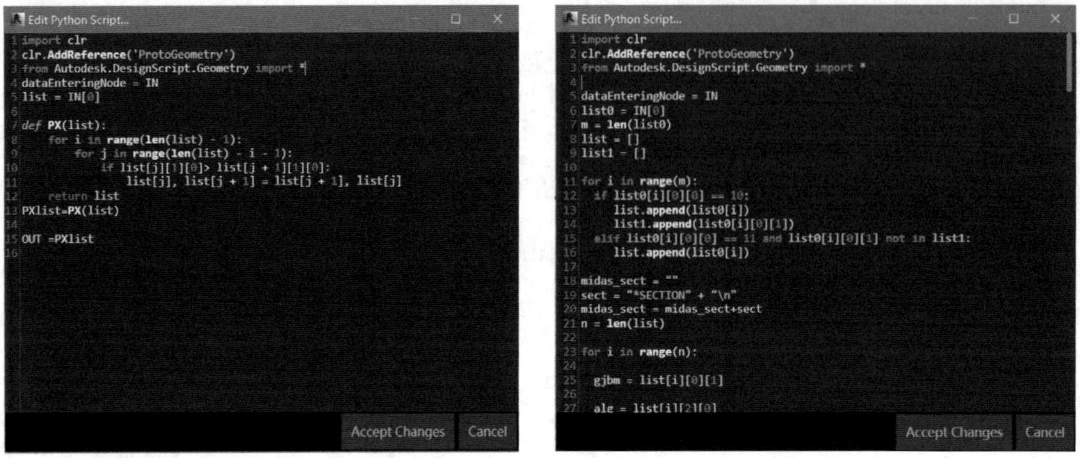

(a) (b)

Figure 7. The example of data conversion code.

3.3. Finite element simulation results based on BIM

In terms of model calculation parameter selection, relevant calculation and design parameter information in BIM model can be extracted. C55 concrete is adopted for the project, with elastic modulus of 3.45E4Mpa and linear expansion coefficient of concrete of 1.0E-5. Q345D steel is used for corrugated steel web and its connecting steel plate with box girder concrete top plate, diaphragm and steering block, with elastic modulus of 2.06E5mpa, thermal expansion coefficient of 1.2E-5 and calculated unit weight of 78.5kN/m3. The allowable axial stress is 200MPa and the allowable shear stress is 120MPa. The prestressed reinforcement adopts low relaxation 1860 steel strand. The diameter of a single steel strand is 15.2mm, the area is 139mm^2, the elastic modulus is 1.95E5Mpa, the standard strength is 1860MPa, and the coefficient of thermal expansion is 1.2E-5.

Combined with the division of construction stages, the axial force, bending moment, stress, dead load displacement, shrinkage and creep displacement and live load displacement of the main beam, as well as the cumulative deformation of the main beam in the construction stage and the manufacturing alignment of corrugated steel web can be simulated to provide technical parameters for construction monitoring.

4. Conclusion and Prospect

In this project, the main bridge model of Zhaojun Yellow River Super Large Bridge is established with the help of the powerful parametric modeling and data management ability of BIM core modeling software Revit. On this basis, the data extraction of BIM model and data conversion with Midas/Civil finite element model are realized with the help of visual programming software Dynamo, and the finite element structure simulation is carried out, which improves the efficiency and accuracy of modeling.

The integration of BIM and finite element analysis is a difficulty in the application of BIM technology. The in-depth application of BIM technology is inseparable from the integration with traditional structural analysis. Promoting the combination of BIM modeling and finite element analysis and endowing BIM model with the highest mechanical properties is not only an innovative breakthrough in BIM application, but also a key breakthrough in BIM forward design and BIM + monitoring.

Acknowledgments

The research is sponsored by the transportation science and technology project of Inner Mongolia (Grant No. NJ-2018-10) and the academician special research funding projects of CCCC (Grant No. YSZX-02-2020-01-B).

Reference

[1] Bhusar A A and Akhare A R 2014 Application of BIM in structural engineering. *SSRC International Journal of Civil Engineering(SSRC-IJCE)* **1(5)** 11-17.

[2] Kim, Yu-Ri, Cho and Young-Sang 2009 The study on application of korean bim process in the structural design phase based on a case study. *JOURNAL OF THE ARCHITECTURAL INSTITUTE OF KOREA Structure & Construction* **25(10)** 19-26.

[3] Aladag H , DemirdoGenb Gokhan and Isikc Zeynep 2016 Building information modeling (BIM) use in turkish construction industry. *Procedia Engineering* **161** 174-179.

[4] Song Jie, Zhang Yadong, Wang Mengjin, Zhang Jinhua and Fan Junyu 2015 Research of model exchang interface between Revit and ANSYS. *Journal of Civil Engineering and Management* **33(01)** 79-84.

[5] He Xiangping, Wang Hao, Zhang Yiming, Wang Feiqiu, Mao Jianxiao and Xie Yishun 2021 Revit-Midas/Civil model conversion approach and its application. *Journal of Southeast University(Natural Science Edition)* **51(05)** 813-818.

[6] Zhang Zheyuan 2021 Research on parametric modeling and structural analysis conversion interface of curved continuous beam bridge based on BIM technology. *Lanzhou Jiaotong University.*

[7] Wang xiongjue 2016 Finite element analysis on bridge reinforcementand reconstruction project BIM technology. *Hubei University of technology.*

[8] Chen Zhiwei, Wu Kun, Huang Ying and Huang Menggang 2018 Generation of refined FE model of complex structures base on BIM. *Journal of Civil Engineering and Management* **35(05)** 60-64

[9] Chen Hao 2017 Construction control and simulation analysis of the cable-stayed bridge based on BIM. *Beijing Jiaotong University.*

Study on High-precision Grid Outage Region Based on Distribution Network Topology Analysis

Weikang Yan[1, a*], Bing He[2, b], Tianxiang Xie[3, c]

[1]State Grid Sichuan Electric Power Company Tianfu New District Power Supply Company, Chengdu, Sichuan, China

[2]State Grid Sichuan Electric Power Company Tianfu New District Power Supply Company, Chengdu, Sichuan, China

[3]State Grid Sichuan Electric Power Company Tianfu New District Power Supply Company

[a]e-mail: 120028379@qq.com

[b]e-mail: b41823305@qq.com

[c]e-mail: c182453445@qq.com

ABSTRACT - Based on the recursion method and the Java language, this paper develops a program for analyzing the distribution network topology diagram and explores various scenarios of power outages. The power line topology analysis can pinpoint the power outage range, providing a solid foundation for power outage information release and distribution network reliability calculation.

1. Introduction

With the vigorous advancement of the new energy security strategy of "energy consumption revolution, energy supply revolution, energy technology revolution and energy system revolution", in 2020, my country put forward the new energy transition goal of "carbon neutrality and carbon peaking" for the first time.[1] Under the blueprint of the strategic goal of energy transition, my country's power system is gradually developing in the direction of a new power system with renewable energy as the main body. It is worth noting that the distribution network, as the main form of distributed energy access and a key link in improving the consumption of renewable energy, is a strong guarantee for achieving the "dual carbon" goal.

But with the access of distributed new energy sources, higher requirements are put forward for the reliability of the distribution network. The possibility of failure of the low-voltage distribution network during operation is relatively high, and timely and accurate determination of the faulty line section is of great significance to reduce the user's excessive power outage and improve the reliability of power supply

In order to accurately determine the fault area in time Section, it is necessary to accurately analyze the topology of the distribution network.[2 3]

2. Distribution network topological structure

The distribution network topology diagram is a file describing the topological structure of a 10kV distribution line. It contains the essential information and topological locations of the power device, such as the line name, line number, outgoing switch, overhead line, cable, ring network cabinet, switch cabinet, line switch, isolator and distribution transformer.

2.1 Format of distribution network topology diagram

Distribution network topology is stored in XML format. As a markup language used to mark electronic files to make them structural, the XML language is mainly applied in structural data storage.

2.2 Basic structure of distribution network topology diagram

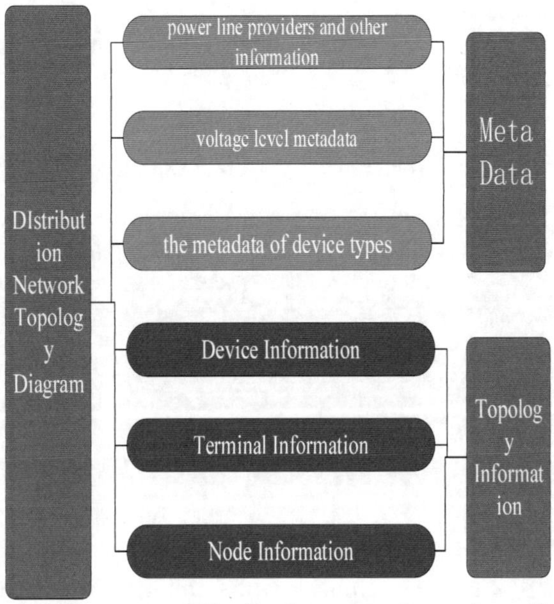

Figure 1 Structure of distribution network topology diagram

1） *Line attribution*

Describe the power line providers and other information.

2） *Voltage level metadata*

Describe the voltage level metadata

3） *Device type metadata*

Describe the metadata of device types connected with each line.

4） *Device information*

Describe the basic attributes of each device on the line. Among them, the voltage level and device type all refer to metadata. It is notable that the ID of the device is critical in subsequent analysis. As Figure 2.

```
<cim:Breaker rdf:ID = "PD_30500000_719497">
  <cim:IdentifiedObject.mRID>30500000_719497</cim:IdentifiedObject.mRID>
  <cim:IdentifiedObject.name>Shenyang Road West Section at Tianqi Station along Line 2 # Ring network cabinet 1 # Switch 2 </cim:IdentifiedObject.name>
  <cim:Switch.open>false</cim:Switch.open>
  <cim:Switch.normalOpen>false</cim:Switch.normalOpen>
  <cim:PowerSystemResource.PSRType rdf:resource="#PD_30500000"/>
  <cim:ConductingEquipment.BaseVoltage rdf:resource="#PD_BaseVoltage_22"/>
  <cim:PowerSystemResource.Assets rdf:resource="#PD_e41075ef2f8a30829f72a9f2c50172e40fcca021a1"/>
  <cim:Equipment.EquipmentContainer rdf:resource="#PD_32400000_1258833"/>
</cim:Breaker>
```

Figure 2 the basic attributes of each device on the line

5) Terminal information

Describe the ConnectivityNode atrribute, like, PD_xxxx, refers to the ID number of the Node, and multiple terminals correspond to one Node. The ConductingDevice corresponds to the ID number in the device information; the corresponding relationship may be one-on-one or two-on-one. As Figure 3.

```
<cim:Terminal rdf:ID = "PD_30500000_719502_1">
  <cim:IdentifiedObject.mRID>30500000_719502_1</cim:IdentifiedObject.mRID>
  <cim:IdentifiedObject.name>BRK_MV_30500000_719502__TO__CNODE_HYPERNODE_431429105</cim:IdentifiedObject.name>
  <cim:Terminal.ConductingEquipment rdf:resource="#PD_30500000_719502"/>
  <cim:Terminal.ConnectivityNode rdf:resource="#PD_431429105"/>
  <cim:Terminal.phases rdf:resource="http://iec.ch/TC57/2016/CIM-schema-cim17#PhaseCode.ABC"/>
</cim:Terminal>
```

Figure 3 ConnectivityNode atrributes

6) Node information

In Node information, only the Node ID number, i.e., PD_xxxx, is useful. As Figure 4.

```
<cim:ConnectivityNode rdf:ID = "PD_431435919">
  <cim:IdentifiedObject.mRID>431435919</cim:IdentifiedObject.mRID>
  <cim:IdentifiedObject.name>CNODE_JCT_NODE_MV_431435919</cim:IdentifiedObject.name>
</cim:ConnectivityNode>
```

Figure 4 Node information

2.3 Basic analysis of distribution network topology diagram

The XML file can be analyzed by the Jsoup Library in Java. The Nodes and Terminals are mainly used to describe the line topology.

The device information includes the name, status, type, voltage level and other attributes of all devices on the line.

The Terminals part consists of multiple terminals with Terminal ID, name, type, ConductingDevice, and ConnectivityNode. Among them, ConductingDevice is the ID of the device, through which the device corresponding to the terminal can be found, and ConnectivityNode is the ID of Node

The Nodes part comprises multiple Nodes with ID numbers, which is in many-for-one correspondence with the ConnectivityNode in Terminals.

2.4 Analysis of distribution network topology diagram

The Nodes and Terminals are matched according to the ConnectivityNode. Some devices and Terminals are in a one-on-one relationship, while others are in many-for-one correspondence, as shown in the following figure:

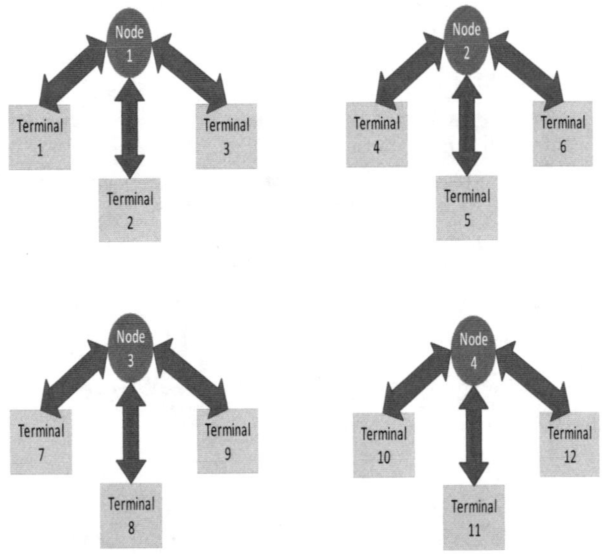

Figure 5 The relationship of *terminals and node*

1) *The Terminals are connected to devices in accordance with the ConductingDevice*, as shown in the figure below.

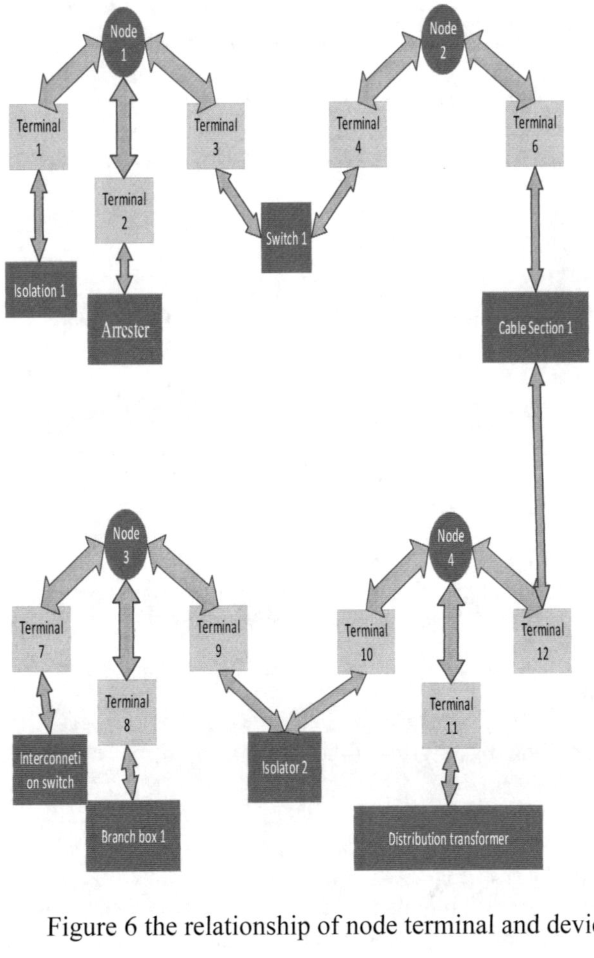

Figure 6 the relationship of node terminal and devices

3. Analysis of distribution network topology diagram

3.1 Analyzing the lines into a multiple-way linked list

Nodes with the same Terminal ID are connected to yield a multi-way linked list, representing this line's basic topology, as shown in the following figure.[4]

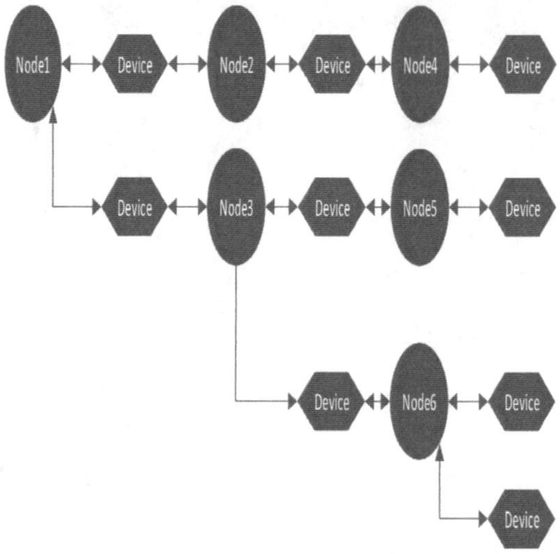

Figure 7 the relationship of devices and nodes

3.2 Analyzing the lines into a tree diagram according to the power supply point[5 6 7]

After choosing a power supply point, the multi-way linked list can be analyzed into a tree diagram by the tail recursion algorithm, just like lifting any place of a fishing net to turn it into an umbrella structure, as can be seen below:

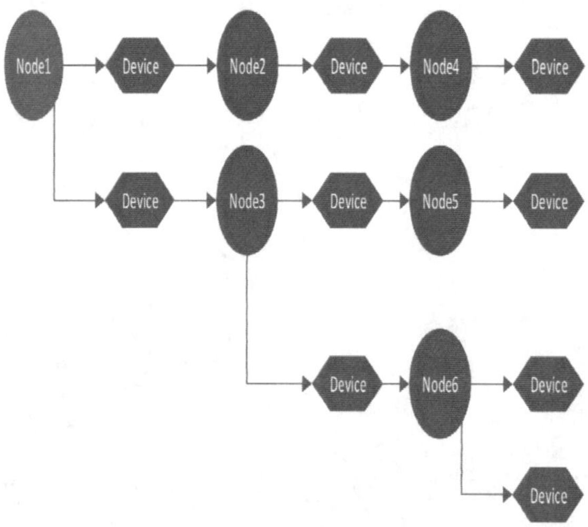

Figure 8 the tree figure of the relationship of node and devices

4. Precise analysis of various outage ranges

4.1 Analysis of a single line power outage

Whether it is a fault power outage or a planned power outage, in most cases, the dispatching department will only tell the power outage time, lines, and positions rather than the accurate power outage range.

Therefore, the precise power outage range can be calculated by analyzing the distribution network topology diagram, thus providing a solid guarantee for releasing power outage information and calculating the reliability of the distribution network.

1) Power supply analysis under the single-break scenario

For a single-break power supply, the root node of the tree diagram can be set as the station switch to generate the tree diagram, and the power outage range is below the leaf node where the breakpoint is located.

2) Power supply analysis of series feeding of interconnection switch through other lines under the single-break scenario

For the series feeding of interconnection switch through other lines under the single-break scenario, the root node of the tree diagram can be set as the interconnection switch to generate the tree diagram, and the power outage range is below the leaf node where the breakpoint is located.

3) Power supply analysis under the multiple-break scenario

For a multi-break power supply, the root node of the tree diagram can be set as the station switch to generate the tree diagram. The power outage range can be calculated by regarding each breakpoint as a single breakpoint and then unifying the results.

4) Power supply analysis under the scenario of breakpoints supplied by multiple power supply points

For breakpoints supplied by multiple power supply points, the tree diagram can be generated with each power supply point as the root node. The power outage range can be calculated according to the power supply under the multiple-break scenario, and the final power outage range is the intersection of the results.

4.2 Analysis of the multiple line outage

1) Power outage analysis of two separate lines

For the power outage analysis of two separate lines, the power outage range can be calculated following the steps mentioned in 3.1 and then unifying the results.

2) Power outage analysis of multiple separate lines

The calculating steps of power outage range for multiple lines are the same as those of two separate lines.

5. Precise releasing of the power outage and supply information

After the power outage range is calculated, the distribution transformers in the faulty components can be screened out. Then users' information under each transformer can be obtained through the ledger. The accurate release of power outage information can be realized by mass texting through the contact information.

6. Distribution network reliability calculation

6.1 Calculating method for distribution network reliability

1) Annual reliability calculation of a single user[8]

$$\frac{\text{ASPSD} - \textit{Annual total power outage}\ \text{duration}}{\text{Annual}\ \textit{supposed power supply duration}} \times 100\%$$

ASPSD: Annual *supposed power supply duration*

2) Annual reliability calculation of multiple users

$$\frac{NOU * \text{ASPSD} - \sum \textit{Annual power outage}\ \text{duration}\ \textit{of each user}}{\textit{Number of users} * \text{Annual}\ \textit{supposed power supply duration}} \times 100\%$$

NOU: Number of users

ASPSD: Annual supposed power supply duration

Where, the total annual power outage duration of each user, ∑, is called the number of power outage households. Generally speaking, the number of users is constant, and the annual power supply duration is 8,760 hours. Only when there is a power outage, the number of households is difficult to determine. Therefore, it is necessary to calculate the number of power outage households for each power outage within one year. The number of users involved in each power outage is not as easy to determine as the starting and ending times of power interruption. As a result, the dispatcher will only report the approximate power failure range when a line is partially cut off, including the power supply points and breakpoints, without targeting every user. Usually, the reliability is calculated according to the simultaneous power failure of the whole line. However, this is equivalent to expanding the power outage range, and the calculated reliability is lower than the actual. Therefore, as long as the users involved in each power outage can be pinpointed, the number of households during each power outage can be precisely calculated, thus improving the accuracy of power supply reliability calculation.

6.2 Improving the calculation accuracy of power outage households by analyzing the distribution topology diagram

To accurately calculate the number of power outage households, the information of lines involved in the power outage, the starting and ending times of power outage, the power supply points and breakpoints are obtained from the dispatching department. Then, the power outage duration is calculated using the starting and ending times. Next, through the distribution network topology diagram, the power supply points and breakpoints, the accurate power outage range is calculated, and the distribution transformers are screened out. Then, through the equipment ledger and users' ledger information, the number of users involved in the power outage transformer, users' nature, and other information are obtained. Finally, the number of users involved in this power outage is calculated by adding users under each power outage transformer. The power outage duration is obtained by multiplying the figure with the power outage duration.

With the number of households during each power outage, the reliability of each line, each power supply station, each power supply company in each county and other information, even the monthly and annual reliability can be calculated. It provides a solid foundation for accurately calculating the power supply reliability of the distribution network.

7. Conclusion

This article introduces the basic structure of the distribution network topology diagram, and gives the analysis method of the topology relationship, which can parse the topology diagram into a tree diagram. Based on the tree diagram, an accurate analysis of the range of power outages under various conditions is done, based on the power outage Scope analysis provides a method for accurate push of power outage and transmission information. Based on outage scope analysis and distribution network reliability calculation formula, a method to improve reliability calculation accuracy is given. In order to improve the accuracy of power outage and transmission information push A solution is proposed to improve the reliability of the distribution network.

References

[1] Li X, Gao L, Liu JT. The approach to carbon emission quotas of road transportation: a carbon emission intensity perspective[J]. Journal of advanced transportation, 2020, (3): 1-18.

[2] HU Funian, SUN Shoujuan. Fault location of distribution network by applying matrix algorithm based on graph theory[J]. Electric Power, 2016, 49(3): 94-98.

[3] XU Xinmin, MA Zhenliang. A novel algorithm for fault location in a distribution network based on topology division[J]. Electric Power, 2012, 45(5): 20-24.

[4] WEI T, MA Z, SU J, et al. Development and application of distribution network planning calculation and analysis software[J]. Southern Power System Technology, 2016, 10(5): 38-44

[5] WEI Ran, ZHANG Pan, GAO Qiangwei, et al. Simulation analysis of low-voltage distribution network fault diagnosis based on tree structure diagram[J]. Power System Protection and Control, 2021, 49(13): 167-173.
[6] HU Funian, ZHANG Ren, BIAN Xiaoliang. Fault location method of active distribution network based on graph theory[J]. Control Engineering of China, 2021, 28(5): 833-838.
[7] LI Jiawei, WANG Xiaojun, HE Jinghan, et al. Distribution network fault location based on graph attention network[J]. Power System Technology, 2021, 45(6): 2113-2121.
[8] XIAO B, ZHAO D P, JIANG Z, et al. Power supply capability evaluation of urban distribution network [J]. Power Generation Technology, 2018, 39(3): 213-219.

Research on BIM model lightweighting methods and IoT technology application in the context of WebGL

Yuhe Yuan[1], Jing Yang[1,2,*], Ruping Shao[1,2] , Jiahuan Mao[1] and Chengwei Zhang[1]

[1]College of Electrical Engineering and Control Science, Nanjing Tech University, 211816 Nanjing, China

[2]Engineering Research Center of BIM, Nanjing Tech University, 211816 Nanjing, China

muyi_qing@njtech.edu.cn*

Abstract. In view of the fact that most of the existing building information modeling (BIM) platforms are based on the Client/Server (C/S) mode of client-side applications, which brings about problems such as weak data interoperability, poor model semantic integrity and bad user-friendliness, this paper proposed an IFC-JSON and IFC-glTF model file format conversion process based on WebGL and an improved Draco algorithm that can further lighten glTF files. The experimental results showed that this option can effectively reduce data redundancy, then combined IoT (Internet of Things) as well as sensor technology in project practice to collect and transmit building-related O&M data, and finally visualised the equipment status of the building on the Web based on WebGL technology, providing a new application way for the combination of BIM technology and IOT technology in the browser/server mode.

1. Introduction

Building information modeling (BIM) technology is widely used for data sharing and information integration management in all phases of projects in the architecture, engineering & construction and facilities management (AEC/FM) industry[1], with its characteristics of serving the whole life cycle of buildings. The core value of BIM technology is to use the BIM server as the storage carrier for information, so as to interact with multiple clients of the project[2]. With the volume of engineering projects increasing year by year, the difficulty of managing large volume 3D models and data gradually increases, so it is especially important to ensure the complete information display of BIM models and meet the user's convenient access under different operating devices.

Classified according to the different forms of clients, they can be divided into client/server (C/S) architecture and browser/server (B/S) architecture. Most of the existing BIM platforms are based on the client-based C/S architecture, which brings about problems such as weak data interoperability, poor model semantic integrity and bad user-friendliness, limiting the integration of BIM with other technologies[3-5]. In contrast, the application of BIM technology based on B/S architecture effectively reduces the hardware and software requirements of the web front-end and improves the compatibility with mobile terminals, among which the efficient rendering of BIM models based on WebGL technology is a current research hotspot.

In view of this, this paper proposed a file format conversion process and a 3D model lightening method based on Draco algorithm for large-volume building models for web-based display. The glTF file was selected as the final output format to achieve efficient compression of the model to guarantee the display effect, while ensuring the retention of building attribute data. In order to broaden the application scenario of the operation and maintenance phase, the IoT (Internet of Things) and sensor technologies were combined to simulate the collection and transmission of building-related operation and maintenance data, combined the dynamic building operation and maintenance data with the lightened BIM model, and finally realised the visualisation of the field equipment status of the building on the web side based on WebGL technology.

2. Research methodology

With WebGL technology as the background, this paper proposed an improved solution for the lightweight presentation of BIM models on the web side in response to the problems of semantic data loss and geometric data redundancy under the current technical approach, and practiced the real-time interaction function of building operation and maintenance data in conjunction with IoT technology.

The solution addressed the problems of lack of semantic integrity and weak information interoperability in the current research, and started with the common IFC file in the field of BIM technology[6], carried out decoupling of semantic and geometric information, stored them as JSON and glTF files respectively, and realised the dual relationship mapping from IFC geometric space and JSON semantic attributes to glTF. To address the problem of stalling in displaying large volume models, a Draco algorithm with improved parameters was proposed to recode the glTF file of the geometry part so as to achieve further lightweighting; to address the problem of displaying static data and dynamic data in the application scenario of operation and maintenance of BIM technology, IoT technology was used to monitor sensor information to achieve 3D model display and O&M function development on the web side. The technical route of the research is shown in Figure 1.

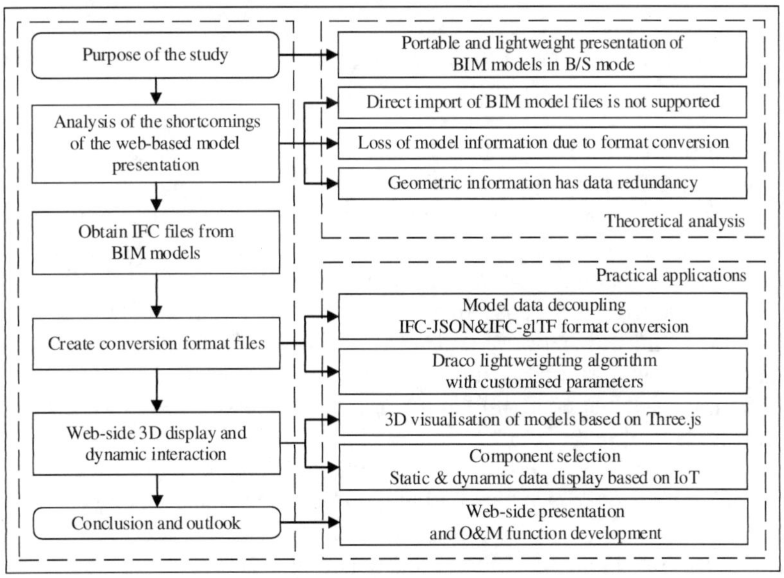

Figure 1. Research technology route.

3. BIM model processing

3.1. IFC-JSON and IFC-glTF file format conversion

IFC files are a common data file format used by the AEC/FM industry for data interaction and sharing, but they are not compatible with 3D formats used for web display, so they need to be converted to a document supported by WebGL for BIM model visualisation. The proposed approach took the common IFC file as input and the JSON and glTF files supported by the Three.js platform as output, splitting the IFC file into semantic and geometric files for processing, which will ultimately retain the complete component information and enable preliminary file classification and lightweighting. To ensure the completeness of the BIM application, two common GIS data: CityGML files, GeoJSON files were also given for processing, the specific conversion process and the open source tools used are shown in Figure 2.

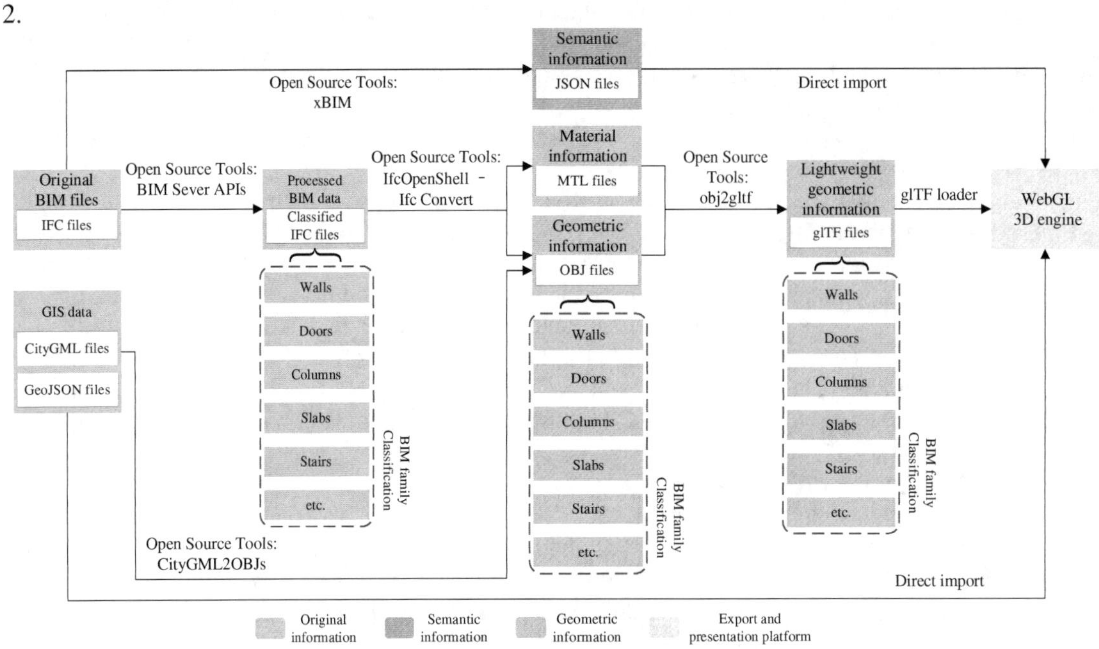

Figure 2. Conversion process of BIM model file (contains geographical information).

As shown in Figure 2, for semantic information extraction, the content under the IFC semantic parameters can be extracted in a targeted manner using Xbim[7] and stored in the newly generated JSON file.

For geometric information, the original IFC file as input was decomposed into several small IFC files by component type using BIM Server APIS[8] before format conversion, in order to solve the problem of not being able to identify the model components individually after model conversion in the previous study[9]. The next step was to use Ifc Convert[10] to extract and store the geometry information in the IFC file as an OBJ file and a separate MTL file to store the material information. The OBJ format was chosen as the first step in the conversion because there are many tools built around it for geometry format conversion; finally, the open source tool obj2glTF[11] was used to obtain the glTF format file, which is a graphics language interchange format designed for real-time rendering applications, combining scene description information such as node hierarchy and model geometry. The glTF file stores data information such as node hierarchy and model geometry as well as animation images in chunks, while the model is usually loaded with a glTF file in text format only, and other block files can be introduced externally and passed directly to the graphics API without additional parsing and conversion. As a result, the glTF format allows for the maximum compression of the original IFC file while ensuring the semantic integrity of the model and data interaction, enabling the web visualisation of lightweight BIM models.

3.2. Improved Draco algorithm for lightweighting

Previous model file lightweighting has mostly stopped at the change in data structure due to the conversion of file types. The improved Draco lightweighting algorithm was used, specifically by adding the control file compression ratio parameter draco.compressionLevel to the build type, to achieve further lightweighting of the final glTF file generated by the above format conversion process.

The Draco algorithm is a technical tool for compressing and decompressing geometric mesh data and point cloud data, encoding the connectivity and geometric information of the mesh separately and storing the code, supporting the compression of points, connectivity information, texture coordinates, colour information, normals and other geometry-related attributes. The design features are well suited for processing glTF geometry files with point and surface structures, and can improve the storage and transmission of 3D images, allowing for a significant reduction in the number of applications using 3D graphics while maintaining visual fidelity.

The core idea of the algorithm is the classical 3D mesh compression algorithm Edgebreaker, and in particular the Corner Table structure of the algorithm[12], which essentially encodes and compresses the mesh graph by traversing it in a breadth-first principle, compressing it to reduce duplicate vertex information and replace it with lighter index information. By describing the partition path of the mesh through operators, the 3D scene can then be created from the transformed encoding, thus achieving compression with much less visual bias using the Draco compression method as opposed to direct surface reduction. As an example of re-encoding glTF files, the encoding process is shown in Figure 3.

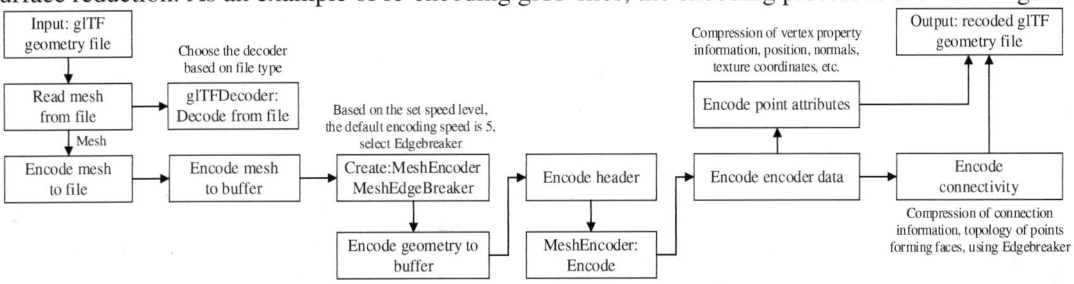

Figure 3. Coding process of Draco.

For the application of the Draco algorithm, the open source tool glTF Pipeline is available as support[13]. In order to further strengthen the compression capability of the Draco algorithm, this study set the parameter draco.compressionLevel that controls the compression ratio of the file on the basis of recoding, and set separate compression parameters for different types of component glTF files for different application scenarios, and compressed the grid data index on the basis of retaining the original grid order, the specific steps of the application are as follows.

Step1: Serialisation of glTF mesh data. Traversing component information, generate binary indexes, insert relevant nodes and generate scene-node-mesh objects.

Step2: Mesh model data compression. Read the serialised glTF file as input to the draco_encoder and determines the component type based on the component id, adds a draco.compressionLevel parameter ranging from 0 to 10 and outputs it to the algorithm's custom .drc format file.

Step3: Mesh model data decompression. Read the .drc file as draco_decoder input and re-exports it as a glTF file after decompression.

Step4: glTF file reading display. Use GLTFLoader and DracoLoader in three.js as data interfaces for compressed model output display.

4. Experimental analysis and web-side development

4.1. Model file conversion comparison

In order to verify the effectiveness of the geometric part IFC-glTF file format conversion strategy and the lightness of the Draco algorithm, this paper used Revit software as the drawing tool for BIM models, and selected six groups of building model files with similar building structures but different scales to compare and analysed the compression rate and web-side rendering effect of the models.

Applying the above proposed process to convert the file format of the geometric profiles and recording the file size data for IFC-OBJ, IFC-glTF and IFC-DracoglTF, Figure 4 is used to show the above files used for geometric presentation, which clearly showed the change in file size before and after compression.The final decoupled JSON semantic file and lightweight DracoglTF geometry files sizes obtained under this approach are shown in Table 1.

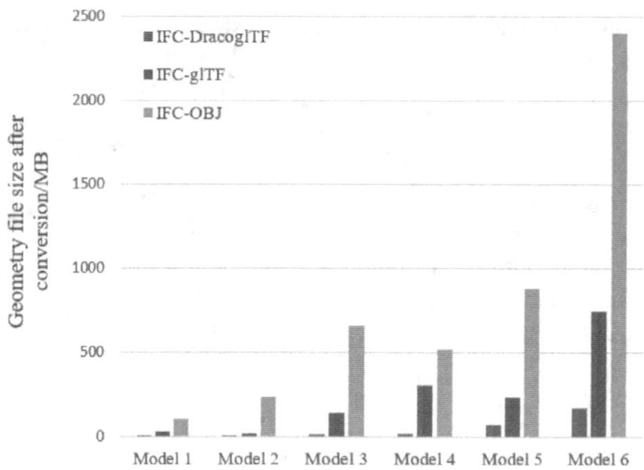

Figure 4. Geometry file size after conversion.

Table 1. Compression results for model files.

Serial number	Original model size /MB	Semantic file size (JSON) /MB	Geometric file size （DracoglTF）/MB	Geometric compression ratio
1	16.5	0.4	2.9	82.42%
2	29.6	4.3	4.9	83.45%
3	82.1	10	15.4	81.24%
4	203.5	15.8	20.7	89.83%
5	507.2	63	75.3	85.15%
6	1551.7	72.5	172.5	88.89%

Analysis of the data in Table 1 showed that decoupling the IFC file into a JSON file for the semantic part and a glTF file for the geometric part, and invoking the Draco algorithm to compress the glTF file, had a significant compression effect, with compression rates above 80% and up to 89.83%. Model 2, a BIM model file of a school building with an original size of 29.6MB, was selected and the converted intermediate file was rendered under the common 3D engine Three.js. The rendering results and frame rates for the three formats are shown in Table 2. The analysis showed that after the conversion of DracoglTF format, only the geometric part of the glTF file was extracted for rendering, which can effectively reduce the file size of the part to be rendered, and the frame rate was greatly improved.

Table 2. Rendering and file information of BIM model.

File format	OBJ file	glTF file	DracoglTF file
Web side display	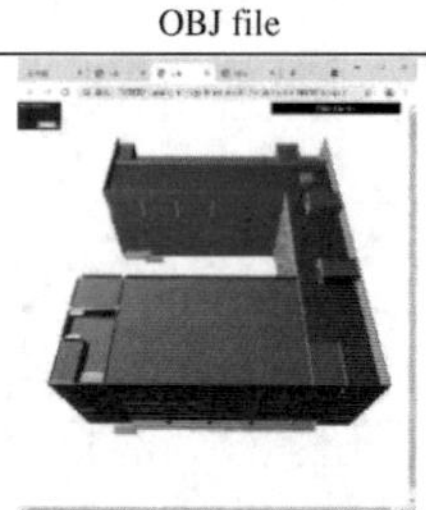	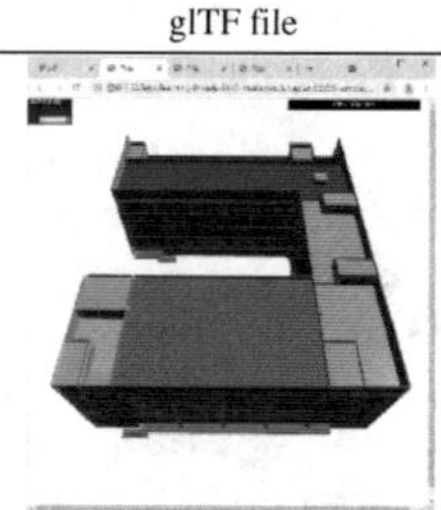	
File size/ MB	236.5	74.3	4.9
Frame rate/ fps	13	35	58

4.2. *Web-side functional development*

The Three.js framework allows for the development of 3D interactive functionality, allowing users to query component property information through the browser. Taking a structural column of this building as an example, the properties of the Revit scene and the web page are shown in Figure 5 and Figure 6 respectively.

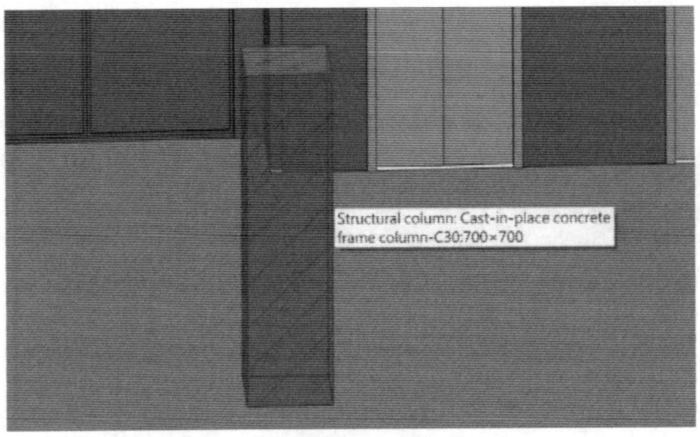

Figure 5. Revit model scene.

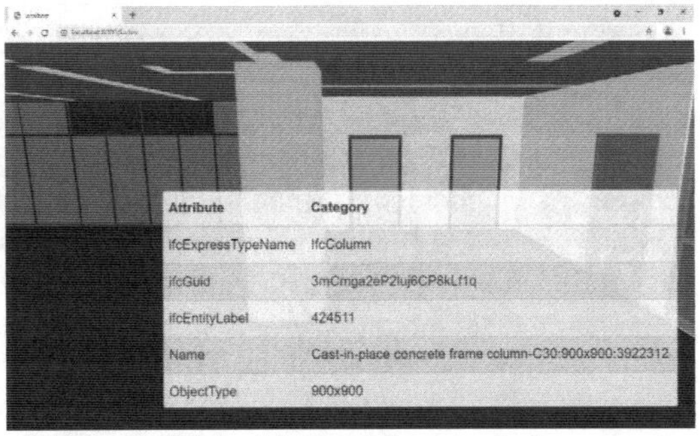

Figure 6. Web-side model scene and property display.

This study introduced IoT technology into the lightweighted BIM model, aiming at visualisation of the model and real-time display of equipment status data, which was exported via open database connectivity in the modeling software Revit.

Dynamic data transmission in the IoT is realised by an IoT program consisting of a node driver and a server-side program. The Zigbee technology is suitable for scenarios in buildings with low network bandwidth and unstable networks due to its low cost, low power consumption and no device limitations, as well as the MQTT protocol's ability to provide an orderly and reliable, bi-directional connection with guaranteed network connectivity that is open, simple and easy to implement. Therefore, this study used Zigbee protocol and MQTT protocol as the core to achieve protocol interfacing and designed a high degree of freedom IoT device control module. After the device was started, the raw data collected by the sensor was switched on and off by the controller, and the raw data was verified, then the verified data was transmitted to the message middleware with the help of the Zigbee network, which sifted and analysed the data and sent it to the MQTT server, completing the interface between the Zigbee and MQTT protocols. Finally, the data was received, parsed and stored by the Web server to the MQTT server in a publish or subscribe communication. The specific transmission process of dynamic IoT data is shown in Figure 7. The sensor application and data presentation is shown in Figure 8.

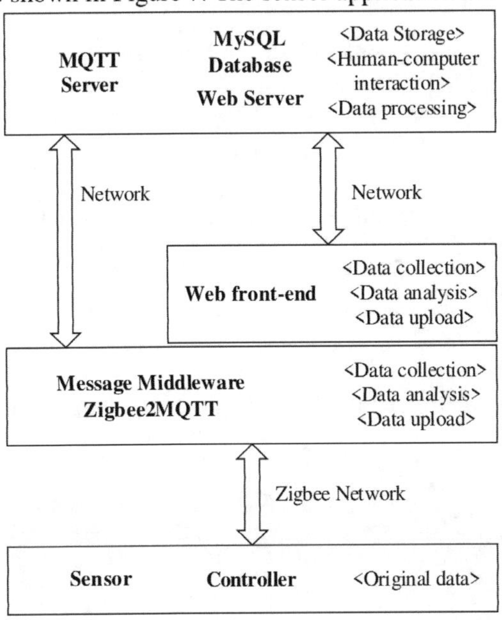

Figure 7. IoT data transfer flow chart.

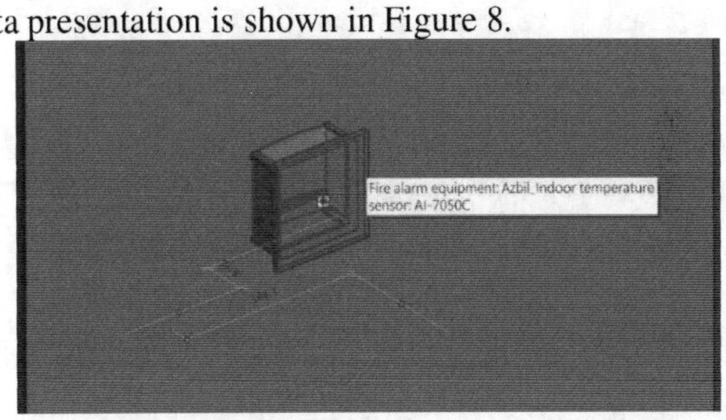

(a)

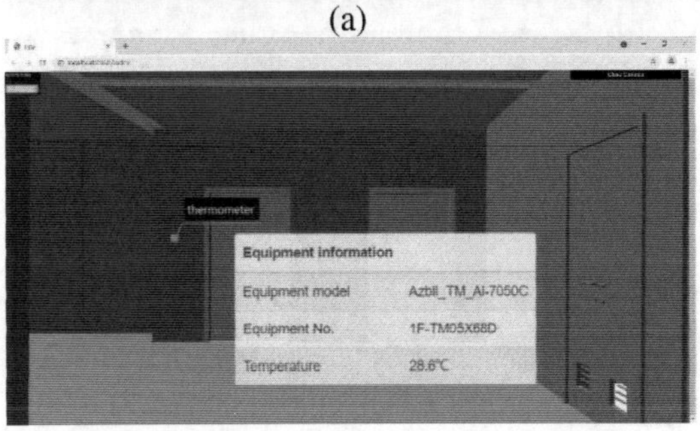

(b)

Figure 8. Sensor display (temperature sensor as an example).

5. Conclusion

This paper proposed a file conversion process for B/S model development and a model lightweighting method to address the problems of weak data interoperability, poor semantic integrity and bad user-friendliness under the existing C/S model and finally realized model information display and the development of interactive functions based on IoT technology. By decoupling the semantic part and geometric part of the IFC format files common to BIM platforms, the JSON file and glTF file supported by the 3D engine were obtained respectively, then by introducing the Draco algorithm and compression parameters, the 3D models were lightened while retaining the information of building components.

The experimental results showed that the method has a significant compression function for models, the compression ratio of the geometric part can be guaranteed to be above 80%, and can complete the interactive development of IoT technology based on Three.js platform while retaining the component information, which not only conforms to the current development process recommended by the latest 3D engine, but also well extends the application scenario, which has positive significance for the promotion of BIM technology.

References

[1] Ozturk G B Interoperability in building information modeling for AECO/FM industry 2020 *Automation in Construction* 113:103122.

[2] Xu Z, Zhang Y, Xu X 3D visualization for building information models based upon IFC and WebGL integration 2016 *Multimedia Tools and Applications* 75(24):17421-17441.

[3] Du Z, Wang H J, Tan Z and Huang H D Research on the lightweight application of water conservancy project BIM based on WebGL and IFC 2020 *China Rural Water and Hydropower* 199-203.

[4] Wang C. P Large architectural models real-time display system based on WebGL 2018 South China University of Technology.

[5] Zhang H Y Research on BIM visualization methods based on WebGL 2017 Xi'an University of Architecture and Technology.

[6] Laako M and Kiviniemi AO The IFC standard: A review of History, development, and standardization 2012 *Information Technology*.

[7] Černý M, Ward A, Benghi C and Lockley S xBIM project https://github.com/xBimTeam.

[8] Laat R D, Lindeque R and Nijs B D BIMSever https://github.com/opensourceBIM/BIMserver.

[9] Chen Y, Shooraj E, Rajabifard A and Sabri S From IFC to 3D Tiles: An Integrated Open-Source Solution for Visualising BIMs on Cesium 2018 *ISPRS International Journal of Geo-Information* 7(10):393.

[10] Moult D, Krijnen T, Kämäräinen A and Hahnebach B IfcOpenShell-IfcConvert https://github.com/IfcOpenShell/IfcOpenShell.

[11] Lilley S, Cozzi P and Amato M obj2glTF https://github.com/CesiumGS/obj2gltf.

[12] Rossignac J, Safonova A and Szymczak A 3D Compression Made Simple: Edgebreaker on a Corner-Table 2001 *IEEE Shape Modeling International Conference. Genoa: Georgia Institute of Technology* 278-283.

[13] Lilley S, Taglang R, Li K N and Cozzi P gltf-pipeline https://github.com/CesiumGS/gltf-pipeline.

Intelligent Blockchain System Based on Differential Privacy

Z J WU[1], H DAI[1], Y K YUAN[1] and D MENG[1]

1 School of Computer and Software Engineering,University of Science and Technology Liaoning, Anshan 114051,China

Z J WU Email: 939674778@qq.com

Abstract. Due to the advantages of decentralization, traceability and non deletion, blockchain technology can be well combined with differential privacy algorithm to protect data privacy. Therefore, it effectively ensures the privacy security of data stored in the system in terms of technical structure and algorithm. This paper proposes a micro service system architecture based on super ledger fabric, which carries out data privacy noise processing combined with smart contract, and carries out different noise processing according to the security level recognized by the system at the time of user registration. The implementation and test results of the system show that the system architecture has good scalability, traceability and security, can meet the query control of users for data privacy, and effectively prevent the tampering and destruction of data by malicious attackers.

1. Introduction

In the era of big data, the storage and use of personal information is a pain point faced by the current Internet system technical architecture. The traditional system architecture adopts micro service or distributed architecture, which is stored in the central server or the third-party cloud disk architecture[1]. The system management mostly adopts the way of human management. In this way, there will be potential code vulnerabilities, the possibility of human subjective tampering and destruction of data and human theft of data. Because the traditional system does not record the query records of the thieves, the difficulty coefficient of supervision and investigation of information theft is higher[2]. Due to the advantages of decentralization, traceability and non deletion, blockchain technology can be well combined with differential privacy model in the field of data privacy security protection. Therefore, it can effectively ensure the privacy and security of data stored in the system in terms of technology and algorithm. This chapter proposes an architecture model of micro service system based on super ledger fabric, which is combined with smart contract for data privacy and noise processing. Finally, the test results of hyperledger caliper, a blockchain testing framework, show that the framework has better interaction performance than Ethereum blockchain technology architecture[3].

2. System architecture

Based on the noise processing algorithm of differential privacy and combined with blockchain super ledger technology, this paper designs a blockchain protection framework model based on differential privacy[4]. This structure has the characteristics of micro service architecture, which is convenient for the scheduling and management of blockchain business, and also reduces the attack risk faced by the traditional architecture. The differential privacy protection framework model based on blockchain is shown in Figure 1. The architecture is mainly divided into five modules: Micro service module, local database, cloud cluster and source information collection module.

The crowd gathered. Cloud clusters periodically collect data in the network. The data interaction between cloud clusters will trigger some transactions in the blockchain, which will be uploaded to the blockchain network periodically. Cloud clusters need to interact with the source information collection module.

Local database. The local database is responsible for storing some unnecessary storage information obtained from the source information collection module[5]. Cooperating with the blockchain network can greatly reduce the pressure of the blockchain system and effectively guarantee the data. Once the data is leaked, it is convenient for traceability and query.

Blockchain network. The blockchain network stores the data collected from the source information module through the smart contract, protects the integrity of the data, and stores it in a distributed manner, which can ensure stability and security. Due to the non erasable characteristics of the blockchain, every transaction on the blockchain can be followed.

Microservice module. As the core part of the whole system, the microservice module provides a one-stop solution for service flow limiting degradation, service registration and discovery, message driven, object storage, distributed read scheduling and distributed transactions in the microservice system architecture[6]. The common problems easily encountered in the system architecture are effectively avoided.

Source information collection. As the hub of the blockchain protection framework, the source information collection module interacts with the blockchain network through smart contracts, and has the function of collecting and commanding clusters. The real data uploaded to the blockchain is disturbed by differential privacy injection[7]. When the data from the cloud cluster is received, you can choose whether to submit it to the blockchain. If the information is judged to be non essential storage type, the data will be submitted to the local database for storage.

3.Process analysis of protection framework

In the blockchain protection framework model based on differential privacy, there are three stages of data interactive transmission between modules. It includes data receiving and block packaging of the blockchain system[8], disturbance forwarding and audit decision-making of data. The specific steps are as follows.

Blockchain data packaging. The data of the blockchain is uploaded to the blockchain through the smart contract through the source information collection module. Once the smart contract is triggered, the transaction will be triggered. The blockchain packs the transaction information into the node of the Merkle tree through sha256 hash value calculation.The root node of Merkle tree is calculated recursively and stored in the blockchain header of the block. The root node, timestamp, random number, hash value of the previous block, difficulty coefficient, etc. in the Merkle tree form a new block[9].

Disturbance forwarding of data. For the audited data, the default data is real and secure. In order to prevent malicious attackers from damaging the blockchain architecture and malicious reading of data. Adding differential privacy noise to the data stored in the blockchain and disturbing the real value can effectively prevent malicious attackers from damaging the blockchain.

Review decisions. The data is collected through the cloud cluster, and the time sequence is recorded, and the cluster operation of the cloud cluster is monitored for abnormalities. After passing the audit, corresponding decisions are made for the results.

For the blockchain protection framework model based on differential privacy in this chapter, the following process is designed. Firstly, at the beginning of the operation of the model framework, the operation state of the current blockchain network will be detected. If it is in the initialization state, the current blockchain network needs to be created through the configuration file of the blockchain network, and the digital ID card will be issued by the CA organization of the blockchain consensus node to join the node into the network. First, the response data information is obtained from the source data acquisition module, and the data information is sent to the endorsement node in the blockchain consensus node for endorsement. If the endorsement is passed, the data information is broadcast and recorded in the blockchain network, and the hash value and transaction timestamp of the block are generated at the same time,

4. Framework workflow

The microservice framework model is used to distribute and process blockchain requests. These functions are mainly undertaken by microservices. The internal workflow of the module is as follows.

The blockchain protection framework based on differential privacy combines the springcloud Alibaba microservice framework and blockchain technology. The system adopts micro service architecture to ensure the fine-grained division of services and service process control. The micro service architecture mode can divide the functional requirements of privacy protection applications into a group of small service groups, so that each service can run in its own process, and the communication between services adopts lightweight communication mechanism. Combined with the characteristics of the framework, the system uses Nacos for service registration and discovery component, ribbon for internal service node load balancing, sentinel for service fault tolerance component and gateway for internal gateway forwarding component. The blockchain adopts fabric super ledger technology as the blockchain technology platform. The platform is equipped with smart contracts, which can realize the automatic operation of functional contracts, so as to avoid artificial third-party tampering. Since there is a consensus node in the fabric super ledger, and the blockchain is interacted in the form of non smart contract, the request cannot be filed by the consensus node. Therefore, the platform architecture can only interact with the blockchain platform through the call of smart contract. The operation status of the smart contract and the data stored in the blockchain cannot be tampered with. Even if an attacker wants to carry out malicious attacks and data coverage against the smart contract and data, even if the forged data is submitted to the blockchain platform by means of authentication bypass, the blockchain will record the malicious operations of the user, Thus, by reviewing these submitted information and tracing the hash value of the address requested by the attacker, and searching and querying the hash address, we can quickly obtain the real identity of the attacker. In the blockchain protection framework of differential privacy, the overall architecture of combining springcloud Alibaba and blockchain fabric server is shown in figure 1.

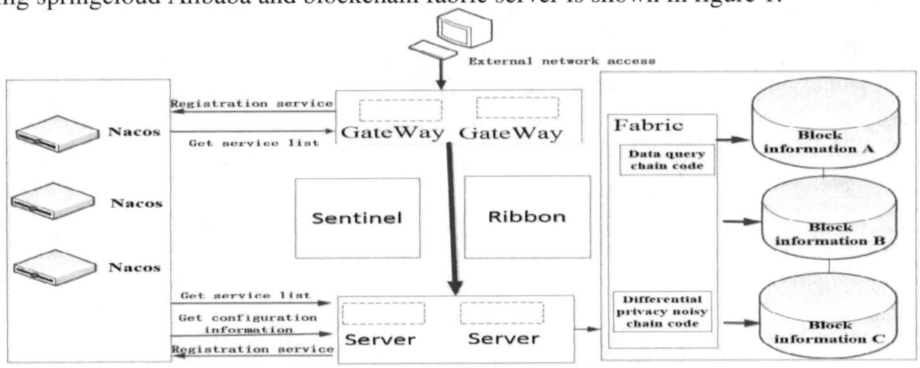

Figure 1. Blockchain system architecture.

5. Analysis of test results

Firstly, relevant function tests are conducted for the user query smart contract, differential privacy noise smart contract and data uplink smart contract submitted to the blockchain environment.The chain code function test results are shown in Table 1.

Table 1. Function test of blockchain code.

Chain code function name	Expected output	test result
Query_cc	Data can be queried according to query criteria	Pass
DPrivacy_cc	It can disturb the data	Pass
Insert_cc	The blockchain browser enables you to query the changes of blocks	Pass

Hyperledger caliper is a general blockchain performance testing framework with excellent performance. It allows users to test different blockchain solutions with customized use cases and get a set of performance test results. Caliper can test the performance of hyperledger fabric's blockchain platform. Caliper currently supports performance indicators including throughput, transaction delay, blockchain resource consumption, CPU, memory, and network io. Caliper is a service that generates a workload for the system under test and continuously monitors its response. After the test, caliper can generate a test report based on the observed kernel SUT response.

As the blockchain protection framework based on differential privacy proposed in this chapter is a blockchain architecture based on super ledger, Ethereum blockchain private chain architecture is used as a comparative test when verifying the fabric framework structure. Ethereum is selected as the comparison architecture. On the one hand, because Ethereum officials provide friendly open source solutions to help Ethereum developers better carry out the development and testing of smart contracts and run their own private chain network for testing, it has the same technical architecture as this experiment. On the other hand, if Ethereum public chain environment is selected as the comparison experiment, Submitting smart contracts on the public chain requires a certain amount of ether, and submitting smart contracts with various functions in the public chain environment will have some uncertain impact on the public chain environment. Therefore, this comparative test uses remix, an online smart contract development tool of Ethereum, to write smart contracts with the same functions and submit them to the private chain environment.

In the experiment of this chapter, 100 data submission queries were conducted for fabric super ledger architecture and Ethereum private chain architecture respectively, and the average time response statistics were conducted for the response time of the two environmental architectures. The experiment used different submitted data sizes for stress test, and the data transmission sizes were set to 1KB, 2KB, 4KB, 8KB, 18KB, 32KB and 64KB respectively, 128KB。 The test results obtained are shown in figure 2.

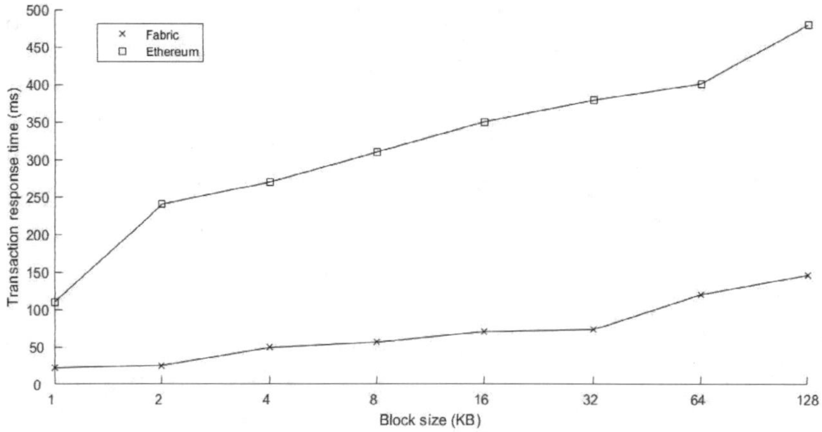

Figure 2. Comparison diagram of Ethereum and Fabric.

From the response time analysis of fabric system architecture and Ethereum private architecture, the architecture based on fabric super ledger tends to be more stable. The average response time will be about 47ms when transmitting 32KB data to 64KB data, while the average response time of the compared Ethereum environment will be higher than that of fabic system architecture, However, from the perspective of average response time, the overall response of fabric meets the request requirements of most blockchains.

6. Conclusion

This chapter integrates the differential privacy algorithm into the blockchain technology architecture, faces the application scenario of the non interactive publishing framework, ensures that the final protection framework model meets the constraints of differential privacy, and greatly ensures the decentralization, traceability and tamperability of the system architecture, The blockchain based differential privacy protection framework constructed in this chapter can not only ensure data quality, but also prevent users' information from being leaked. It provides a new technical architecture and new ideas for solving a series of problems arising from the technical system in the field of information security.

References

[1] Cynthia,Dwork.Aaron The Algorithmic Foundations of Differential Privacy(Foundations & Trends in Theoretical Computer Science) pp 211–407.

[2] Acs G Castelluccia C.I have a dream!differentially private smart metering(.Springer Berlin He'idelberg)pp 118–132.

[3] Jia J J and Huang H.A trajectory(k,e)-anonymous algorithm against trajectory similarity attacks(Computer Engineering &Science)pp 828-834.

[6] Wei Yinzhen and Deng Zhonghua. A scientific data security Traceability Method Based on blockchain and smart contract pp 32-38

[7] Zhu Liehuang, Gao Feng, Shen Meng, et al. Overview of blockchain privacy protection [J]. Computer research and development,(Modern intelligence) pp 2170-2186

[8] Zhao Gang and Liu Tao. Trusted storage technology of unmanned cooperative system based on blockchain [J].(Firepower and command and control) pp 141-144

[9] Weng Xiaoyong. Research on cloud computing data sharing system based on blockchain (Computer engineering and application)pp 120-124

Computing offloading method with low power consumption constraint for smart grid

Yuebin WU[1,a], Pengcheng LU[1,b], Xiande BU[2,c], Shidong LIU[2,d], Cheng ZHONG[1,e], Chuan LIU[2,f]

[1]Xiongan New Area Power Supply Company, State Grid Hebei Electric Power Co. LTD, Xiongan, China

[2]Global Energy Interconnection Research Institute, State Grid Laboratory of Electric Power Communication Network, Nanjing, China

[a]xa_wuyb@he.sgcc.com.cn; [b]13398619196@163.com; [c]buxiande@geiri.sgcc.com.cn; [d]liushidong@geiri.sgcc.com.cn; [e]hebeizhongc@163.com; [f]liuchuan@geiri.sgcc.com.cn

Abstract—With the development of the Power Internet of Things, power transmission and transformation business have a wide range of operation, large regional differences, a wide variety of equipment and other problems, making the access and deployment of mass sensors with low power consumption crucial. In order to realize the cloud-side collaborative computing of the Internet of Things, it is necessary to build a low-power information acquisition and transmission network, and realize the low-power deployment of the low-power sensor network and equipment. The embedded node device can be used as the basis of low-power transmission network to realize low-power transmission of cloud-side collaborative computing information, combined with the existing low-power transmission and transformation equipment.

1. INTRODUCTION

Internet of Things technology has been widely used in smart transmission networks. Edge computing devices have been deployed on transmission lines to realize intelligent perception and local intelligent processing of line status. However, the application of high-performance equipment improves the intelligent level of line monitoring, but also raises the requirements on the power supply of equipment. Only improving the capacity of battery power supply cannot fully meet the power supply requirements of the equipment. Therefore, it is necessary to reduce the requirements on the power supply of the equipment based on the working mechanism of the equipment.

2. LOW POWER CONSUMPTION REALIZATION MECHANISM OF EMBEDDED DEVICE

An important characteristic of embedded system is the non-uniformity and dynamic variation of workload. Since the workload of an embedded system typically changes over time, a balance between system performance and power consumption can be achieved by switching off the device or dynamically adjusting the operating voltage of the processor. After years of research and development, these two technologies have now been proposed at many levels of the system and have become the dominant techniques in the dynamic low power technology design process [1-2].

2.1Dynamic Power Management (DPM)

The basic prerequisites for DPM application are as follows. A system, or system unit, will be under an uneven workload during normal operating hours. The non-uniformity of workload is a very common phenomenon in embedded systems and most interactive systems. The essence of DPM technology is to selectively set system resources to low-power mode according to the changes of system workload, so as to achieve the purpose of reducing system energy consumption. System resources can be used to construct a corresponding model using an abstract diagram of working state, in which each state is a compromise between performance and power consumption. For example, a system resource may have two working modes: normal and sleep. Among them, sleep state has low power consumption, but it takes some time and energy cost to return to Normal state. Switching behavior between states is controlled by commands sent from the Power Management (PM) unit, which observes the workload to determine when and how to switch modes of work. The power minimization (or maximization) strategy model under performance constraints is a constrained optimization problem.

The basic idea of DPM is to view a workload as a collection of multiple task requests. For example, for hard disks, task requests are read and write commands. For the network adapter, the task request contains two parts of packet sending and receiving. The device is in busy state when there are requests. Otherwise, the device is in idle state. Based on this concept, during T1-T4 period, the device is in idle state, in which it is possible to enter sleep low-power working mode. The device is turned off at point T1 and woken up at point T4 when a task request is received. It takes time for this state to change. In the case of hard drives or displays, it can take a few seconds to wake up these devices. It also takes extra power to wake up a device in sleep mode. In other words, changes in the operating mode of the device will inevitably incur additional overhead. Without this additional overhead (both time and energy), DPM itself would not be necessary. Any device enters idle once, status immediately turns it off. Therefore, a device should go into Sleep mode only if the energy savings justify the extra cost. The rules that determine whether a device is worth shutting down are called policies. In the process of power management, we generally only consider the power consumption of the device in Idle state, but not in Busy state.

2.2Dynamic Voltage Regulation (DVS)

Recently, with the development of commercial CMOS chip power supply technology, it is possible to adjust the operating voltage of processor core in real time during operation. The emergence of efficient DC-DC voltage converter also provides conditions for the dynamic adjustment of the operating voltage of the processor core.

In addition, in soft real-time system, tasks only need to be completed before the specified deadline to meet the performance requirements of the system, but there is no requirement for immediate response of the system. DVS technology dynamically adjusts the processor operating voltage according to the urgency of the task to achieve a balance between task response time and low power consumption of the system.

DPM technology can provide significant energy savings for non-real-time systems. However, DPM cannot be applied to real-time system due to its inherent probabilistic characteristics and uncertainty. DVS technology is a good solution to the performance and power consumption requirements of embedded real-time system, which can adjust the processor operating voltage according to the performance requirements of the current running task. DVS technology is primarily based on the fact that the energy consumption of the processor is square proportional to the operating voltage. If only the frequency of the processor is adjusted, the energy savings will be limited. This is because power consumption is inversely proportional to cycle time, and power consumption is proportional to execution time and power consumption. The early DVS principles set the speed of the processor based on its utilization and did not take into account the different requirements of running tasks. Some voltage regulation strategies have been proposed for real-time systems. For example, processors with continuous and discrete voltage variations are discussed. To reduce the voltage that continuously changes processor power consumption, a specific voltage needs to be found for each task to extend the entire execution time to the corresponding deadline. For processors with discrete voltage variations, at least two execution voltages need to be found for each task. The problem of joint online scheduling of periodic and non-periodic tasks is considered. This principle can ensure that the deadlines of all periodic tasks are met and the number of non-periodic tasks accepted is maximized [3-5].

3 Low Power Node Devices Adapted to Power Transmission and Transformation IoT

3.1Hardware Architecture

In view of the growing power business of the IoT, such as overhead line monitoring, cable multidimensional perception and other services, node equipment is required to have higher processing capacity. The high-intensity work of node equipment inevitably brings power consumption problem. In this section, the hardware module circuit design and low power scheme design are processed.

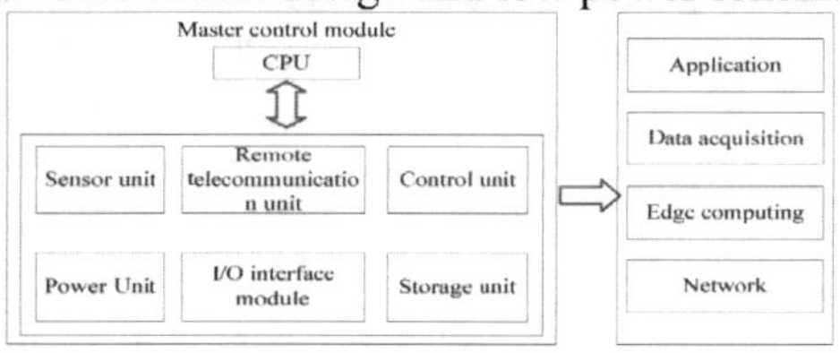

Fig.1. Hardware architecture design of node devices adapted to power transmission and transformation Internet of Things.

Edge computing node devices are mainly used in transmission lines and substation multi-service scenarios. The hardware of the device includes the CPU module, power module, and sensor unit, as shown in Figure 4-1. This section studies the hardware architecture design of node devices adapted to the IoT.

3.1.1 Power supply and management unit: In the transmission and transformation environment, it is difficult to get power, and more battery power is used. The power supply mainly uses lithium battery, and has battery management unit, which mainly completes battery charge and discharge management, electric quantity detection, temperature monitoring, system power supply management.

3.1.2 Sensor unit: Real-time information interaction with sensor nodes. According to the business type of the IoT, a variety of sensors are applied in different scenarios, including temperature, humidity, tension and other sensors.

3.1.3 Control unit: With low-power MCU as the core, on the one hand, edge computing can be carried out, and at the same time, local communication can be carried out through wireless modules. Cloud-side collaborative information interaction can also be carried out with the main station of power transmission and transformation business through remote communication units.

3.1.4 Telecommunication unit: According to the real-time demand of power transmission and transformation service, communication unit should have strong data transmission capability. The communication network of power-saving equipment should be equipped with the functions of special wireless power networks, 4G or 5G public network communication modules.

3.2Low Power Circuit Method

In view of the low power performance requirements of power transmission and transformation edge node equipment, this section studies the low power of hardware units from the aspects of circuits of each module. The devices include digital CMOS circuits, sensor unit circuits, and CPU circuits to reduce hardware complexity, miniaturize and integrate devices, and reduce low power consumption, as shown in Figure 2.

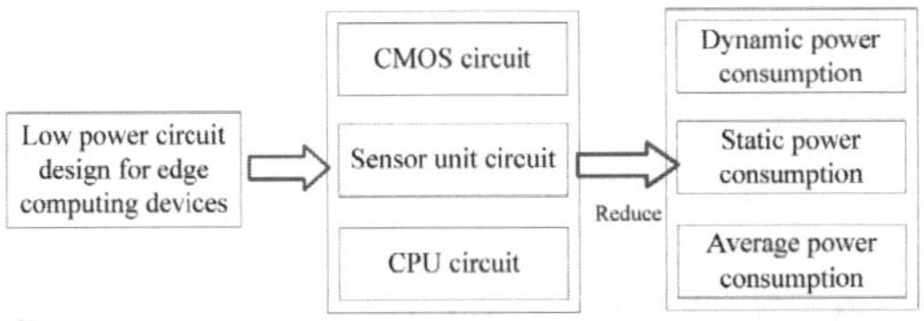

Fig.2. Low power circuit design.

For low power circuit design, the principle of simplifying circuit should be followed, to reduce useless power consumption. Firstly, the power consumption can be effectively reduced by reducing the length of the wire and the load capacitance and the working frequency. When the signal activity is zero, even if the load capacitance is large, the circuit does not consume energy. Therefore, in practical work, when a certain system or module of the circuit does not work and is in a dormant state, the clock of these systems can be shielded, so that part of the circuit can stop working and flipping, so as to reduce the power consumption of the circuit.

3.2.1 Digital CMOS circuit: Digital CMOS circuit is a common low power circuit, which should be selected first for hardware modules and units in node devices. Because short-circuit current power consumption and driving voltage, switching current power consumption is mainly related to load capacitance, driving voltage, turnover frequency. The main source of CMOS circuit power is dynamic power, which consists of two parts: switching current and short-circuit current. Dynamic power consumption is proportional to the square of the operating voltage, and proportional to the switching frequency and load capacitance.

3.2.2 Power design: In power supply design of transmission and transformation equipment, reducing working voltage is an effective way to reduce power consumption. If the rest of the device's power circuit is exactly the same, reducing the operating voltage to 3V can save 75% of power consumption, which indicates the effect of reducing voltage to save energy is of great significance. Therefore, a careful design of the operating voltage of the circuit and the design of an efficient power supply to reduce the power loss in the voltage conversion can reduce the power consumption of the device.

3.2.3 Component selection: The selection of low-power components of the equipment should meet the logic circuit that saves the number of devices and select small devices to reduce the load capacitance, so as to effectively reduce the power consumption of the equipment. In order to give consideration to low power consumption and high efficiency at the system level, the requirements of practical applications must be considered. For example, in special power transmission and transformation environment, the device sensor can turn off the MCU master clock and CPU, and only turn on the low-frequency clock and wake up the peripheral circuit for detection. When an event that meets the specified conditions occurs, the CPU to process the event should be quickly started, so that the CPU can maintain the standby state of the sensor network on the node device.

In view of the special application scenarios of the Power transmission and transformation Internet of Things, the hardware circuits of edge computing node devices are analyzed in detail. According to the actual situation, the design satisfies the different hardware circuit low power consumption scheme. The design of low power circuit follows the principle of digital CMOS circuit. Low power circuits are enhanced according to specific functions between different cell modules [6-7].

4 Offloading and Offloading Strategies of Mobile Edge Computing in the Internet of Things

Mobile edge computing offloading technology refers to a technology that can unload part or all of the tasks to the cloud with more superior and sufficient resources when the resource capacity of the device cannot meet the requirements of the tasks with limited resources and high service demands. This approach addresses many of the shortcomings of traditional networks, such as allocating user resources, improving service performance and reducing power consumption.

4.1 Off-loading Process

The task unloading process of mobile edge computing is mainly composed of six steps, which are described as environment perception, task allocation, unloading decision, unloading task transmission, MEC task execution and task processing result return. The decision to unload a task is critical throughout the process. That is, the user chooses the appropriate decision-making mode according to the task characteristics, which is the key factor to determine the user service experience in the system. Figure 3 shows the process for off-loading tasks

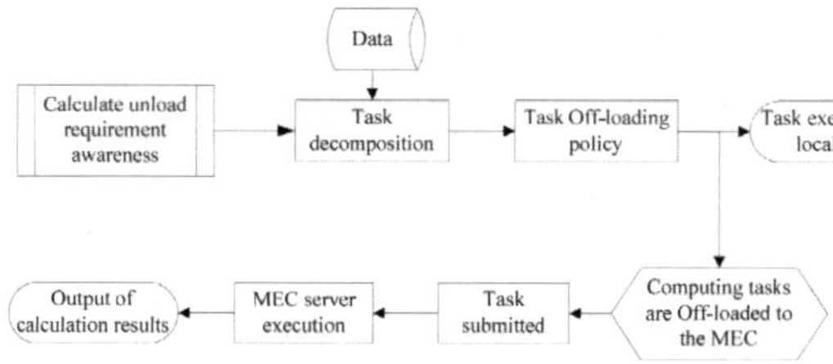

Fig.3. Task off-loading process.

According to the previous introduction of the system framework for moving Edge Computing (MEC), MEC is an infrastructure architecture that operates as a virtual machine. In the MEC system, when a user unloads a task, the task will be first unloaded to the MEC within the communication range. The edge server develops a VM environment corresponding to the task submitted by the user, and then provides the corresponding service to the user. When the MEC server completes the computing task off-loaded by the user, the processing result of the task is fed back to the client through the backhaul link. The MEC server then destroys the virtual machine, along with some reclaiming of available resources. In general, the MEC paradigm can be divided into two categories: binary unload and partial unload. With binary off-loading, computing tasks at the user cannot be partitioned, but must be executed at the user or at the edge cloud. If partial unload is used, tasks can be divided at the user's office so that local computation and unload can be performed.

4.1.1 Off-loading environment awareness: When users in the system have tasks to arrive, UE will first perceive the current network environment. Such as detecting the current wireless channel environment and the working status of edge computing servers that are expected to provide services for their own tasks. These factors will determine the user's off-loading decision to some extent. Therefore, this stage can be regarded as the preparation stage before task off-loading.

4.1.2 Task assignment: When a task is request from the user, it can be divided to some extent before unload. This division can make it easier for users to make decisions on tasks. If the processing capabilities of the MEC in the system are inconsistent, users can select the MEC based on the task requirements. For example, if a computationally intensive task is performed, users may be willing to offload to a server with better resources in order to achieve lower latency.

4.1.3 Off-loading decision: The essence of the user's off-loading decision is to decide whether the current task needs to be off-loaded, the appropriate proportion of off-loading, and the appropriate power to submit the off-loaded data, etc. In the system, when the user makes the decision of task unloading, it usually adopts the appropriate decision algorithm, along with various indicators of task requirements. For example, time constraint of task processing, task energy consumption, service experience, user cost and many other factors are taken to weigh it and then make the best unloading plan for itself.

4.1.4 Task transfer: According to the steps mentioned above, after the user has made the off-loading decision, the task can be delivered with the MEC and submitted to the edge cloud server for execution. Edge computing servers have more processing power and superior resources than end users, and are geographically closer to users than traditional cloud centers, which leads to very direct benefits in terms of low latency and low energy consumption. Therefore, it has more advantages than traditional cloud offloading, so it can bring better and more convenient service to users.

4.1.5 MEC server execution: After the user makes the decision about the task and submits it to the MEC server, the edge server assigns virtual machines to the off-loaded task and uses its own resources, such as computing resources, to process the task submitted by the user.

4.1.6 Results back: This is the last step in the unloading process. After the edge cloud completes the task delivered by the user by using its own resources, the result of task processing will be sent back to the user through the backhaul link. Users receive the feedback results of MEC, and then process and use them according to their own needs. The user at this stage may unload the delivery again in response to the received volume processing results, or terminate the request and disconnect from the edge server.

4.2 Off-loading Model

The unloading model of tasks directly affects the decision making of unloading. For the user's computing task model, there are mainly two unloading models, namely the two-state unloading model and the partial unloading model of the task.

4.2.1 Binary off-loading: The dimorphic unloading model considers each computing task as an indivisible and independent unit. An independent computing task can be processed locally (without unload) or unloaded to the MEC server (unload). Typically in this type of unloading model, the task is represented by a triplet λ (D,C,T). D represents the number of input bits of the task. C represents the computing resources required by the task, the number of available CPU cycles. T represents the time limit for completing the task.

*4.2.2 Partial off-loading:*For partial off-loading, computing tasks can be regarded as further subdivided into sub-task modules. For each subtask, either execute locally or off-loading can be processed. Thanks to the development of code decomposition and parallel processing technology, this model has been paid more and more attention. Due to the refinement of unloading granularity, through the parallel execution of each sub-task, computing tasks can be completed more effectively and faster and computing resources can be utilized more rationally, thus improving unloading efficiency. In addition, there are dependencies among subtask modules in the partial unload model (The output of one module may be the input of another module). The order in which each subtask is executed should be considered. Although the partial unloading model further refines the unloading granularity and improves the unloading efficiency, the partial unloading model must consider the execution order between sub-tasks. However, in multi-user scenarios, it is too complex to consider the execution order of each user's tasks, unload decisions and resource allocation at the same time. Therefore, the two-state unloading model of tasks is generally considered in multi-user scenarios, while the partial unloading model is more common in single-user scenarios.

4.3 Formulation of Task Unload Decision

Edge computing technology is applied to transmission Hub Net, which aims to realize local data processing and analysis and transmission station operation and maintenance control. However, due to the limitations of computing, storage and power consumption, Hub Net can unload the monitoring video analysis, electricity analysis and other computation-intensive tasks to the master station server for operation. Only sending computing tasks and receiving results are needed. There is no need to take up local computing and storage resources, which can effectively solve the problem of local resource constraints. The computational unload technology actually solves the joint optimization problem of computing resources and communication resources allocation in the technical support system of transmission stations.

The decision of task unloading mainly considers the execution location (where), execution time (when), which tasks (what) to off-loading, and how much wireless resources and computing resources to allocate to each task. At present, there are two main decision-making modes: centralized control and distributed control.

4.3.1 Central station control

Centralized computational unload control is a centralized optimal control based on a global perspective by integrating the wireless channel state in the network, the number of resources in the network, the number of tasks requested at the same time and the requirements of each task. The objective is to obtain the optimal results of scheduling, resource allocation and unloading decisions. Generally, the role of the centralized control unit is assumed by the base station or MEC server. In the multi-cell collaboration scenario, the centralized control unit can be a macro station, a unified aggregation point, or an SDN controller with a global perspective to achieve the separation of control signaling and specific execution.

4.3.2 Distributed control

Distributed computing offload control, which focuses on the specific preferences of each user/computing task, is an equilibrium rather than optimal result for multiple users to maximize their utility function. Each user is not only the initiator but also the decision maker of task unloading, which is usually modeled by game theory or auction model. In specific research, users can make the choice according to the focus and scene of the research problem. Centralized control can obtain optimal (or sub-optimal) task scheduling results from the global perspective, but the complexity is large. Distributed control is widely used in the Internet of Things. Because of the massive device connection, the distributed control method can get the decision result quickly with low complexity.

4.4 Calculation unloading politics

The main factors affecting the decision of off-loading are network transmission technology, basic resources of Hub Net, off-loading of application program and basic resources of host server. The most important factors to be considered are delay and energy consumption of Hub Net transmission.

4.4.1 Unloading decisions aimed at reducing delay

For computing tasks requiring high real-time performance, the offloading decision aiming at reducing delay should be adopted so as not to affect the quality of service when the computing tasks are offloaded to the master station. If the computing task is executed locally, the time taken is the computing time, i.e.,

$$T_E = \frac{m}{v_E} \tag{1}$$

Where, T_E is the local execution time of a computing task, m is the computing resources required to complete the computing task, and v_E is the local execution speed.

If the computing task is unloaded to the primary server, the total time spent is composed of three components.

That is, the time for transmitting the data required for computing to the master site, the time for the master site to perform the computing task, and the time for the master site to receive the data returned from the master site. That is,

$$T_C = \frac{b_1}{B_1} + \frac{m}{v_C} + \frac{b_2}{B_2} \quad (2)$$

In the formula, T_C is the time required for unloading the calculation task to the master station, and b_1 is the size of data transmitted to the master station; B_1 is the current network bandwidth; v_c calculates the speed of the master station; B_2 is the size of the returned data; B_2 is the network bandwidth at return time.

If $T_E > T_C$, that is, the local execution time exceeds the execution time when the computing task is off-loaded to the primary site, the computing task is off-loaded to the primary site to meet the minimum delay requirement.

If $T_E > T_C$, that is, the local execution time is less than the off-loading time to the primary site, the computing task is executed locally to meet the minimum delay requirement.

4.4.2 Unloading decisions aimed at reducing energy consumption

Low power consumption is an important index of Hub Net. Unloading some computing tasks to the master station will greatly reduce the energy consumption of Hub Net transmission. Calculate the power consumption of a locally executed task.

$$E_E = P_E \cdot \frac{m}{v_E} \quad (3)$$

In the formula, E_C is the energy consumption of the local execution of the task. P_E is the operating power of Hub Net.

The energy consumption of a computing task unloaded to the master station consists of two parts: the energy consumption of unloading data transferred to the cloud and the energy consumption of receiving returned data.

$$E_C = P_1 \cdot \frac{b_1}{B_1} + P_2 \cdot \frac{b_2}{B_2} \quad (4)$$

In the formula, E_C is the energy consumption of the unloading task to the main station. P_1 is the transmission power during data upload. P_2 is the transmitted power when data is received.

If $E_E > E_C$, that is, the energy consumption of the local computing task exceeds that of the local computing task being off-loaded to the primary site, off-load the computing task to the primary site to meet the minimum power consumption requirement.

If $E_E > E_C$, that is, the energy consumption of the local computing task is less than that of the off-loading to the primary site, the local computing task is executed to meet the minimum power consumption requirement.

4.4.3 Unloading decision aiming at tradeoff delay and energy consumption

Different application scenarios have different requirements for delay and energy consumption. How to take two optimization objectives into consideration is an important factor in unloading decision. The total transmission cost can be set as $Q = \alpha T + \beta E$, $\alpha + \beta = 1$, α, β are the weight coefficients of delay and energy consumption respectively. Therefore, unloading decision is based on joint optimization of computing and communication resource allocation.

5 Tests and Results

Figure 4 shows the relationship between the total energy consumption of the system and the calculation demand. The results show that the energy consumption of the three schemes increases with the increase of computing requirements. Similarly, when only local calculation is performed, the energy consumption is the largest, and the MEC scheme has the middle performance, and the complete unloading scheme has the lowest power consumption. The performance of the scheme in this paper is significantly better than other schemes. With the increasing of computing tasks, the gap between the proposed scheme and other schemes is basically expanding. In order to meet the requirement of task delay, when computing tasks increase, more tasks will be unloaded to optimize the system energy efficiency.

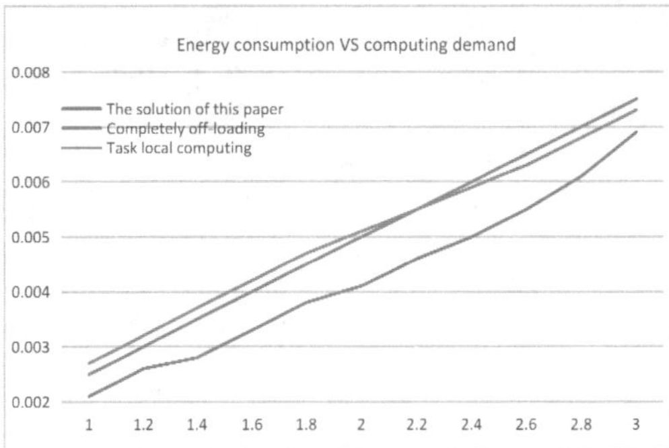

Fig.4. Power consumption in different task off-loading modes

6 Conclusion

This paper proposes a low-power implementation method for the edge devices of the Power Internet of Things. It uses computing offloading technology to solve the problem of the distribution and joint optimization of computing and communication resources when energy is limited, and occupies less local computing and storage resources, so as to effectively solve the problem of local resource limitation. At present, the computational unloading technology has been widely used in the transmission lines of urban distribution network. In the state of no charge, the equipment can be extended twice the working time. In the future, the hibernation technology can be combined to further extend the working time of the equipment and effectively reduce the power consumption of the equipment.

Acknowledgment

This work is supported by the Science and Technology Project of State Grid Corporation of China: Research and application of key technologies for wireless communication network coverage of urban power underground pipe gallery (Grant No. 5700-202113189A-0-0-00).

References

[1] Y. Chen, L. Wei, and Q. Jiao, "A Low Power Communication Technology for Power Line Status Surveillance," Automation of Electric Power System, vol. 36, No.24, pp. 69-73, December 2012.
[2] Y. Xie, "Design of Low Power Consumption and Real-Time Monitoring System of Micro-meteorological for Power Transmission Line", Process Automation Instrumentation, Vol.39, No.4, pp.33-36. April 2018.
[3] L. Gou, J.Q. Jia and S. Cao, "Analysis of Special Network of Energy and Power IoT Based on LPWAN Technology," Telecom Power Technology, Vol.37, No.15, PP.127-129, August 2020.
[4] L. Gou, J.Q. Jia and S. Cao, "Analysis of Special Network of Energy and Power IoT Based on LPWAN Technology," Telecom Power Technology, Vol.37, No.15, PP.127-129, August 2020.
[5] Internet of Things in 2020, Roadmap for the future. http://www.smart-systems-integration.org /public/internet-ofthings, 2010.
[6] Z. Nie, J.M. Zhang and H.W. Fu, "Key Technologies and Application Scenario Design for Making Distribution Transformer Terminal Unit Being a Containerized Edge Node," Au
[7] M. Young, The Technical Writer's Handbook. Mill Valley, CA: University Science, 1989

Intelligent computing and analysis of breakdown voltage of rod-rod long air gap

Xiuyuan Yao[1], Bingxue Yang[2], Jianghai Geng[2]*, Yu Su[1], Hua Yu[3], Guohua Yang, Xinwei Wang[3]

[1]China Electric Power Research Institute Co., Ltd., Beijing, China

[2]North China Electric Power University, Beijing, China

[3]Electric Power Research Institute of State Grid Shanxi Electric Power Company, Taiyuan, China

[4]Pinggao Group Co., Ltd., Pingdingshan, China

[*] Corresponding author: gengjianghai@163. com

***Abstract*—**The discharge characteristics of air gap is one of the important factors to determine the insulation level which is crucial for electrical engineering construction. In order to calculate the 50% breakdown voltage of rod-rod air gap under different meteorological conditions and different gap distances, we take the positive standard switching impulse test data of rod-rod air gap under different meteorological conditions as samples. Based on training set, a breakdown voltage prediction model of rod-rod air gap under different meteorological conditions was established by cuckoo search optimized weighted support vector regression (CS-ω-SVR). The results show that the average absolute percentage error of the test set is 3.6%, which verifies the validity of this computing method. The method can provide reference for the computation of air gap breakdown voltage under different altitudes and extreme meteorological conditions.

1. INTRODUCTION

The discharge characteristic of long air gap play a significance role in the external insulation design of high-voltage electrical equipment. And changes of gap structures and meteorological conditions will have a greater impact on it [1]. As one of the typical electrode structures, the effective calculation method for the breakdown voltage of rod-rod air gap is beneficial to research the discharge characteristics of typical air gap under various conditions.

Many factors such as gap structures, voltage wave forms, atmospheric conditions and so on can affect air gap breakdown voltage. Due to the intricate non-linear relationship between them, the breakdown voltage usually cannot be accurately calculated by empirical formulas based on traditional statistical methods [2]. Under extreme atmospheric conditions, the accuracy of these traditional calculation methods is further reduced. However, the machine learning algorithms can deal with the complex relationship. Appropriate machine learning algorithms can be used to establish the rod-rod long air gap breakdown voltage prediction model. And this kind of prediction model can solve the shortcomings of traditional calculation methods effectively.

Early researchers mainly used artificial neural network algorithms to construct the air gap breakdown voltage prediction model. And they achieved good prediction results in a small range of test sets [3]. However, the number of parameters of neural network is larger, and the performance on small sample data is often not as good as SVM. The SVM is a training model basis for the principle of structural risk minimization, that can avoid overfitting effectively by introducing regularization item. The air gap discharge data is multi-dimensional, non-linear and small sample data, and SVM has good regression performance for this kind of data, which can establish a prediction model with high accuracy for such data [4]. At present, the intelligent calculation model of air gap is mostly aimed at the short gap, while the long air gap is the usual insulation form in real engineering application. And most of the researches on breakdown voltage prediction of long air gap only focus on the rod-plane gap, and other typical electrode structure gaps are rarely involved.

In order to obtain the required samples for modeling, a series of rod-rod air gap switching impulse test are conducted in UHV DC test base of State Grid Corporation of China. And 70 samples were obtained from spring to winter in Beijing. In this article, four features are selected as the input variables, and the 50% breakdown voltage of rod-rod long air gap is taken as the forecast target. Firstly, the grey relation degree between input variables and the breakdown voltage is calculated. The breakdown voltage prediction model rod-rod long air gap under positive standard switching impulse voltage is established on account of a weighted support vector regression algorithm optimized by cuckoo search. And the reserved test set is used to verify the validity of the model. Last, the trained model is used to research the discharge characteristics under extreme temperature and humidity.

2. METHODOLOGY

2.1 Grey Relation Analysis

The grey relation analysis is an algorithm to determine the relation between factors through the geometry similarity of the change curves of various factors in the system, and the relation between different factors is quantitatively described by grey relation degree [5]. The basic steps are as follows:

(1) Normalize the original data (or standardization, etc.).

(2) Determine the reference sequence: $x_0 = \{x_0(k)|k = 1,2, \dots, n\}$; comparison sequence: $x_i = \{x_i(k)|k = 1,2, \dots, n\}(i = 1,2, \dots, m)$

(3) Calculate the relation coefficient $\gamma(x_0(k), x_i(k))$ of each corresponding point of reference sequence and comparison sequence:

$$\gamma(x_0(k), x_i(k)) = \frac{\min_i \min_k |x_0(k)-x_i(k)| + \xi \max_i \max_k |x_0(k)-x_i(k)|}{|x_0(k)-x_i(k)| + \xi \max_i \max_k |x_0(k)-x_i(k)|} \quad (1)$$

(4) Calculate the grey relation degree $\gamma(x_0, x_i)$ of reference sequence and comparison sequence:

$$\gamma(x_0, x_i) = \frac{1}{n}\sum_{k=1}^{n} \gamma(x_0(k), x_i(k)) \quad (2)$$

2.2 CS-ω-SVR Model

The SVR is a supervised learning method. Its main idea is to map input variables to a high-dimensional space through the kernel function, and then the simple multiple linear regression is carried out in the high-dimensional feature space [6].

Give the training sample $(\boldsymbol{x}_1, y_1), (\boldsymbol{x}_2, y_2), \cdots, (\boldsymbol{x}_n, y_n)$. We hope to learn a hyperplane to make $f(\boldsymbol{x})$ and y as close as possible:

$$f(\boldsymbol{x}) = \boldsymbol{\omega}^{\mathrm{T}} \boldsymbol{x} + b \quad (3)$$

In (3), $\boldsymbol{\omega}$ is the regression coefficient and b is the bias term. Give a small positive number ϵ. If there is the following formula:

$$|f(\boldsymbol{x}_i) - y_i| \le \epsilon \quad (4)$$

Then the SVR problem can be formalized as:

$$\min_{\omega,b} \frac{1}{2}\|\boldsymbol{\omega}\|^2 + C\sum_{i=1}^{n} l_\epsilon(f(\boldsymbol{x}_i) - y_i) \quad (5)$$

Where, C is the regularization constant, and l_ϵ is ϵ- insensitive loss function:

$$l_\epsilon = \begin{cases} 0 & , \quad if\ |z| \le \epsilon \\ |z| - \epsilon, & otherwise \end{cases} \quad (6)$$

Introduce the non-negative slack variable: ξ_i, ξ_i' and Lagrange multiplier $\mu_i, \mu_i', \alpha_i, \alpha_i' \ge 0$:

$$L(\boldsymbol{\omega}, b, \boldsymbol{\alpha}, \boldsymbol{\alpha}', \boldsymbol{\xi}, \boldsymbol{\xi}', \boldsymbol{\mu}, \boldsymbol{\mu}') = \frac{1}{2}\|\boldsymbol{\omega}\|^2 C\sum_{i=1}^{n}(\xi_i + \xi_i') - \sum_{i=1}^{n}\mu_i\xi_i - \sum_{i=1}^{n}\mu_i'\xi_i' + \sum_{i=1}^{n}\alpha_i(f(\boldsymbol{x}_i) - y_i - \epsilon - \xi_i) + \sum_{i=1}^{n}\alpha_i'(y_i - f(\boldsymbol{x}_i) - \epsilon - \xi_i') \quad (7)$$

Calculate the partial derivatives of $\boldsymbol{\omega}$, b, ξ_i and ξ_i', make them equal to 0, and transform the original problem into its dual problem. Then introduce the kernel function to map the original data to a high-dimensional plane, and obtain the hyperplane expression after solving:

$$f(\boldsymbol{x}) = \sum_{i=1}^{n}(\alpha_i' - \alpha_i)\,\kappa(\boldsymbol{x}, \boldsymbol{x}_i) + b \quad (8)$$

The role of kernel function is to avoid the difficulty in calculating the inner product of vectors in high-dimensional space and RBF is selected as the kernel function of SVR in this paper:

$$\kappa(\boldsymbol{x}_i, \boldsymbol{x}_j) = exp\,(-\gamma\|\boldsymbol{x}_i - \boldsymbol{x}_j\|^2) \quad (9)$$

From the above derivation, it is not difficult to find that the values of two parameters cannot be directly obtained from the sample. And we use Cuckoo search (CS) algorithm to search the optimal C and γ because it has a good global optimization ability [7].

To improve the prediction precision of models, the grey relation degree of each input variable and breakdown voltage analyzed by grey relation analysis is used as the weight of variables in this paper. And it is applied in the calculation of kernel function to construct a weighted SVR model optimized by CS algorithm.

2.3 *Error Analysis Method*

In this paper, we select two error indexes to quantitative the accuracy of predicted results.

(1) Absolute percentage error (APE)

$$e_{APE} = |(U_p - U_t)/U_t| \quad (10)$$

(2) Mean absolute percentage error (MAPE)

$$e_{MAPE} = \frac{1}{N}\sum_{i=1}^{N}|(U_p - U_t)/U_t| \quad (11)$$

Among the tow formulas, U_t is the test values, U_p is the predicted values, and N stands for the number of samples. The smaller the two indexes are, the more accurate the prediction results are.

3. TEST RESULTS AND DISCUSSIONS

3.1 Intelligent Computation and Analysis of the prediction model

In this paper, 70 groups of long rod-rod air gap test data under different meteorological conditions are taken as the samples. Four features of rod-rod gap distance (d/m), temperature (t/℃), air pressure (p/kPa) and absolute humidity (h/g/m^3) are selected as the input variables to construct the prediction model of 50% breakdown voltage of long rod-rod air gap based on CS-ω-SVR. According to the sample distribution of samples, respectively select 2 groups of samples under high temperature, 7 groups of samples under low temperature and 5 groups of samples at about 20℃ as the test set, and others as the training set.

First, all samples after normalization are analyzed based on the grey relation analysis method. And the parameter of ξ=0.5 is set to obtain the grey relation degree between four features and the 50% breakdown voltage of rod-rod gap, as shown in Table I.

TABLE I. THE GREY CORRELATION DEGREE BETWEEN BREAKDOWN VOLTAGE AND FEATURES

Features	Grey correlation degree
d	0.7991
t	0.6147
p	0.6673
h	0.5760

Take the grey relation degree in Table I as the weight of each input variable to calculate the RBF kernel function, and take the mean square error of 5-fold cross-validation as the fitness function to search the optimal parameter of SVR by CS. The parameter setting of CS is: iteration = 100, number of nests = 20 and random elimination probability = 0.25. Input the training set to obtain the optimal parameter: C = 50.00 and γ = 0.34.

Under the condition that the optimal model is obtained, the prediction result of the training set is shown in Table II. It can be indicated that the maximum of e_{APE} of the test sample is 8.8% and the e_{MAPE} is 3.6%. From the comparison between the prediction and test results of Figure 1, it can also be directly seen that the prediction results are pretty close to the test values, which verifies the effectiveness of breakdown voltage intelligent calculation method of long rod-rod air gap proposed in this article.

TABLE II. THE COMPUTATION RESULTS OF CS-ω-SVR MODEL

	d/m	U_t/kV	U_p/kV	e_{APE}/%
Near standard atmospheric conditions	2	1075	988.0	8.1
	3	1339	1261.4	5.8
	5.5	1599	1654.4	3.5
	6.5	1770	1819.5	2.8
	7.24	1877	1863.7	0.7
Low temperature	1	630	658.0	4.5
	2	1045	1033.3	1.1
	3	1282	1271.7	0.8
	4	1490	1621.8	8.8
	5	1865	1788.0	4.1
	6	2054	2058.1	0.2
	7	2078	2103.5	1.2
High temperature	3	1341	1353.6	0.9
	5	1724	1850.4	7.3

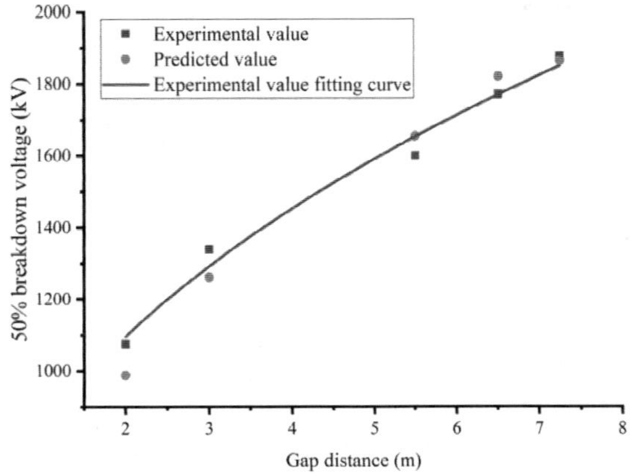

(a) Near standard atmospheric conditions

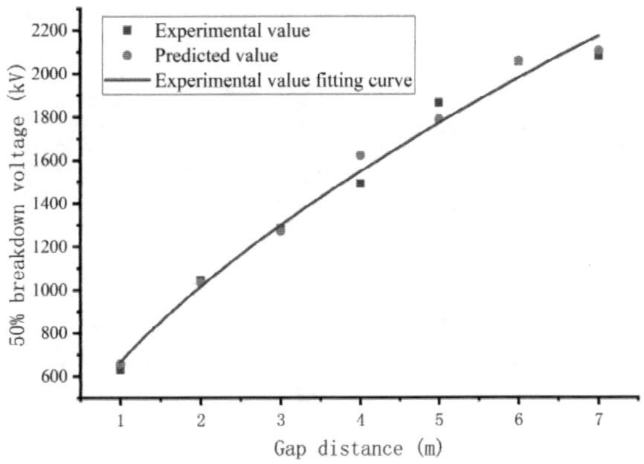

(b) Low temperature

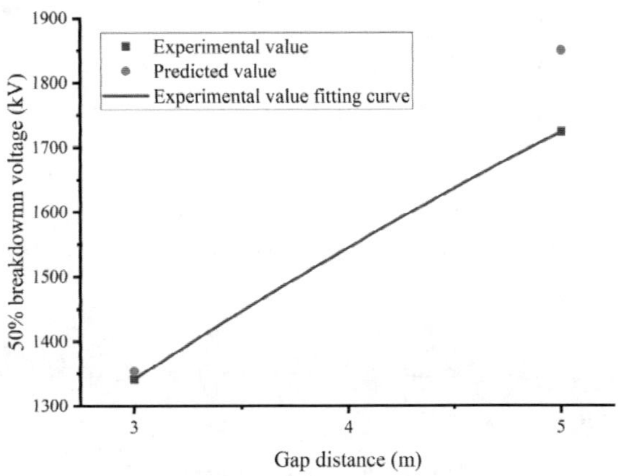

(b) High temperature

Figure 1. Comparison of predicted and experimental value

Journal of Physics: Conference Series **2215** (2022) 012011 doi:10.1088/1742-6596/2215/1/012011

3.2 Calculation of Breakdown Voltage of Rod-rod Gap at the Extreme Atmospheric Conditions

The 50% breakdown voltage prediction model of rod-rod long air gap constructed above can be used to calculate the breakdown voltage at various conditions. Subject to laboratory conditions, it is difficult to obtain discharge characteristics of rod-rod gap at some extreme atmospheric conditions by real test. In this section, we calculate the breakdown voltage of rod-rod air gap under three extreme atmospheric conditions based on the intelligent model and draw the corresponding discharge curve.

Taking the test condition of 70 samples into consideration, we select three extreme atmospheric conditions listed in Table III according to the climate characteristics of Beijing. In the table, Rh (%) is the relative humidity. By fitting of the predicted voltage computed by trained model, the discharge curve under three conditions is drawn as Figure 2.

Through the analysis of discharge curve of rod-rod gap under 3 extreme atmospheric conditions, the results show that the 50% breakdown voltage under high temperature and high humidity is the minimum. And the difference on 50% breakdown voltage under low-temperature drying and low-temperature wetting is not large, which is consistent with the rule obtained from the test. The larger the gap distance, the less obvious the influence of meteorological conditions on discharge characteristics. In other words, the difference of discharge characteristics under different conditions is gradually reduced as gap distance grows. When the gap distance is 2 m, the maximum difference on 50% breakdown voltage under high temperature and humidity condition and 50% breakdown voltage predicted under the other three meteorological conditions is about 13.1%, and when the gap distance increases to 6 m, the maximum difference is about 1.9%.

TABLE III. THE PREDICTED RESULTS OF CS-ω-SVR MODEL

	p/kPa	t/℃	h/g/m³	Rh/%
High temperature and high humidity	101.3	30	15.2	50
Low temperature and high humidity	101.3	-6	2.2	70
Low temperature and dry	101.3	-6	0.4	10

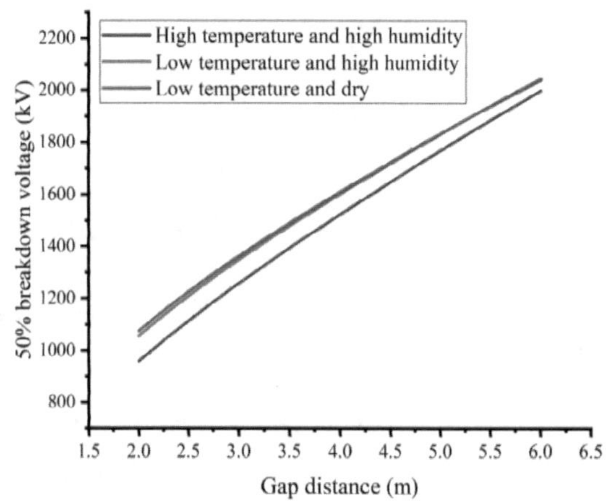

Figure 2. Switching impulse discharge curve of rod-rod gap at three extreme atmospheric conditions

4. CONCLUSION

In this paper, the rod-rod gap distance, temperature, air pressure and absolute humidity are taken as the input variables to construct the prediction model of 50% positive switching impulse breakdown voltage based on CS-ω-SVR algorithm, with the following conclusions obtained:

(1) The ranking of grey relation degree between 50% breakdown voltage of rod-rod gap and its influencing factors from high to low is: gap distance, air pressure, temperature and humidity.

(2) The maximum e_{APE} of the CS-ω-SVR model on test set is 8.8% and the e_{MAPE} is 3.6%. In the 14 test samples, there are 10 test samples with the absolute percentage error predicted by the model under 5%, which verifies the effectiveness of this algorithm.

(3) According to the constructed CS-ω-SVR prediction model of long rod-rod air gap which can be used to calculate 50% breakdown voltage, the discharge characteristics of rod-rod gap under 3 extreme atmospheric conditions are researched, with the results obtained being basically consistent with the rule obtained from the test.

Acknowledgment

This paper was supported by National Natural Science Foundation project of China (51977198, 51861135310) and technology project of State Grid Corporation of China (5200-201955108A-0-0-00).

References

[1] Weijiang Cheng, Rong Zeng, Hengxin He: 'Research Progress of Long Air Gap Discharges', High Voltage Engineering, 2013, 39, (06), pp. 1281-1295. (in Chinese)

[2] Zhibin Qiu, Jiangjun Ruan, Daochun Huang, et al: 'Breakdown Voltage Prediction Method of Rod-Plane Air Gaps Based on Support Vector Machine and Its Applications', Transactions of China Electrotechnical Society, 2017, 32, (19), pp. 220-228. (in Chinese)

[3] Mokhnache L, Boubakcur A, Feliachi A: 'Breakdown voltage prediction in a point-barrier-plane air gap arrangement using self-organization neural networks', IEEE Power Engineering Society General Meeting, IEEE, 2004.

[4] Cortes C, Vapnik V N: 'Support-Vector Networks', Machine Learning, 1995, 20, (3), pp. 273-297.

[5] Julong Deng: 'The relational space of grey system theory', Fuzzy Mathematics,1985, (2), pp. 1-10. (in Chinese)

[6] Lloyd, Brereton Gavin R: 'Support Vector Machines for classification and regression', Analyst, 2010, 135, (2), pp. 230-267.

[7] Yang X S, Deb S: 'Cuckoo Search via Levy Flights', Proc. of World Congress on Nature & Biologically Inspired Computing (NaBIC 2009), December 2009, India, IEEE Publications, USA, pp. 210-214.

Research on Time Delay Feedback Control of Lateral Vibration Axially Moving Viscoelastic Beam

Bo Wang [1], YouQi tang[2]

[1]Mechanical Engineering Institute, Shanghai Institute of Technology, SN: 201400, China

[2]Mechanical Engineering Institute, Shanghai Institute of Technology, SN: 201400, China

196091101@mail.sit.edu.cn

Abstract. The nonlinear dynamic behavior of a viscoelastic beam in axially variable motion under time delay is studied. Considering that the axial velocity pulsation and radial tension change periodically with time under the action of time delay, the viscous damping and finite support stiffness are taken into account, and the viscoelastic constitutive relationship adopts Kelvin model. A mathematical model of lateral vibration of an axially moving viscoelastic beam with changes in axial velocity and tension under time delay is established, and a partial differential-integral control equation describing the lateral nonlinear vibration of an axially moving beam is given. Based on the Galerkin truncation method, the control equations established based on the mechanical model are discretized, so that the direct multi-scale method obtains the numerical solution of the lateral nonlinear vibration of the axially moving beam under the action of time delay, and determines the nonlinear dynamic behavior of the system under the action of time delay. By analyzing the numerical solution of the vibration displacement and velocity at the midpoint of the beam, the whole process of the average velocity of the axially moving structure along the axis, the amplitude of the disturbance tension and the change of the viscoelastic coefficient is simulated.It provides a numerical theoretical basis for the study of the nonlinear dynamic behavior of axially moving beams under time delay.
Through the research in this article, the introduction of time delay is studied, and the time delay is used to analyze the axial force that changes along the radial direction of the beam caused by the speed pulsation, and the theoretical framework for nonlinear vibration analysis of the axially moving viscoelastic beam is constructed to expand the time delay. The application scope of the nonlinear vibration theory and nonlinear dynamics theory, the development and improvement of the approximate analysis and numerical solution of the nonlinear continuum (especially the gyro body) under the action of time delay, to provide a theoretical basis for the analysis and design of the corresponding engineering system and technical reserves.

1. Introduction

With the development of science and technology, the active control of vibration has been received in aerospace, machinery, shipbuilding and other engineering fields..More and more attention has been paid and it has been widely used in practical applications. However, the active control of vibration is inevitable.Bring time delay, although in practical applications, the amount of time delay is very small, but it often affects the stability of the entire system.And system performance[1-4]. Ignoring the time delay may lead to erroneous results. In most cases, the time delay is.It is considered to be an unfavorable factor in the control of the vibration system, and it has a great impact on the stability of the system, and even causes Very complex dynamic behavior[4-8]. Even if it is small, it cannot be ignored, so the time delay needs to be taken into consideration.Time-delay control system to achieve time-delay control of vibration. Aiming at typical nonlinear models, through a lot of researchStudies have shown that time delay can be used as a parameter to suppress nonlinear vibrations[9-11].

At present, the ways to suppress vibration are divided into active control and passive control. Passive control does not require external force. It uses damping devices inside the structure to absorb the energy during vibration to reduce vibration. However, passive control is subject to the characteristics of external excitation, and generally has a good effect on high-frequency vibration, and it has a poor control effect on low-frequency vibration. As the theoretical system of modern control theory matures, the advantages of active control are gradually being discovered. Active control requires the input of external energy. It uses the real-time acquisition of the state quantity for feedback adjustment, and then achieves the suppression of structural vibration. Because the active control does not depend on the characteristics of external excitation, and the control effect is significantly better than passive control, Therefore, the active control of flexible structures has received more and more attention, and many research results have been published.

However, active control inevitably has time delay, which is an inherent property of the system, such as file transmission delay, satellite signal delay, image delay when watching live broadcast, and image shooting and transmission delay in engineering. In the past, people always control the convenience and always ignore the delay. However, even a very small time delay may have a great impact on the system, and even be caused by essential errors.

2. Research on the time-delay stability of axial viscoelastic variable-speed motion

2.1 dynamic model

Consider a uniform Euler beam whose density is ρ, cross-sectional area is A, the length between the two ends of the support is L, the moment of inertia of the section around the neutral axis is I, the coefficient of viscoelasticity is α, and the modulus of elasticity is E to change with time The speed Γ(t) moves along the axial direction. The analysis of the bending vibration of Euler beam is based on the plane assumption of the cross section, that is, the cross section always remains a plane during the bending deformation of the Euler beam. We use the mixed Eulerian-Lagrangian description. **Figure 1** shows the physical model of a viscoelastic Euler beam with axially variable motion. Where a is the location of the speed sensor and b is the location of the exciter.

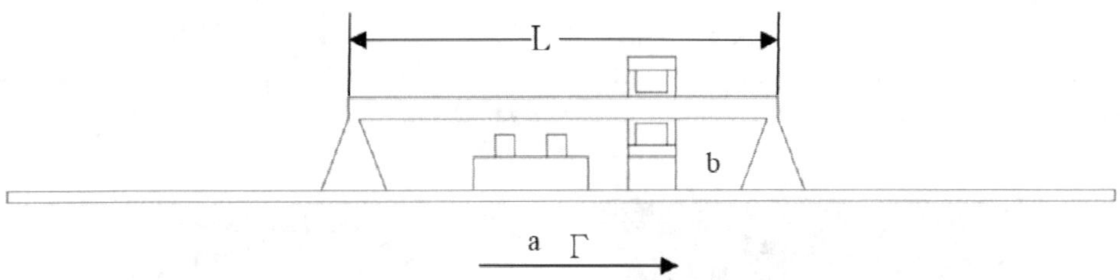

Fig.1 Physical model of viscoelastic Euler beam with axially variable motion

2.1.1 Dynamic equation of axial viscoelastic variable-speed motion

The dimensionless dynamic equation[12] is obtained by applying Hamilton's principle to the axially variable-speed movement Euler beam shown in Figure 2.1 as

$$\begin{aligned} & v_{,tt} + 2\gamma v_{,xt} + [\kappa\gamma^2 - (x-1)\dot{\gamma} - 1]v_{,xx} + k_f^2 v_{,xxxx} \\ & + [\alpha(v_{,xxxxt} + \gamma v_{,xxxxx})] + \frac{1}{2}k_1^2 v_{,xx}\int_0^1 v_{,x}^2 \mathrm{d}x + f(x_0,\tau) = 0 \end{aligned} \tag{1}$$

v(x,t) is the lateral displacement and radial displacement of the Euler beam at the axial coordinate x and time t, the comma "," in the subscript represents partial differential, γ is the initial axial velocity, and γt is the velocity Dynamic, α is the coefficient of viscoelasticity, kf and k1 support stiffness and shear deformation, f(x0,τ)=βvτ,t(x,t)δ(x-x0), x0 is the position of the actuator, and later No further explanation.

2.1.2 Truncated system of time-delay dynamic equations

The Galerkin method is used to discretize the equation (2.1). For the sake of simplicity, considering the gyro effect, N=2 can be obtained.

$$\begin{cases} \ddot{q}_1-\frac{16}{3}\gamma\dot{q}_2+\alpha\pi^4\dot{q}_1-[c\gamma^2+\frac{1}{2}\dot{\gamma}-1]q_1\pi^2-\frac{64}{9}\gamma_t q_2-\frac{128}{3}\gamma\alpha\pi^5 q_2+ \\ k_f^2\pi^4 q_1+\frac{1}{4}k_1^2\pi^4 q_1(q_1^2+4q_2^2)+\beta_1 q_{1\tau}+\beta_2 q_{2\tau}=0 \\ \ddot{q}_2+\frac{16}{3}\gamma\dot{q}_1+16\alpha\dot{q}_2\pi^4-4[c\gamma^2+\frac{1}{2}\dot{\gamma}-1]q_2\pi^2-\frac{16}{9}\gamma_t q_1+\frac{8}{3}\gamma\alpha q_1\pi^5+ \\ 16k_f^2\pi^4 q_2+k_1^2\pi^4 q_1(q_1^2+4q_2^2)+\beta_2 q_{2\tau}+\beta_3 q_{2\tau}=0 \end{cases} \quad (2)$$

Further reduced to

$$\begin{cases} \ddot{q}_1+A_1\dot{q}_2+B_1\dot{q}_1+C_1q_1+D_1q_2+E_1g_1(q_1^2,q_2^2)+\alpha_1\dot{q}_{1\tau}+\alpha_2\dot{q}_{2\tau}=0 \\ \ddot{q}_2+A_2\dot{q}_1+B_2\dot{q}_2+C_2q_1+D_2q_2+E_2g_1(q_1^2,q_2^2)+\alpha_1\dot{q}_{2\tau}+\alpha_3\dot{q}_{2\tau}=0 \end{cases} \quad (2)$$

Where

$$\begin{gathered} q_{i\tau}=q_i(t-\tau), C_1=-\left(c\gamma^2+\frac{1}{2}\dot{\gamma}-1-k_f^2\pi^2\right)\pi^2,\ A_1=-\frac{16}{3}\gamma, B_1=\alpha\pi^4, \\ D_1=-\left(\frac{64}{9}\gamma_t+\frac{128}{3}\gamma\alpha\pi^5\right), E_1=\frac{1}{4}k_1^2\pi^4 q_1, D_2=-\left(c\gamma^2+\frac{1}{2}\dot{\gamma}-1-16k_f^2\pi^2\right)\pi^2, \\ A_2=\frac{16}{3}\gamma, B_2=16\alpha\pi^4, C_2=\frac{8}{3}\gamma\alpha\pi^5-\frac{16}{9}\gamma_t, E_2=k_1^2\pi^4 q_2, g_1(q_1^2,q_2^2)=(q_1^2+4q_2^2), \\ \beta_1=-2\alpha\sin^2(\pi x_0), \beta_2=-2\alpha\sin(\pi x_0)\sin(2\pi x_0), \beta_3=-2\alpha\sin^2(2\pi\dot{q}_1 x_0) \end{gathered} \quad (4)$$

2.2 Steady-state dynamic response

2.2.1 Uncontrolled system

Given the parameters γ=2, α=0.0002, kf=0.8, k1=71.28, β=0, τ=0, the Runge-Kutta method can be used to obtain the time history diagram and phase diagram of the speed feedback control without time delay. As shown in **Figure 2**and **Figure 3**

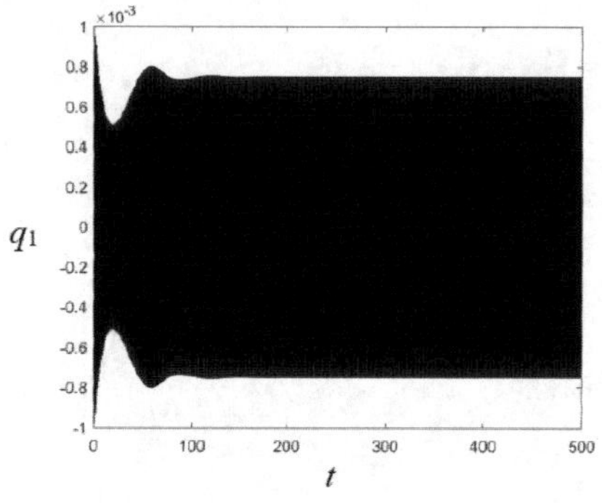

Fig. 2 Time course

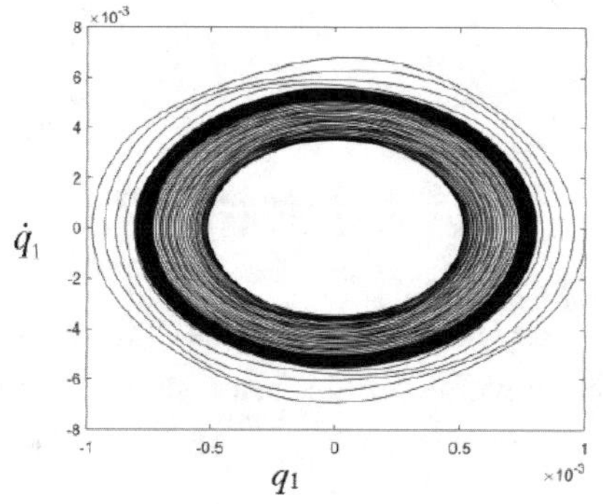

Figure.3General phase

2.2.2 Controlled system

Given the parameters γ=2, α=0.0002, kf=0.8, k1=71.28, β=0.2, τ=0.3, the Runge-Kutta method can be used to obtain the time history diagram and phase diagram of the time-delay speed feedback control. As

shown in **Figure 4** and **Figure 5**

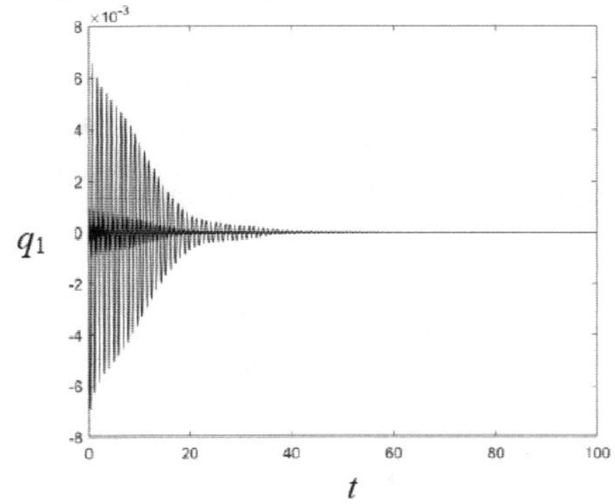

Fig. 4 Time course

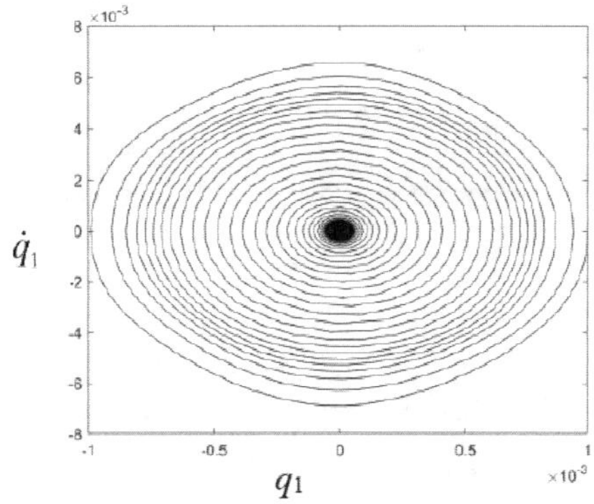

Figure.5 General phase diagram

3. Conclusion

The relationship between the vibration response and the phase of the second-order truncated axially moving viscoelastic Euler beam is analyzed without time delay. By applying the speed delay control, the axially moving viscoelastic Euler beam is obtained. The beam vibration response and phase diagram show that by adding time delay control, the vibration can be effectively reduced. In summary, the research in this article can provide a solid theoretical basis for the vibration of a wide range of axially moving beams. The author also hopes that this research can widely promote the axially moving beam system to move forward in a higher speed and more stable direction.

References

[1] Stepan G. Retarded Dynamical Systems: Stability and Characteristic Functions. Essex: Longman Scientific &Technical, 1989.

[2] Zhang L, Yang C Y, Chajes M J, et al. Stability of Active-Tendon Structural Control with Time Delay[J]. Journal of Engineering Mechanics, 1993, 119(5):1017-1024. Hu H Y, Wang Z H. Dynamics of Controlled MechanicalSystems with Delayed Feedback. Berlin: Springer, 2002.

[3] Niculescu S I. Delay Effects on Stability: A Robust Control Approach. London: Springer, 2001.

[4] Erneux T. Applied Delay Differential Equations. Berlin Springer, 2009 501-512 17-30

[5] Wei, J, Ruan. Stability and bifurcation in a neural network model with two delays[J]. PHYSICA D, 1999.

[6] Cai G, Huang J. Optimal control method with time delay in control[J]. Journal of Sound & Vibration, 2002, 251(3):383-394.

[7] Chung K W, Xu J. An efficient method for studying weak resonant double hopf bifurcation in delay differential equations[C].

[8] Yu P, Yuan Y, Xu J. Study of double Hopf bifurcation and chaos for an oscillator with time delayed feedback[J]. Communications in Nonlinear ence & Numerical Simulation, 2002, 7(1):69-91.

[9] Reddy D V R, Sen A, Johnston G L. Time delay effects on coupled limit cycle oscillators at Hopf bifurcation[J]. Physica D: Nonlinear Phenomena, 1999, 129(1– 2):15-34.

[10] Campbell S A, BeLair J, Ohira T, et al. Complex dynamics and multistability in a damped harmonic oscillator with delayed negative feedback[J] Chaos: An Interdiplinary Journal of Nonlinear ence, 1995, 5(4):640-645.

[11] J, Xu, and, et al. Effects of time delayed position feedback on a van der Pol– Duffing oscillator[J]. Physica D Nonlinear Phenomena, 2003.

[12] Y.Q.,Tang,L.Q.,Chen,X.D.,Yang.Naturalfrequencies,modesandcriticalspeedsofaxiallymovingTimoshenkobeamswithdifferentboundaryconditions,InternationalJournalofMechanicalSciences,2008,50,1448-1458.

Reinforcement Learning Optimized Intelligent Electricity Dispatching System

Xinhang Xiao

Sichuan Tianfu New Area Electric Power Supply Company, State Grid Corporation of China, Chengdu, China

ceongshau@gmail.com

Abstract—With the rapid development of artificial intelligence, new changes are coming to all walks of life including traditional manufacturing industry, bio-pharmaceutical industry, electric-power industry, etc.. For electric-power industry, many machine learning algorithms have been used to achieve intelligent dispatch of electricity, intelligent electricity equipment fault diagnosis and more optimized customer management, among which the automation of power network is one of the most important issue ensuring safe operation of electricity grid. In the recent years, intelligent dispatch method based on big data analysis has been used in intellectual scheduling of electricity and has got significant promotion. However, the intelligent dispatch on big data analysis needs tremendous history data that is sometime not that easy to get. Moreover, once a small probability event happens, the big data based intelligent dispatch method may lose efficacy, thus in this paper we proposed a new intelligent dispatch model based on the reinforcement learning that is more robust, safer and with higher efficiency. Finally, according to the empirical study result, our model outperforms the traditional methods, where the average economic benefit in the test area increases more than 25%, fluctuation of distribution is more stable than before and the carbon emission decreases about 30%.

1. INTRODUCTION

In the recent years, with the development of the energy crisis, safety and stable issues have become a crucial point in the electricity grid. One important and necessary measure to realize power network safe operation is making regional power network dispatching management better. New technologies developed from the artificial intelligence domain offers traditional dispatching management better way to achieve a safer, more stable and more efficient automatic manage system.

The power grid dispatching and control department is a manage center that contains numerous data, rules and experts' experience. The traditional methods for the dispatching system normally rely on the dispatching experience and manual analysis. However, with the significant increasing of the scale of data and data categories, the relation between data and dispatching goal is becoming much more complicated than before, leading to a much more tedious work for the dispatching managers. Meanwhile, higher requirement for dispatching managers of their official knowledge and experience is needed. Moreover, the traditional experience related dispatching method is inefficient and cannot take the long-term benefits into account. Under this background, the using of artificial intelligent methods to solve electricity dispatching problem becomes a necessary the most obvious thing.

Machine learning algorithm is somehow the core part of artificial intelligence domain that can be used to analysis grid feature, get user portrait, electricity equipment fault diagnosis and intelligent dispatch for power grid. Reinforcement learning (RL) is an area of machine learning that concern with how intelligent agents ought to take actions in an environment in order to maximize the notion of cumulative reward. Similar with the learning process that our human beings do, reinforcement learning rewards desired behaviors and punishes undesired ones so that the agent can be better and better.

Reinforcement learning is a young research domain that is mainly being studied in the recent ten years, but it developed in a tremendous speed and has been used in many domains such as resources management in computer clustering, traffic light control, robotics, games, web system configuration etc.. In [1], the authors showed how to use RL to automatically learn to allocate and schedule computer resources to waiting jobs, with the objective to minimize the average job slowdown. In paper [2], researcher tried to achieve automatic control of traffic lights more intelligent and solve the congestion problem by using the RL method. In robotics domain, RL is now playing a significant role because of the excellent learning style of RL and the extreme complexity of intelligent robots, as shown in [3][4][5]. The reason that RL is so famous these days can be owed a lot to its success in games such as the well-known AlphaGo by Google [6] and Dota2 by OpenAI [7], where the computer completely defeated the best human beings. Especially the success of AlphaGo, it gave us human beings an important enlightenment that machine can be trained as smart as human and even with better performance. So if we put its training method, the reinforcement learning method, into medical domain, electricity grid domain and those that closely linked with human life, it is obvious that there will be a tremendous progress.

In this paper, we will introduce the using of RL into electricity dispatching. By considering lots of aspects that can influence the usage of electricity, we defined and trained a RL network to learn the most efficient electricity dispatching decision. Then, the empirical study result shows that the using of RL method in power dispatching can significantly improve the power dispatch efficiency and reduce the waste of electricity resources.

2. RELATED WORK

In this section, we will introduce the basic knowledge of RL, including the definition of RL and how it works.

Reinforcement learning is the science of decision making. It is about learning the optimal behavior in an environment to obtain maximum reward. This optimal behavior is learned through interactions with the environment and observations of how it responds, similar to children exploring the world around them and learning the actions that help them achieve a goal.

Reinforcement learning focuses on training an algorithm following the cut-and-try approach, its working process shows in Figure 1 bellow. The agent refers to algorithm that evaluates a current situation (state). Then the agent takes an action, the action will go through the environment and lead to a certain result, after this interaction the environment will output a feedback which is exactly the reward. The feedback can be positive and negative so the reward can be a positive reward or negative reward respectively. When the reward is a positive one, it means that the agent had a right decision in the last step and the environment encourages it to learn in this direction, while the negative reward means the opposite one. Finally, after many rounds of study iterations, the agent can get a maximized reward and get close to reach its best. The working process of RL is just like us human beings, we learn again and again to be the best we can be. In conclusion, agent is an assumed entity which performs actions in an environment to gain some reward; environment ($\boldsymbol{e}$) is a scenario that an agent has to face; reward ($\boldsymbol{R}$) is an immediate return given to an agent when he or she performs specific action or task; state ($\boldsymbol{s}$) refers to the current situation returned by the environment; policy ($\boldsymbol{\pi}$) means a strategy which applies by the agent to decide the next action based on the current state; value ($\boldsymbol{V}$) is expected long-term return with discount, as compared to the short-term reward; value function specifies the value of a state that is the total amount of reward and can be understood as an agent which should be expected beginning from that state.

Journal of Physics: Conference Series **2215** (2022) 012013 doi:10.1088/1742-6596/2215/1/012013

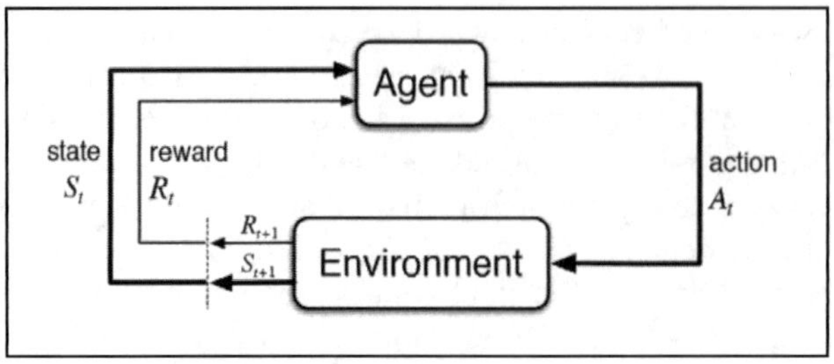

Figure 1. Working process of reinforcement learning

Reinforcement learning can be implemented into three approaches, value-based, policy-based and model-based. In a value-based RL method, we try to maximize the value function ***V(s)***, in which the agent is expecting a long-term return of the current states under policy π. In a policy-based RL method, we try to come up with such a policy that the action performed in every state helps to gain maximum reward in the future. While in model-based method, it is needed to create a virtual model for each environment and the agent learns to perform in that specific environment, which is modelled as a tuple $< S, A, P, R, r >$ of Markov Decision Process.

3. ALGORITHM

3.1 Markov Decision Process

There are two important algorithms in RL that are the fundamental ones, Markov Decision Process (MDP) and Q-learning. Basic reinforcement is modeled as an MDP. An MDP is a discrete time stochastic control process. It provides a mathematical framework for modeling decision making in situations where outcomes are partly random and partly under the control of a decision maker [8]. Additionally, the MDP is the foundation of Q-learning.

An MDP is a tuple $< S, A, P, R, r >$, where:

- S is a finite set of states,
- A is a finite set of actions,
- P is a state transition probability matrix, and
  $P_{ss'} = P[S_{t+1} = s' | S_t = s, A_t = a]$,
- R is a reward function, $R_s^a = E[R_{t+1} | S_t = s, A_t = a]$,
- γ is a discount factor $\gamma \in [0,1]$.

3.2 Q-learning

The other important learning model in RL is the Q-learning, which is a model-free algorithm with the purpose to learn the value of an action in a particular state. The so called model-free is because it does not require a model of the environment and can handle problems with stochastic transitions and rewards without requiring adaptations[9]. To introduce Q-learning we first need to introduce the Q-table. Q-table is a fancy name for a simple lookup table where we calculate the maximum expected future rewards for action at each state, basically, this table could guide people to the best action at each state. In Q-learning, the final purpose is to find out the best Q-table. Given an action a_t and state s_t , the Q value function is:

$$Q^{\pi}(s_t, a_t) = E[R_{t+1} + \gamma R_{t+2} + \gamma^2 R_{t+3} + \cdots | s_t, a_t], \tag{1}$$

where γ variates from 0 to 1. When γ is close to 0, the cumulative reward considers more about immediate benefits, while γ approximates 1, the long term benefit will be thought more. Equation (1) can be then wrote to:

$$Q^{\pi}(s_t, a_t) = E[R_{t+1} + \gamma Q^{\pi}(S_{t+1}, A_{t+1}) | s_t, a_t]. \tag{2}$$

The optimal state action value function is $Q^*(s, a) = \max_{\pi} Q^*(s, a)$ and the expansion formula on expectation is:

$$Q^*(s,a) = \Sigma_{s'} P(s'|s,a)\left(R(s,a,s') + \gamma \max_a Q^*(s',a')\right) \quad (3)$$

After k times of iterations we will get the final Q-table showing in equation 4 below. In Q-learning, the core of learning process is looking for the optimal Q-table, where each action and state of Q-learning corresponds to a Q value that is the exact corresponding reward value and all these values make up the Q-table. In other words the Q-table is the map of Q-learning.

$$Q^*_{k+1}(s,a) = \Sigma_{s'} P(s'|s,a)\left(R(s,a,s') + \gamma \max_a Q^*_k(s',a')\right) \quad (4)$$

Generally, the Q-table uses the Bellman equation to update $Q(s,a)$ and get the best one, the process can be summarized as follows:

1. Initialize Q-values arbitrarily for all state-action pairs.
2. For life or until learning is stopped.
3. Choose an action (a) in the current world state (s) based on current Q-value estimates ($Q(s,\cdot)$).
4. Take the action (a) and observe the outcome state (s') and reward (r).
5. Update $Q(s,a) := Q(s,a) + \alpha[r + \gamma \max_a Q(s',a') - Q(s,a)]$.

4. MODEL

In this section we will introduce our model, including its mathematical structure, parameters, input and output.

Our model is based on the Q-learning method. As introduced in the last section we first need to determine what are the states, actions and design the reward. We choose a large residential quarter level electricity dispatching system as example, discretize the dispatching process that is originally a continuous process. Energy storage in distribution network, load rate of electricity, time and battery capacity are simulated as the states of the system; charge, discharge, electric energy conversion and the increase and decrease of dispatching are set as the actions. Once one of the actions (a_t) happens, the related reward will be transmitted from the environment to the formal state (s_t) that will lead the state update to s_{t+1}.

As we have time as one of the system state, we can discretize the dispatching process into 24 parts, of which one hour is a segment. Assume the energy storage in distribution network, load rate of electricity and battery capacity have L, M, N segments respectively, then we have $lmn * 24$ states, which means that our state set S is $S = \{s_1, s_2, \dots, s_{lmn*24}\}$.

After setting the set of states in our model, we need to determine action set, where each action a_t has a interaction with the environment and will determine the state's transformation $s_t \rightarrow s_{t+1}$. The design of actions is shown in the table bellow:

TABLE 1 POLICY SETTING

Policy	Value a
Charge by thermal power generation	+1
Charge by clean energy (wind, sun)	+2
Energy storage system→distribution	+1
Distribution → energy storage system	-1
Distribution network → ordinary use	-1
Distribution network → commercial use	-2
Leave unused	-1

From the table we can see that we give positive feedback to those policies that increase the amount of distribution network and negative feedback to actions that output electricity to users. When electricity is generated by clean energy like wind, sun light and hydropower, we give larger action feedbacks in order to encourage the use of clean energy.

Journal of Physics: Conference Series **2215** (2022) 012013 doi:10.1088/1742-6596/2215/1/012013

Once the states and actions are determined, we begin the training of our model, based on the Q-learning method, we continuously update the Q-table after every iteration and finally the Q values can infinitely approaches convergence after thousands of training episodes. The training process is showing bellow:

1. Initialize environment.
2. Initialize the original Q-table according to the current environment.
3. Based on the environment including number of residents, season, current electricity prize, transmission and distribution cost and many other variables to calculate the final reward and economic benefits. Then the reward or economic benefits will be put into the *R*-table which refers to the reward that the certain action caused.
4. Under the current state S_t, for example current time, storage in distribution network, determine which action can be chose.
5. Using Q-learning algorithm, combined with the greedy method, choose the corresponding policy and then update Q-table.
6. If Q- I s convergent, end the iteration and get the final Q-table; if not, then continue step 3.

Pseudo-code of the training process of our model is showing bellow:

Algorithm 1

Algorithm parameters: $\alpha \in (0,1]$, **small** $\epsilon > 0$
Initialize $Q(s,a)$, **for all** $s \in S, a \in A$
for each episode **do**
  Initialize S
  for each step of episode **do**
    Choose a from s using policy derived from Q (e.g., ϵ -greedy)
    Take action a, observe r, s'
  end for
  $Q(s,a) \leftarrow Q(s,a) + \alpha[r + \gamma \max_{a'} Q(s',a') - Q(s,a)]$
  $s \leftarrow s'$
  Until S is terminal
end for

5. EMPIRICAL STUDY

In the simulation experiment, we choose a distribution network with the capacity of 5000 kWh, we set energy storage in distribution network, load rate of electricity and battery capacity have 5 segments, which means that we have $5^3 * 24 = 3000$ actions. Policies for each action obeys the policy setting in the last section. We set iteration upper limit as $8 * 10^4$.

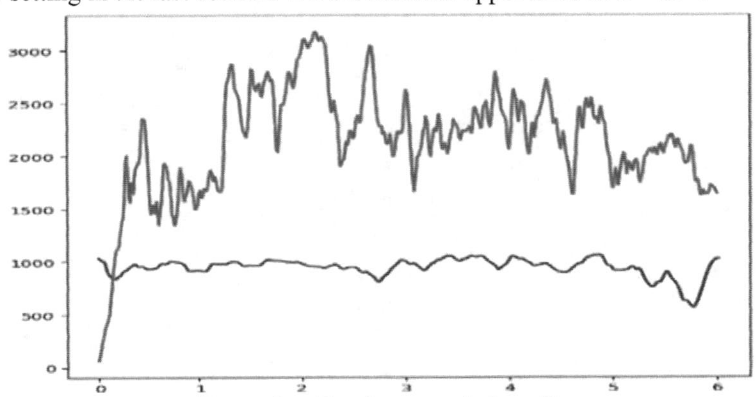

Figure 2. Total economic benefit

In Figure 2 and Figure 3, the green line is the variation trend of economic benefit using the reinforcement learning based model

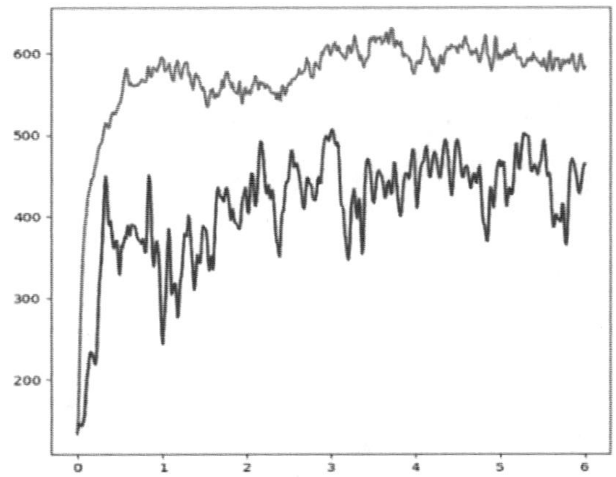

Figure 3. Average economic benefit

for dispatching electricity, while the blue line shows the trend of original experience related methods. It is obvious that the RL method based model can significantly improve the economic benefit for the dispatching system.

Figure 4 and Figure 5 shows the fluctuation of the distribution network before and after optimization by our RL based model. From the two figure we can see that the fluctuation of the distribution network is much more stable than before

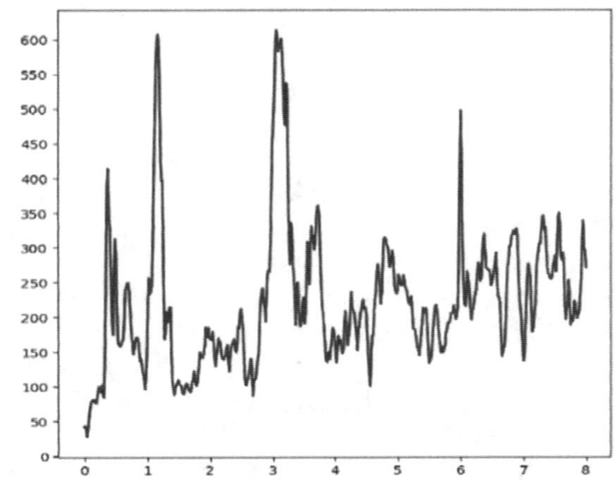

Figure 4. Original fluctuation of distribution network

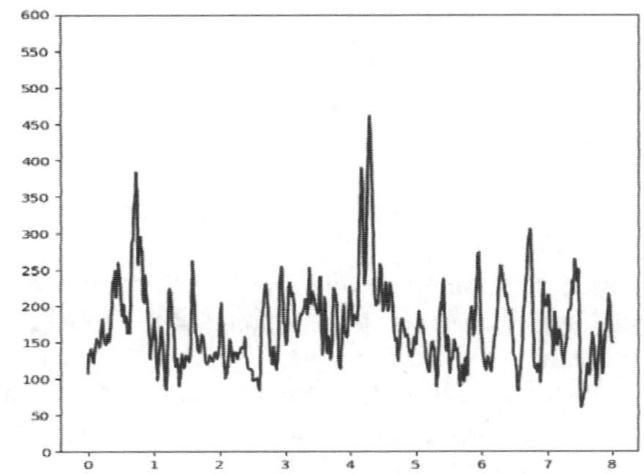

Figure 5. Fluctuation of distribution network after optimization

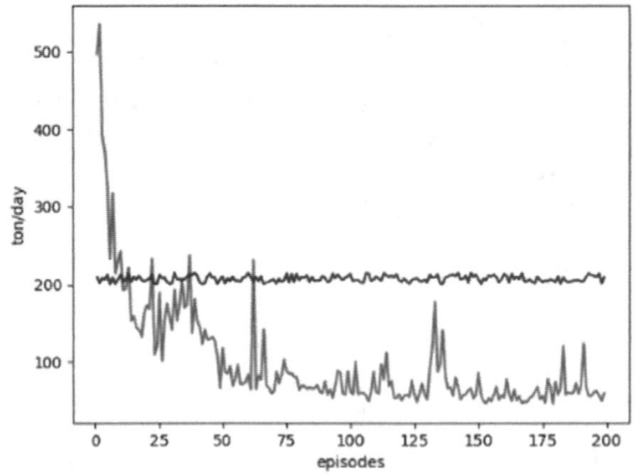

Figure 6. Carbon emission index before and after optimization

Finally, Figure 6 bellow shows the carbon emission after 200 episodes of training. The blue line is the carbon emission per day in the test area variation for original method and the green one shows the variation after optimization, in which we can see that our model has a significant improvement with an improvement of about 30 % in average.

6. CONCLUSION

With the rapid development of artificial intelligence, many astonishing changes have appeared around our usual lives. Even in some traditional industry areas that have developed for years, new artificial intelligence measures are making them work more sensible, reasonable and much more efficiency with higher economic benefits. Reinforcement learning is a new research branch of artificial intelligence that has been widely used in automatic control domain. In this paper, we discuss the using of reinforcement learning into electricity dispatching system that is relevant to the society and our usual lives. From the empirical study we can see that our reinforcement learning based dispatching method can rational allocate resources, allocate power generation tasks reasonably and improve the economic benefit for power grid, etc.. In the next step for further study, we will do more research including design more complicated training networks with deep reinforcement learning methods and use larger electrical network model to simulate our training environment, in order to improve the efficiency of power grid dispatching system further more.

REFERENCES

[1] Mao, Hongzi and Alizadeh, Mohammad and Menache, Ishai and Kandula, Srikanth, "Resource management with deep reinforcement learning," P Proceedings of the 15th ACM workshop on hot topics in networks, pp. 50–56, 2016.
[2] Arel, Itamar and Liu, Cong and Urbanik, Tom and Kohls, Airton G, "Reinforcement learning-based multi-agent system for network traffic signal control," IET Intelligent Transport Systems, vol. 4, pp. 128–135, 2010.
[3] Lin, Long-Ji, "T Reinforcement learning for robots using neural networks" 1992
[4] Mahadevan, Sridhar and Connell, Jonathan, "T Automatic programming of behavior-based robots using reinforcement learning," J Artificial intelligence, vol. 55, pp. 311-365, 1992.
[5] Mahmood, A Rupam and Korenkevych, Dmytro and Vasan, Gautham and Ma, William and Bergstra, James, "Benchmarking reinforcement learning algorithms on real-world robots," Conference on robot learning, pp. 561–591, 2018.
[6] Silver, David and Schrittwieser, Julian and Simonyan, Karen and Antonoglou, Ioannis and Huang, Aja and Guez, Arthur and Hubert, Thomas and Baker, Lucas and Lai, Matthew and Bolton,

Adrian, et al, "Mastering the game of go without human knowledge," nature, vol. 550, pp. 354–359, 2017.
[7] Berner, Christopher, et al, "Dota 2 with large scale deep reinforcement learning," arXiv preprint arXiv:1912.06680, 2019.
[8] Song, Haili and Liu, C-C and Lawarr, Jacques and Dahlgren, Robert W, "Optimal electricity supply bidding by Markov decision process," IEEE transactions on power systems, vol 15, pp. 618–624, 2000.
[9] Watkins, Christopher JCH and Dayan, Peter, "Q-learning," Machine learning, vol 8 (3-4), pp. 279–292, 1992.
[10] Song, Haili and Liu, C-C and Lawarr{\'e}e, Jacques and Dahlgren, Robert W, "Optimal electricity supply bidding by Markov decision process," IEEE transactions on power systems, vol 15, pp. 618–624, 2000.

Journal of Physics: Conference Series **2215** (2022) 012014 doi:10.1088/1742-6596/2215/1/012014

Numerical study of RC slabs under two-point contact explosion load

Xinjing Yuan[1*], Jian Song[1]

[1] School of Highway, Chang'an University, Xi'an 710064, China;

Author's brief introduction: Xinjing Yuan (1997-), Male, Master, E-mail address: 2019121031@chd.edu.cn;

*Corresponding author's E-mail address: 2019121031@chd.edu.cn; yuanxinjing315@163.com.

Abstract. As an important bearing component of bridge, bridge deck is the key to bridge passage. In this paper, a solid model of reinforced concrete slab is established in LS-DYNA software, and the parameters are designed as the amount of explosive and the distance between explosives. The numerical simulation of two-point explosion test is carried out, and the damage mechanism and dynamic response of reinforced concrete slab under two-point explosion load are proposed. The results show that there is a phenomenon of energy accumulation in two-point contact explosion. At the same charge, the failure characteristics and dynamic response of RC slabs are related to the distance between the blast points. When the distance is small (30cm), the stress wave encounters and superposition strengthens between the blast points. When the spacing is large (60cm), the stress waves do not meet between the burst points, but reach the back burst surface and the side of the plate at the same time, and the damage is serious near the side of the plate, and the peak acceleration is small.

1. Introduction

As a key and controlling part of transportation, bridges are a guarantee for smooth road traffic. Occasional explosion of flammable and explosive materials during transportation will cause obstacles to transportation and major economic losses. In particular, on bridges with heavy traffic, once an explosion occurs, it is very likely that two or more vehicles will detonate together on the bridge deck. In the explosion incident on the bridge, the bridge deck is the most vulnerable component. Therefore, studying the damage mechanism and dynamic response of reinforced concrete slabs under two-point contact explosion provides a theoretical basis for the anti-blast design of the bridge deck and the safety assessment of the bridge, which has very important academic and engineering significance.

At present, with the development of computer technology and finite element software, the use of numerical simulation methods to study the dynamic response and damage mode of reinforced concrete slabs under explosive loads can shorten the scientific research cycle and reduce the test risk. Yao et al. [1] numerically simulated the blast resistance of reinforced concrete slabs with different reinforcement ratios, and the results showed that with the increase of the reinforcement ratio, the damage degree, deflection and spall radius of RC slabs all decrease. Casted et al. [2] improved its anti-blast performance by placing steel plates on reinforced concrete slabs and arranging steel fibers in the tension zone. Numerical simulation results show that reinforced concrete slabs containing steel fibers or polypropylene fibers in the tensile stress zone have better blast resistance than ordinary reinforced concrete slabs. Li et al. [3] studied the dynamic response and anti-explosion performance of reinforced concrete slabs under explosion load by means of theoretical analysis and numerical simulation, and analyzed the influence of concrete strength, steel yield strength, plate thickness and boundary on the dynamic response and anti-explosion performance of slabs. In addition, many scholars [4-6] have proposed different material models to study and calculate the dynamic response and damage mode of reinforced concrete slabs, and the research methods have been relatively mature. Scholars [7-10] have carried out relevant tests and numerical simulation and studied the theory of two-point and multi-point explosion. It is found that the energy accumulation phenomenon exists in the multi-point explosion shock wave, which will produce high overpressure. In the case of simultaneous initiation of multiple points, the aggregation effect among multiple shock waves will produce strong damage to the structure.

To sum up, there have been some achievements in the research on the explosion load of reinforced concrete slabs, but all of them focus on non-contact explosion or one-point explosion. There are few studies on contact explosion, especially on two-point contact explosion. In this paper, the 3D finite element model of RC slabs are established by using LS-DYNA software, and a numerical model is proposed to simulate the two-point contact explosion process of RC slab based on multi-substance fluid-solid coupling method. The failure characteristics and dynamic response of RC slabs under two-point contact explosion are studied by numerical simulation results.

2. Explosion test simulation

2.1. Experimental descriptions

The size of rectangular reinforced concrete slab is 3000mm×1000mm×300mm. The fixed mode is hinged at both ends, and its net span is 2.6m. The design parameters of the test are explosive charge and blast point spacing. The explosion load condition is shown in Table 1.

Table 1 Explosion Load Cases of Specimens

Specimen number	Number of explosive points	Explosive spacing /cm	Test dosage /kg
RCP1	2	30	0.5×2
RCP2	2	60	0.5×2
RCP3	2	30	0.25×2
RCP4	2	60	0.25×2

2.2. Basic model

Figure 1 is the basic model of two-point contact explosion simulation, including five parts: steel bar, concrete, air, explosive and rigid body. The reinforcement and concrete are described by Lagrange algorithm. Explosive and air are described by Euler algorithm, which can well simulate the formation and propagation of blast wave and the interaction between shock wave and structure. In the numerical model, concrete and fixed rigid body use SOLID164 element with a grid size of 20mm, steel bar use BEAM161 element with a grid size of 20mm, air and explosive use SOLID164 element with a grid size of 40mm and 20mm respectively. The constraint coupling is set between the reinforcement and the concrete. Fluid-structure interaction is set among air, explosive and reinforced concrete structure. Rigid bodies are in contact with concrete definitions and are used to define articulated boundary conditions. Each explosive shall be detonated independently, and each explosive entity shall have a detonation point with the same detonation time. If the detonation interval is required, different initiation times can be defined.

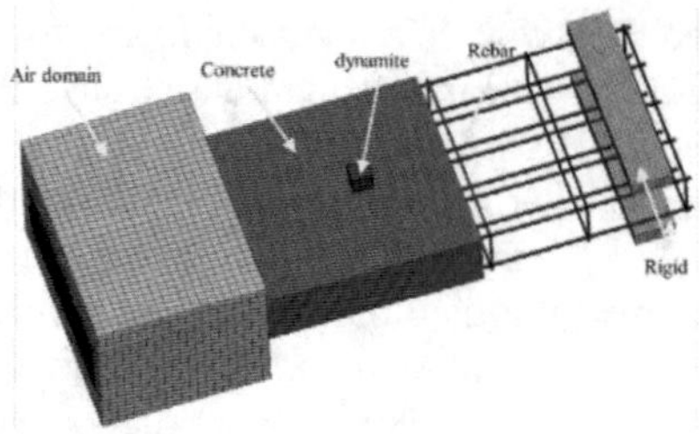

Figure 1 Basic finite element model

2.3. Material model and parameter settings

2.3.1. Detonation products. The explosive material in the model is JH-2. The JWL equation of state, the explosive material model and the *P-V* relationship of the equation of state are used for the calculation of the detonation product action process as follows:

$$P = A(1-\frac{\omega}{R_1 V})\exp(-R_1 V) + B(1-\frac{\omega}{R_2 V})\exp(-R_2 V) + \frac{\omega E_0}{V} \tag{1}$$

Where *P* is the detonation pressure; *V* is the relative volume; E_0 is the initial internal energy per unit volume; ω, *A*, *B*, R_1, and R_2 are the material constants; the value of the parameter in the model is *A*=3.74×1011; *B*= 3.23×109; R_1=4.150001; R_2=0.95; E_0=7.0×109; *V*=1, ω=0.3.

2.3.2. Air. Simplify air as an inviscid ideal gas, assuming that the expansion of the shock wave is an isentropic adiabatic process air material model and linear polynomial equation of state:

$$P = C_0 + C_1\mu + C_2\mu^2 + C_3\mu^3 + (C_4 + C_5\mu + C_6\mu^2)E_0 \tag{2}$$

$$\mu = \frac{1}{V} - 1 \tag{3}$$

Where *E* is the initial specific internal energy of the explosive, that is, the internal energy per unit volume; *V* is the relative volume, that is, the ratio of the volume of the detonation product to the initial volume of the explosive, and ρ_0, ρ, and λ are the initial density, density and adiabaticity of the gas, respectively index. When the linear polynomial equation of state is used for air, the parameter values are $C_0=C_1=C_2=C_3=C_6=0$, $C_4=C_5=0.4$.

2.3.3. Rebar. Since the steel bar is an isotropic material, considering kinematic hardening, the plastic bilinear kinematic hardening model is adopted, and the influence of the material strength on the strain rate is considered. The strain rate effect of the steel bar is calculated using the *Cowper-Symonds* model, that is, when calculating the yield stress of the steel bar, multiply the following coefficient:

$$1 + \left(\frac{\dot{\varepsilon}}{C}\right)\exp(\frac{1}{p}) \tag{4}$$

Where $\dot{\varepsilon}$ is the strain rate, and *C* and *P* are the strain rate parameters of the *Cowper-Symonds* model. When the value of *C* and *P* is zero, the strain rate effect of the material is not considered. For steel bars, the values of these two parameters are: *C*=40, *P*=5.

2.3.4. Concrete. During the explosion, the structure was in a state of high strain, and the concrete material used the CONCRETE_DAMAGE_REL3 model. The model considers the strain rate effect of the material through the dynamic improvement factor DIF, which is the ratio of the dynamic strength of the material to its static strength at a certain strain rate. The dynamic increase coefficient of concrete strength adopts the C&K model, which is improved on the basis of the calculation method recommended by the European standard CEB. The model does not automatically consider material failures, and additional erosion criteria need to be defined to consider the maximum normal strain and shear strain. When any failure criterion is met, these concrete elements can be deleted from the calculation.

2.4. Validation of finite element model

In order to verify the applicability of finite element, 1.5kg single point contact explosion of RC slab was simulated, and the numerical simulation results were compared with the actual explosion test results of the research group. Figure 2 is the damage comparison results. The measured longitudinal damage length of the bursting surface of RC slab is 56cm, the transverse damage length is 64cm, and the maximum depth is 14.2cm. The longitudinal damage length of the back bursting surface is 99cm and the transverse damage length is 60cm. In the numerical simulation, the longitudinal damage length of the bursting surface of RC plate is 58cm, the transverse damage length is 62cm, the maximum depth is 13.8cm, the longitudinal damage length of the back bursting surface is 98cm, the transverse damage length is 56cm. Figure 3 is the actual value and the simulation value of the 1/4 cross section of the RC slab. The peak acceleration value of the actual value is $24176.6m\cdot s^{-2}$, and that of the measured value is $25229.2.0m\cdot s^{-2}$. The error is about 4%. The comparison of damage results and acceleration results shows that the numerical simulation is in line with the reality.

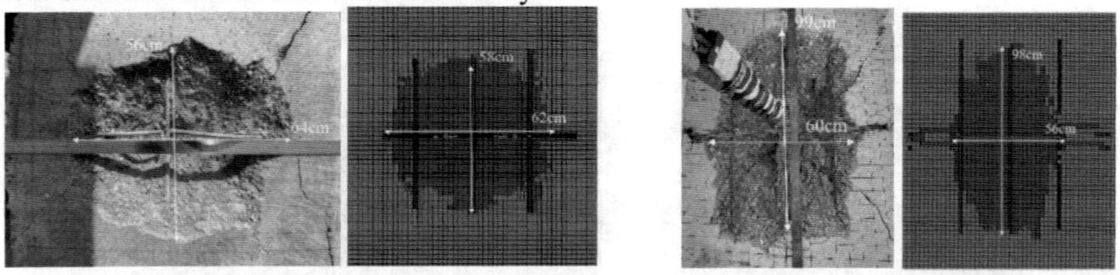

(a)Top damage (b)Bottom damage

Figure 2 concrete damage obtained by numerical calculation with actual test results

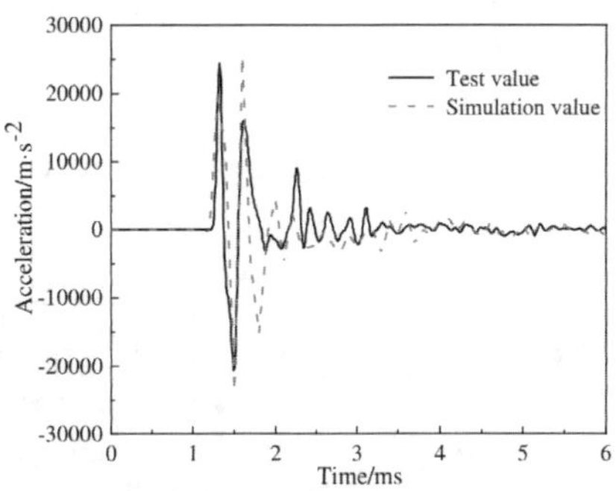

Figure 3 Acceleration time-history curves of 1/4 cross section of RC slab

3. Simulation result

3.1. Result damage

Table 2 shows the damage results of RC slabs under different explosion conditions. Under the condition of large charge (the total charge is 1kg), the entire mid-span section is damaged and the damage is serious. Two blast points on the burst face of RCP1 (spacing 30cm) form a connected oval pit, and the damage on the back burst face is a regular oval pit. The damage in the middle of the two explosive points is the most serious, but also can be seen that the concrete on both sides of the board has cracking; Two interconnected pits were also formed at the two explosive points on the bursting face and back bursting face of RCP2 (spacing 60cm). The concrete on both sides of the slab suffered the most serious damage and the penetrating damage occurred. Under the working condition of small charge (the total charge is 0.5kg), the damage of the slab is light and does not cover the whole mid-span section. The two blast points on the bursting face and back bursting face of RCP3 (spacing 30cm) form an oval pit connected as a whole, and the concrete on both sides of the slab does not crack. Two separate circular pits were formed at the two explosive points on the bursting face and the back bursting face of RCP4 (spacing 60cm), and the concrete on both sides of the slab cracked and spaled.

Table 2 Damage results of RC plates under different explosion conditions

specimen	Top surface	Bottom surface
RCP1		
RCP2	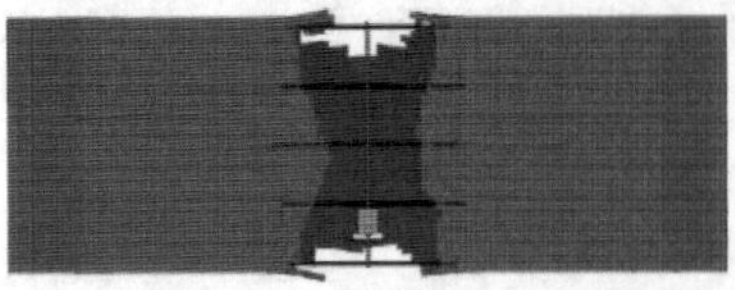	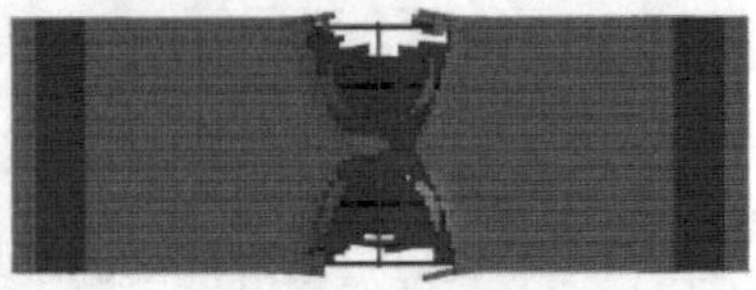
RCP3		
RCP4		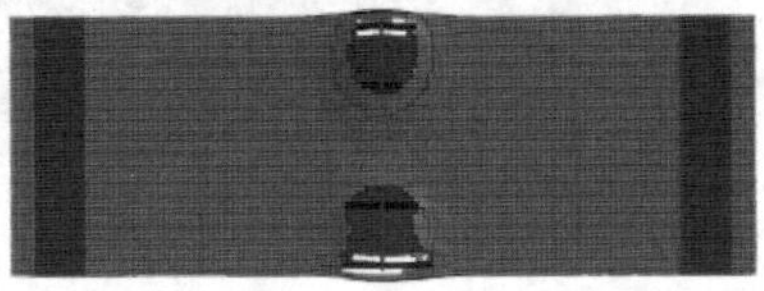

3.2. dynamic response

3.2.1. Vertical stress wave. Table 3 shows the vertical stress waves of the plate during the period of 0-0.09ms after the explosion. It can be seen from the vertical stress waves of the two-point contact explosion that when the explosive spacing is 30cm, the two stress waves meet and superimpose each other in the plate, maintaining a high stress state in a farther range. When the spacing is 60cm, the superposition effect is not obvious. According to the failure characteristics of the RC plate, the damage caused by explosion is that the compressive stress wave is generated on the bursting surface and causes the damage on the bursting surface of RC plate. Then the compressive stress wave propagates from the bursting point to the plate. When it meets other surfaces of the plate, it will reflect and generate strong tensile wave to cause damage. Under the condition of large charge, the stress wave in the plate reaches the back burst surface first when the contact explosion is 0.5kg×2 and the blast point spacing is 30cm. The strong tensile wave will lead to the spalling and collapse of the concrete on the back burst surface and form the pit. However, because the blast point is far away from the side of the slab, the tensile wave generated by the stress wave reflecting on the side of the slab is not enough to cause great damage to the concrete on the side of the slab. When the blast point spacing is 60cm, the stress wave almost simultaneously reaches the side and back burst surface of the plate, and the strong tensile wave generated here causes damage to the side and back burst surface at the same time, and the penetrating failure occurs on both sides of the plate. Under the condition of small charge, the stress waves at two points of 0.25kg×2 contact explosion are just intersecting when the distance between the two points of blast is 30cm, so the damage range of the two points can be connected superimposed, but the damage range caused is small. When the explosive spacing is 60cm, the stress waves of the two blast points do not intersect, the damage range of the two blast points is small, the destructive power is small, and two separate pits are formed, and no penetrating failure occurs on both sides of the plate

Table 3 RC slab mid-span cross section vertical stress waves

specimen	0.03ms	0.06ms	0.09ms

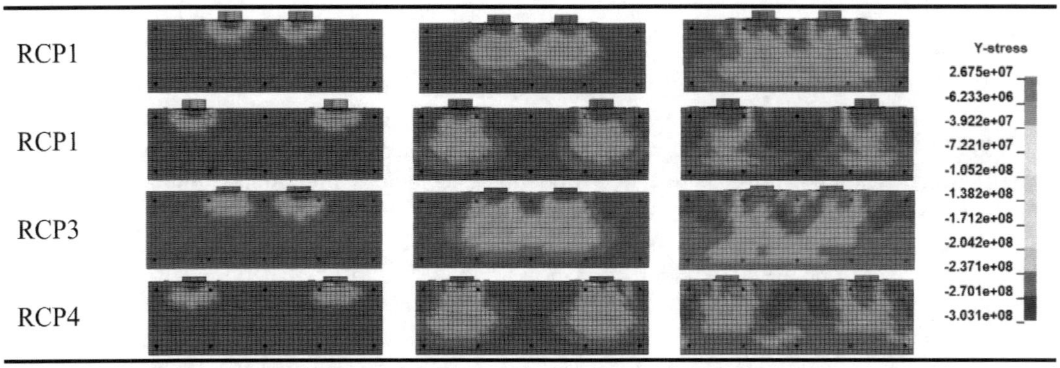

3.2.2. Dynamic response. The Figure 4 shows the acceleration time history curve of RC slabs. As can be seen from the figure, 0.5m away from the explosion center, the peak acceleration of RCP1-RCP4 are 54804.75m·s^{-2}, 48522.92m·s^{-2}, 26132.05·s^{-2} and 18675.72m·s-^{2}, respectively. Among them, the peak acceleration of high charge condition is 2-3 times of that of small charge condition. Under the condition of large charge, the peak acceleration with the blast point spacing of 30cm is 1.13 times that of the peak acceleration with the blast point spacing of 60cm. Under the condition of small charge, the peak acceleration with the blast point spacing of 30cm is only 1.40 times that of the peak acceleration with the blast point spacing of 60cm. From the perspective of the peak acceleration size, the peak acceleration with the blast point spacing of 30cm is larger than that with the blast point spacing of 60cm. Combined with the analysis of the propagation phenomenon of stress wave, the superposition of blast wave is obvious and the peak acceleration is larger when the distance between two explosive points is 30cm. When the distance is 60cm, the phase separation shock waves generated by the two blast points are separated, and the peak acceleration is small.

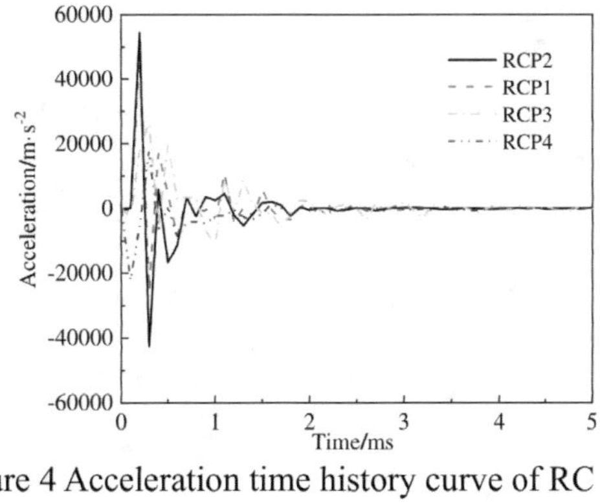

Figure 4 Acceleration time history curve of RC slabs

4. conclusion

Under the two-point contact explosion load, the damage mechanism of RC slabs are as follows: the explosion generates stress waves and thus leads to the damage of the blasting surface of RC slab, and then the stress waves propagate and diffuse from the blasting point to the slab, and when they encounter other surfaces of the slab, they will reflect and generate strong tensile waves to cause damage. Two-point contact explosion energy gathered phenomenon, spacing is small, the explosive stress wave in the encounter between superposition, at this time between the explosive destruction: the worst spacing and larger near the side slab, the blast stress wave at the same time reach the slab back and side, creates strong tensile wave in reflection and back surfaces and side layer crack and collapse, near the location of the slab profile has been disrupted. The dynamic response of RC slabs are as follows: when the blast point spacing is small, the energy accumulation generated by the explosion is obvious, which makes the dynamic response of RC slab more obvious. When the spacing is 30cm, the peak acceleration of RC slab is larger than that of 60cm.

Reference

[1] SHUJIAN Y, DUO Z, XUGUANG C, et al. Experimental and Numerical Study on the Dynamic Response of RC Slabs under Blast Loading[J]. Engineering Structures, 2016, 66:120-129.

[2] CASTEDO R, SEGARRA P, ALAÑON A, et al. Air Blast Resistance of Full-scale Slabs with Different Compositions: Numerical Modeling and Field Validation[J]. International Journal of Impact Engineering, 2015, 86:145-156.

[3] LI Tian-hua, ZHAO Jun-hai, WEI Xue-ying, et al. Dynamic Response and Parameter Analysis of Reinforced Concrete Slab under Explosion Load [J]. Building Structures, 2012, 42(S1): 786-790.

[4] ZHOU X Q, KUZNETSOV V A, HAO H, et al.. Numerical Prediction of Concrete Slab Response to Blast Loading[J]. International Journal of Impact Engineering, 2008, 35(10): 1186-1200.

[5] WANG W, ZHANG D, LU F, et al. Experimental Study on Scaling the Explosion Resistance of a One-Way Square Reinforced Concrete Slab under a Close-in Blast Loading[J]. International Journal of Impact Engineering, 2012, 49: 158-164.

[6] ZHOU X Q, HAO H. Mesoscale Modelling and Analysis of Damage and Fragmentation of Concrete Slab under Contact Detonation[J]. International Journal of Impact Engineering, 2009, 36(12): 1315-1326.

[7] ZHANG Shi-hao, HAN Jing, WANG Hua. Energy Gathering Effect of Multi-point Simultaneous Explosion in Concrete [J]. Blasting, 2014, 31(1): 19-24.

[8] HU Hong-wei, SONG Pu, GUO Wei, et al. Interaction of Shock Waves in Ground Burst [J]. Chinese Journal of High Pressure Physics, 2014, 28(03): 353-357.

[9] HAN Ke-hua, REN Xi, LI Hui, et al. Simulation and Experiment Studies on the Multi-point Synchronization Detonation Overpressure of slapper Detonators [J]. Energetic Materials, 2016, 24(01): 38-44.

[10] LI Xu-dong, LIU Kai-xin, ZHANG Guang-sheng. Focusing of Shock Waves in Cement Mortar Plates [J]. Journal of Tsinghua University. Natural Science, 2008, 48(08): 1272-1275.

Research on global actual measurement of indoor surface flatness and verticality Based on sparse point cloud

Zhongyue Zhang[1,4], Huixing Zhou[1,4], Shun Wang[2,4,] Yannan Lv[1,4], Xiaoyu Zheng[1,4], Langzhao Zeng[3]

[1] School of Mechanical-electronic and Vehicle Engineering, Beijing University of Civil Engineering and Architecture, Beijing 100044, China

[2] School of Civil and Transportation Engineering, Beijing University of Civil Engineering and Architecture, Beijing 100044, China

[3] School of Foreign Studies, Minzu University of China, Beijing 100081, China

[4] Beijing Engineering Research Center of Monitoring for Construction Safety, Beijing 100044, China

Corresponding author's e-mail address: 2108550020058@stu.bucea.edu.cn

Abstract. At present, it has become a trend to realize indoor actual measurement by reconstructing indoor 3D information with dense point cloud. Considering the large time consumption of traditional manual measurement and data overload of 3D reconstruction for a construction, the paper presents a global measurement method based on sparse point cloud for indoor actual measurement of flatness and verticality. First, given the point cloud is present with low density, our data size is much smaller than that in 3D reconstruction, which greatly reduces time consumption for data processing. Second, we calculate surface flatness degree by using a tolerance formula in machinery industry as reference after we conduct fitting of surfaces by least square method. As for the measurement of verticality, making use of the more orderly and effective point cloud clustering plus surface fitting based on sparse point cloud, we are able to calculate the verticality degree easily by a mathematical method. Compared with existing methods, this approach might feature a more tailored or specialized measurement method for actual measurement of flatness and verticality in acceptance of construction work in indoor scenes. To validate this approach, we present our experimental results and examine the systematic error by the mathematical assessment modelling. It is proved that the systematic error of the measurement method based on sparse point cloud is nearly ignorable.

1. Introduction

In a construction project, indoor actual measurement plays an important role in project tolerance management throughout a project planning, where the traditional manual measurement method is still mostly used [1]. However, it can be seen that more advanced measurement methods will give rise to higher efficiency and precision of measurement of quantities like flatness and verticality of surfaces [2]. At present, with the increase of labor cost and the progress of technology, automatic measurement methods are gradually replacing traditional manual measurement in acceptance of construction work [3].

The traditional manual measurement has inherent defects such as the limited measurement range by hand in indoor environment and the low time-efficiency and high cost of pure manual work. Under this background, we need to adopt a more global actual measurement method for flatness and verticality of surfaces in indoor environment, which can get a more comprehensive and large-range measurement result of indoor structures. Therefore, merits of 3D LiDAR are highlighted. For example, 3D laser scanning equipment is able to analyze complicated scenes and realize 3D reconstruction of cultural relics, tunnels, bridges and other civil engineering constructions by extracting point cloud information of them. Thus, the acceptance of construction work is simplified [4].

However, comparatively speaking, for indoor measurement for flatness and verticality, 3D reconstruction method seems to unduly capture and process 3D scanning data from an indoor environment. As a matter of fact, a large number of relevant literatures focus on the indoor measurement method for varied scenes. Y Sun set up framework called patch graph convolution network (PGCNet), which can achieve effective performance equivalent to state-of-the-art for segmenting standard indoor scene dataset [5]. H Wu proposed to use few labelled samples to solve the problem of the requirement for abundant labelled samples for training with deep learning classification methods, which can enhance the efficiency of training [6]. Jung, J proposes a scheme for automated 3D geometric modeling of indoor structures, including regularized boundary by means of the constrained least-squares method to get better positional accuracies [7].

Despite their meaningful attempts to discuss approaches to point cloud clustering and information matching, they may be apt to ignorance of the relatively excessive time consumption of point cloud data collecting, considering that the scanned object is confined to indoor environment. Therefore, we may need to propose a new actual measurement method targeted at the measurement of flatness and verticality surfaces in indoor scenes. With reference to relevant literatures of a group of experts [8-10], we believe that the key goals of the specialized method include the following three points: 1). Searching measurement requirements or standards for flatness and verticality with a smaller size of point cloud data. 2). Achieving more simple and effective point cloud clustering. 3). Matching well the measurement results with standards in traditional and mature manual acceptance of construction work.

To achieve the three main goals of a specialized or tailored measurement of flatness and verticality of indoor surfaces, we propose three main solutions: 1) Referring to the concept of interspace of point clouds mentioned in ACI building tolerance standard [11], we abandon the high-density point cloud collection in 3D indoor reconstruction, but collect sparse point cloud to process and build a set of new standards for measurement of flatness and verticality of indoor surfaces. 2) This paper proves that sparse point cloud will help improve the orderliness and effectiveness of point clouds clustering and reveal the pattern of curve correlation. 3)least squares method is employed to make horizontal and vertical surface fitting so that calculation of flatness and verticality degree is achievable based on sparse point cloud. Based on fitted surfaces, we use for reference the assessment formula for the roughness of 3D surface in machinery industry to calculate the flatness degree of surfaces and calculate verticality degree by formula. By mathematical method rather than relatively time-consuming, less accurate and small-range manual measurement. Base on the consideration as above, we launch a research on the global indoor actual measurement of flatness and verticality based on sparse point cloud and carry an experiment on it.

2.Indoor global measurement method

2.1. Sparse points and the maximum interspace of point clouds

In order to obtain smaller size of point cloud data, we propose a method of extracting sparse point cloud. First, in the idealized construction model of 6 surfaces, we identify the largest fitted surface and calculate the number of points on it, marked as n_s, and the s_{max} is the area of the largest surface. (see Figure 1). Calculate it in the formula s_0 as below. According to ACI building tolerance standard, the maximum interspace of point (s)= 303 mm [13], so the value of s_0 cannot surpass 303mm. s_0 is estimated as the maximum interspace of point clouds captured in our experiment.

$$s_0 = \sqrt{\frac{s_{\max}}{n_s}} \qquad (1)$$

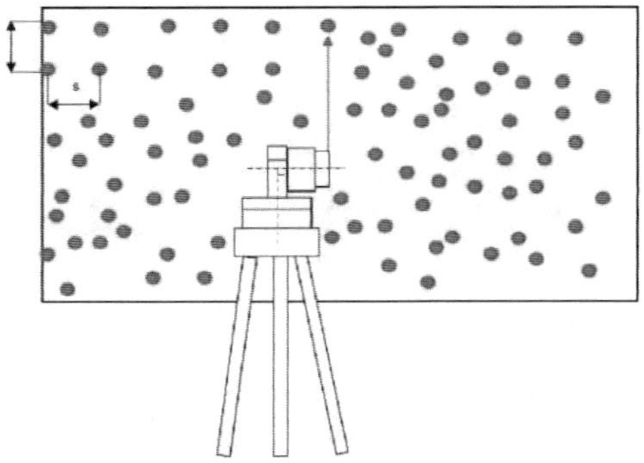

Figure 1. sparse point cloud and interspace model.

2.2 Mathematical method for calculating indoor surface flatness degree based on sparse-point cloud surface fitting

In the whole process of building construction acceptance, flatness information has different acceptance indicators in different stages of construction technology, but from the essence of the measurement information of these indicators, they are the collection of data used to solve flatness related problems in indoor information. Combined with the above analysis and comparison, the flatness problem can be regarded as a large-scale surface roughness. As shown in the figure (see Figure 2), taking the flatness of the wall as an example, the traditional measurement method with a guiding rule is to obtain the wall flatness by measuring the gap between the guiding rule and the wall. However, we measure the flatness of surfaces by sparse point cloud.

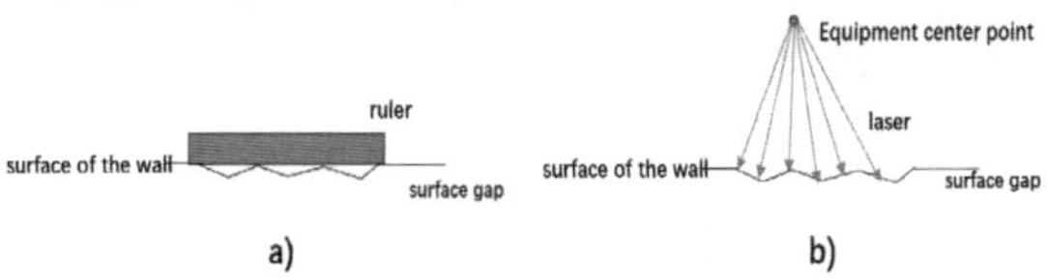

Figure 2. Comparison between traditional flatness measurement and global flatness measurement.

After the completion of the global point cloud data collection, they will be segmented according to the indoor point cloud clustering rules. In the ideal hexahedral indoor space, all the point clouds will be segmented into six clusters. According to 3D quadrant coordinate of point cloud clusters in the 3D coordinate system, they are marked as 6 surfaces. Each points cloud cluster represents a 3D fitted surface.

The wall surface equation is expressed as:

$$Ax + By + Cz + D = 0 \quad (2)$$

After transformation, the fitting plane of the wall surface can also be expressed as:

$$f(x, y, z) = ax + by + cz + 1 = 0 \quad (3)$$

For point set of point cloud cluster $M\{(x_1, y_1, z_1), (x_2, y_2, z_2), ..., (x_n, y_n, z_n)\}$.

The best fitting plane of point cloud cluster M is:

$$\sum_{i=1}^{n} [f(x, y, z) - f(x_i, y_i, z_i)]^2 = \min \quad (4)$$

So:

$$g = \sum_{i=1}^{n} (ax_i + by_i + cz_i + 1)^2 = \min \quad (5)$$

In order to make function g hold, partial derivatives of a, b and c should be obtained: $\frac{\partial g}{\partial a} = 0, \frac{\partial g}{\partial b} = 0, \frac{\partial g}{\partial c} = 0$ The results are as follows:

$$a\sum_{i=0}^{n} x_i^2 + b\sum_{i=0}^{n} x_i y_i + c\sum_{i=0}^{n} x_i z_i = -\sum_{i=0}^{n} x_i$$
$$a\sum_{i=0}^{n} x_i y_i + b\sum_{i=0}^{n} y_i^2 + c\sum_{i=0}^{n} y_i z_i = -\sum_{i=0}^{n} y_i \quad (6)$$
$$a\sum_{i=0}^{n} x_i z_i + b\sum_{i=0}^{n} z_i y_i + c\sum_{i=0}^{n} z_i^2 = -\sum_{i=0}^{n} z_i$$

$$Q = \begin{bmatrix} \sum_{i=0}^{n} x_i^2 & \sum_{i=0}^{n} x_i y_i & \sum_{i=0}^{n} z_i x_i \\ \sum_{i=0}^{n} x_i y_i & \sum_{i=0}^{n} y_i^2 & \sum_{i=0}^{n} y_i z_i \\ \sum_{i=0}^{n} x_i z_i & \sum_{i=0}^{n} z_i y_i & \sum_{i=0}^{n} z_i^2 \end{bmatrix} \quad (7)$$

$$X = [a, b, c]^T \quad (8)$$

$$K = \begin{bmatrix} -\sum_{i=0}^{n} x_i \\ -\sum_{i=0}^{n} y_i \\ -\sum_{i=0}^{n} z_i \end{bmatrix} \quad (9)$$

Coefficient of plane equation:

$$X = Q^{-1}K \quad (10)$$

For a, b, c in X set:

$$a = \frac{A}{D}, b = \frac{B}{D}, c = \frac{C}{D} \quad (11)$$

Deviation set of all point clouds in point cloud cluster {di}:

$$d_i = \frac{|Ax_i + By_i + Cz_i + D|}{\sqrt{A^2 + B^2 + C^2}} \quad (12)$$

$$m_0 = \sqrt{\frac{\sum_{i=1}^{n} d^2}{n-3}} \quad (13)$$

2.3 Mathematical method for calculating surface verticality degree based on sparse-point cloud surface fitting

In the traditional manual measurement of indoor surface verticality, guiding ruler is the key tool. During construction acceptance, the guiding ruler is placed basically vertical to the ground (see Figure 3), specifically, with the bubble in the guiding ruler is located at the middle of the horizontal ruler scale. As shown in the figure, the length of the guiding ruler is known to be γ, the maximum allowable deviation distance is e, and the guiding ruler is vertical to ground at 90 degrees. Therefore, γ (the deviation angle between the wall and the guiding rule) can be figured out.

$$l = \sqrt{r^2 + e^2} \quad (14)$$

$$\gamma = \arccos\left(\frac{r^2 + l^2 - e^2}{2 \times r \times l}\right) \quad (15)$$

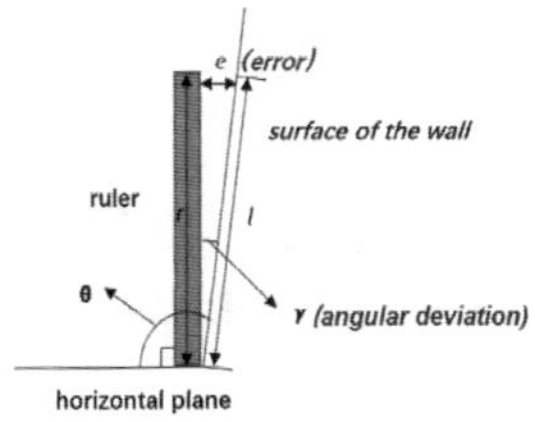

Figure 3. Traditional verticality measurement.

The actual angle between the wall and the ground is θ :

$$\theta = 90^{\circ} + \arccos(\frac{r^2 + l^2 - e^2}{2 \times r \times l}) \quad (16)$$

In the process of global actual measurement of indoor surface verticality, with the indoor surface fitting equation (the plane fitting equation is from least squares), we measure the angel between every couple of surfaces. The fitting equation is as follows.

For two surfaces, their surface fitting equations are as:

$$A_s x + B_s y + C_s z + D_s = 0, A_j x + B_j y + C_j z + D_j = 0 \quad (17)$$

Solving the angle between two surfaces:

$$\theta = \arccos(\frac{A_s A_j + B_s B_j + C_s C_j}{\sqrt{A_s^2 + B_s^2 + C_s^2}\sqrt{A_j^2 + B_j^2 + C_j^2}}) \quad (18)$$

In practical measurement, the surface selection depends on indoor structure. In many cases, ceiling is the alternative to ground.

3. Application of the global actual measurement method based on sparse point cloud in our experiment

3.1 Experiment conditions

In order to examine the effectiveness of the global actual measurement of indoor surface flatness and verticality based on sparse point cloud, we applied the global measurement method in indoor scenes in this experiment. In this experiment, the point cloud data is collected by our innovated 3D LiDAR in an existing home covering a floor area of 17 square meters (see Figure 4) with its structure in conformity to the 3D model mentioned in the paper and then is processed on MATLAB 2016B in Windows10.

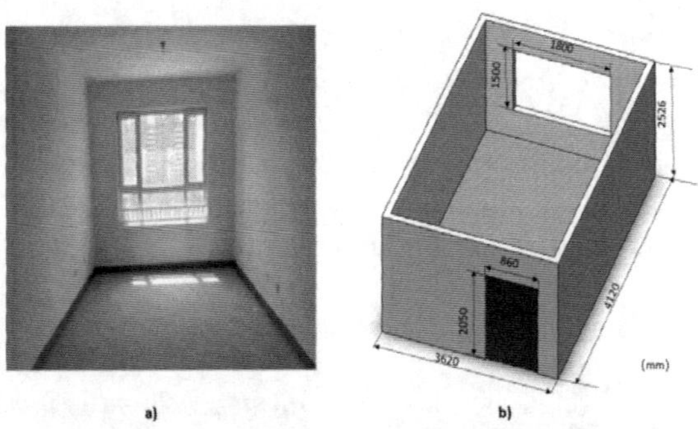

Figure 4. Experiment scene：(a) Indoor photos (b) 3D model of experimental scene.

3.2 The analysis of time consumption in visualization system

In the global actual measurement of surface flatness and verticality, with the growth of the point cloud data size, the indoor conditions will be well revealed by data visualization in MATLAB. However, the growth of data size gives rise to the load of algorithm. (see Figure 5)

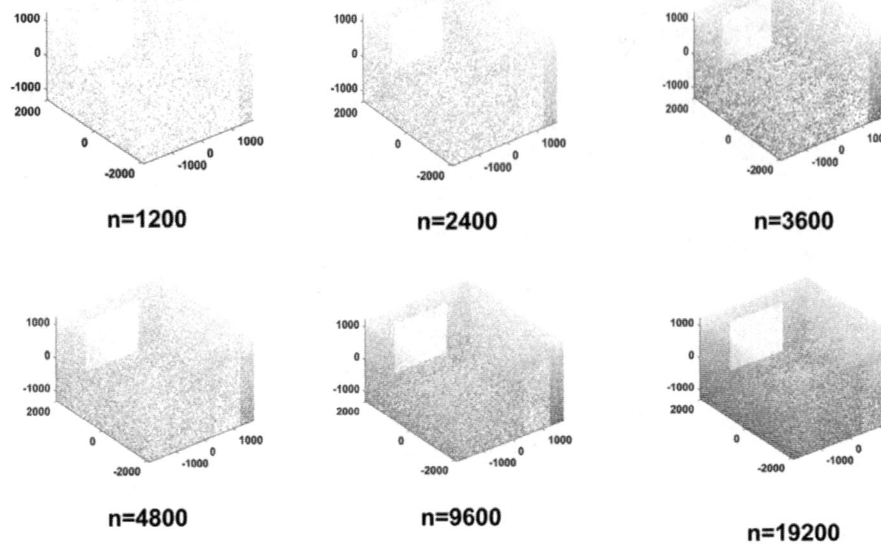

Figure 5. 3D images of visualized indoor points cloud data of different sizes.

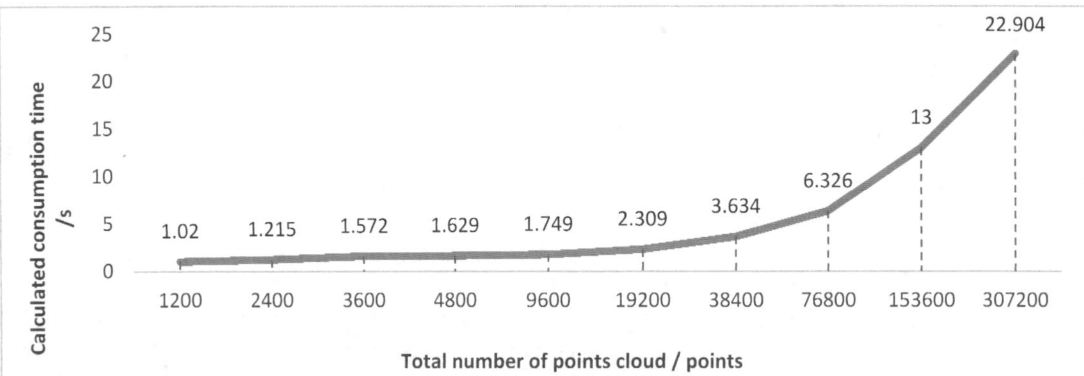

Figure 6. The variation of time consumption of the visualization program with different data sizes.

To explicitly examine the time-efficiency of sparse point clouds in the processing of data, this paper takes the visualization program of extracting point cloud data of varied sizes as examples by calculating the time-consumption of it with MATLAB. (see Figure 6) The results are shown in the graph above-the two times rise of the data size of point cloud will lead to exponential rise of the time consumption of program operation. Thus, based on the premise that measurement is undertaken in circumstances in line with standard acquirements, the smaller size of point cloud, the more time-friendly.

3.3 The experiment results of the global actual measurement of indoor surface flatness and verticality.

3.3.1 The accuracy of the global actual measurement of surface flatness based on sparse point cloud
Weighing the time of processing point cloud and the concept of sparse point cloud model, we choose the indoor point cloud with the number of point clouds ($n = 3000$) for experimental measurement. In this magnitude, the maximum distance of any two points cloud is $s = 114.184$mm, which is below the maximum number of coefficient point cloud model. In this paper, the plane fitting and surface roughness solution calculation are by the equations (2)-(13). In the experiment, the sftool toolbox in MATLAB is used to complete the three-dimensional surface fitting.As shown in Table 1, by the means of the global

measurement proposed, the indoor flatness is m_0, and the flatness result obtained by traditional manual measurement in the same scene is m_c. Taking the traditional manual measurement results as the true value, the measurement error of this method is less than 1.6 mm.

Table 1. Flatness measurement results of different planes.

Measurement objects	The fitting plane	m_0	m_c
Plane A	192.3x-0.009961y-z-330100=0	3.781	2.45
Plane B	0.0001934x-1194y-z+2423000=0	1.3352	2.1
Plane C	$5.237e^{-6}x-7.944e^{-7}y-z+1246=0$	1.435	1.8
Plane D	-216.6x-0.04866y-z-412000=0	3.444	2.28
Plane E	0.09517x-813.8y-z-1697000=0	1.397	1.95
Plane F	$-2.752e^{-5}x-1.002e^{-5}y-z-1280=0$	1.4347	1.64

3.3.2 The measurement of verticality of fitted walls.

After the measurement for indoor horizontal surfaces, we gain a formula for the measurement results of all the fitted surfaces. Based on these equations (14)-(18), we are able to figure out the verticality of fitted surfaces according to the formula, the entire measurement results of verticality in the experiment are revealed in the following Table 2. It proves that the measurement error of the global actual measurement of fitted surfaces verticality do not surpass 0.12 degrees.

Table 2. Errors of global verticality measurement with manual results as standard.

Measurement objects	Measurement results of global measurement method (°)	Measurement results of manual method (°)	Error (°)
A	89.7018	89.8056	0.1038
B	89.9526	89.9654	0.0128
D	89.7018	89.8056	0.1038
E	89.9526	89.9654	0.0128

4. Conclusion

In conclusion, global actual measurement of indoor surface flatness and verticality based on sparse point cloud is a comprehensive and specialized measurement method for indoor measurement in construction works. It has following merits, validated by the experiment results present in this paper: 1). It is able to cover larger measurement range than traditional manual measurement method. 2). Smaller size of point cloud data reduces the load of data processing, making measurement more efficient. For instance, in our experiment to measure a room in size of 4120×3620×2526, we merely need to process 3000 points clouds, superior to 3D LiDAR measurement in the aspect of time consumption. 3). It achieves higher orderliness of point cloud clustering because it is performed in visualization system in MATLAB based on sparse point cloud. 4). Surface fitting based on sparse point cloud allows us to figure out flatness and verticality of indoor surfaces in mathematical method. With surfaces fitting by least square method, the distance deviation of points from the fitted surfaces is able to be used in the calculation of flatness by referring to a tolerance formula. Plus, the verticality of indoor surfaces, typically walls to horizontal surface, is calculated by the angle between two fitted surfaces. Furthermore, the flatness measurement error is less than 1.6mm and verticality error less than 0.12 degrees. Additionally, the systematic error of the method is examined through a mathematical assessment modelling.

Because of the lower cost and efficient method, researching for the global actual measurement based on sparse point cloud is meaningful to realize the automatic measurement in construction. Although this paper proposed the indoor actual measurement method with sparse point cloud and made a simple attempt, there are still a lot of related work needed to do, including: indoor point cloud cluster segmentation of sparse point cloud, influence of cloud density at different points on measurement accuracy and how to reduce the systematic error of measurement equipment etc. Our team will continue to research relevant issues in the future.

Acknowledgments
This study is sponsored by Beijing Natural Science Foundation and Education Commission fund (No.KZ202110016024) and the BUCEA Doctor Graduate Scientific Research Ability Improvement Project (No. DG2021015)

References

[1] Saeed Talebi.,et al. "Tolerance Management in Construction: A Conceptual Framework," Sustainability, vol. 12(3), pages 1-24(2020).
[2] Enshassi, Msa , et al. "Integrated Risk Management Framework for Tolerance-Based Mitigation Strategy Decision Support in Modular Construction Projects." Journal of Management in Engineering 35.4(2019).
[3] More references Kumar, G. A. , et al. "A LiDAR and IMU Integrated Indoor Navigation System for UAVs and Its Application in Real-Time Pipeline Classification." Sensors 17.6(2017):1268-.
[4] Hong, S. , et al. "Indoor 3D Modeling Approach based on Terrestrial LiDAR." Journal of the Korean Society of Civil Engineers 32.5D(2012).
[5] Sun, Y. , et al. "PGCNet: Patch Graph Convolutional Network for point cloud segmentation of indoor scenes." The Visual Computer 36.10(2020).
[6] Wu, H. , et al. "Classification of Point Clouds for Indoor Components Using Few Labeled Samples." Remote Sensing 12.14(2020):2181.
[7] Jung, J. , et al. "Automated 3D Wireframe Modeling of Indoor Structures from Point Clouds Using Constrained Least-Squares Adjustment for As-Built BIM." Journal of Computing in Civil Engineering (2015):04015074.
[8] Alonso, S. , et al. "Analysis of Uncertainty in a Middle-Cost Device for 3D Measurements in BIM Perspective." Sensors 16.10(2016):1557-1574.
[9] Wang, Y. T. , et al. "A Single LiDAR-Based Feature Fusion Indoor Localization Algorithm." Sensors 18.4(2018):1294-.
[10] Liu, S. , et al. "A LiDAR-aided indoor navigation system for UGVs." Journal of Navigation 68.2(2015):253-273.
[11] Birkeland, Philip W., and Westhoff, Leonard J., "Dimensional Tolerances in a Tall Concrete Building", Journal of the American Concrete Institute, August1971, pages 600-607.
[12] Recommended Practice for Concrete Formwork (ACI 347-78), American Concrete Institute, Detroit, Michigan 48219.
[13] Suchocki, C., et al. "Detection of defects in building walls using modified OptD method for down-sampling of point clouds. Building Research and Information" (2020):1-19.

Journal of Physics: Conference Series **2215** (2022) 012016 doi:10.1088/1742-6596/2215/1/012016

Experimental study on head loss coefficients of tailrace surge tank for Baihetan hydropower station

Hongliang Sun[1], Fei Yang[1], Fulin Cai[2], Yimin Chen[1] and Jianrong Xu[1]

[1] Power China Huadong Engineering Corporation Limited, Hangzhou 311122, China

[2] College of Water Conservancy and Hydropower Engineering, Hohai University, Nanjing 210098, China

sun_hl2@hdec.com

Abstract: The scale and hydraulic inertia of the tailrace surge tank of Baihetan hydropower station are huge. Its resistance characteristics are directly related to the accuracy of the calculation results of the hydraulic transient process, the stability of the unit operation, the economy and safety of the project. The influence of split ratio, cross-sectional area of the restricted orifice and unit operation on the head loss coefficient of Baihetan surge tank was studied by model experiment method. The results show that the head loss coefficient increases with the increase of split ratio and decreases with the increase of the cross-sectional area of the restricted orifice. Under the diffluence condition, the head loss coefficient of the restricted orifice in double unit operation is obviously larger than that in single unit operation. Under the confluence condition, the head loss coefficients of the restricted orifice in single and double unit operation are the same. The deviation will be large when using the recommended value in the specification, the experimental results are more reasonable. The research conclusions will provide technical support for the design and operation of Baihetan hydropower station as well as reference for similar projects.

1. Introduction

Baihetan hydropower project is located on Jinsha River, of which the underground powerhouse is set at the head of the water diffluence system. The gross installed capacity of this station is 16 000 MW. There are four hydraulic units arranged on the left and right banks respectively, totally 16 single unit 1 000 MW Francis type water turbine-generator units, and the rated flow is 547.8 m^3/s. For the headrace tunnel, one unit uses one tunnel, and for the tailrace tunnel, two units share one tunnel and one tailrace surge tank.

The form of Baihetan surge tank is cylindrical throttled surge tank with double restricted orifices. It is the largest underground surge tank in the world, of which the diameter of the surge tank is 43m – 48m, and the maximum height exceeds 100 m. Two circular restricted orifices with a diameter of 7.6m are arranged on the upstream side of the restricted orifice plate, at the bottom of which a split piers is arranged in the flow passage with a height of 18m. The typical structural arrangement of the surge tank is as shown in figure 1.

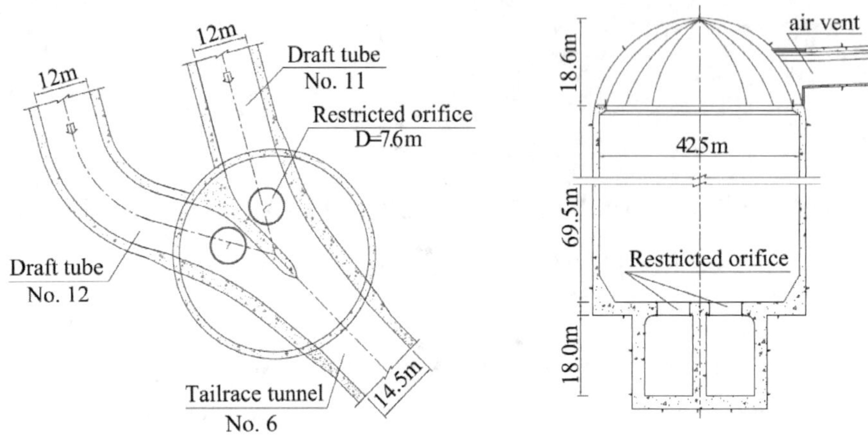

(a) Plan (b) Elevation

Figure 1. Typical structural layout of No. 6 tailrace surge tank.

The resistance characteristics of the tailrace surge tank have significant influence on the amplitude and attenuation of the surge, safety of water conveyance system, stability of power generation and water hammer reflection effect [1]. Therefore, in the calculation of hydraulic transient process, determination of the resistance loss coefficient of surge tank is the key, which is usually obtained based on relevant codes and manuals for small and medium sized projects [2]. In Baihetan hydraulic project, due to the factors of huge scale of surge tank, double restricted orifices, and the interaction of flow at the junction of draft tube and draft tunnel. The deviation will be large when using the value recommended in the specification. In view of this, the hydraulic model experiment method is used in this paper to study the water head loss coefficient of the tailrace surge tank shared by two units, which can provide a basis for accurate calculation of hydraulic transient process of the water conveyance system as well as a reference for the design and research of other similar projects.

2. Model experiment

2.1. Experimental model

The No. 6 tailrace surge tank on the right bank where No.11 and No.12 units are located is taken as the research object. The experimental model is designed according to gravity similarity criterion, and a normal model with the scale of 1:40 is adopted. Simulation scope: the simulated length of the draft tube of No.11 and No.12 is 182m and 150m respectively, and that of the tailrace tunnel is 200m, which is more than 10 times of the tunnel diameter. The simulated range of the surge tank is from the bottom to the top. Experimental model layout is as shown in figure 2. The experimental model is made of PMMA (polymethyl methacrylate) and the auxiliary pipe is made of PVC plastic pipe. The model is shown in figure 3.

In the experiment, orifice flowmeter was used to measure the flow of draft tube, and triangle measuring weir was used to measure the flow of tailrace tunnel. In order to measure the head loss coefficient at the bottom of the surge tank and when the water flows into and out of the tailrace surge tank (hereinafter referred to as diffluence flow and confluence flow), four pressure measuring sections from 1-1 to 4-4 were arranged on the experimental model, as shown in figure 2. The pressure measuring section of draft tube and tailrace tunnel is located three times the tunnel diameter from the surge tank, and that of the surge tank is located one time the surge tank diameter from the bottom. Four pressure taps are arranged equidistant along the circumference of the tunnel section in each measuring section. They are connected to the piezometer to measure the piezometric head on the corresponding section.

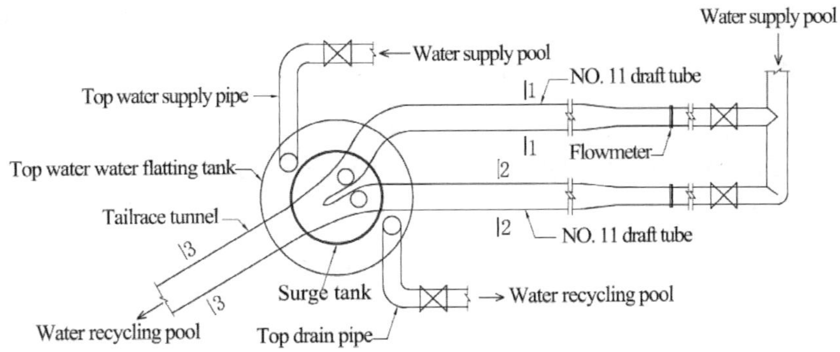

(a) Plan

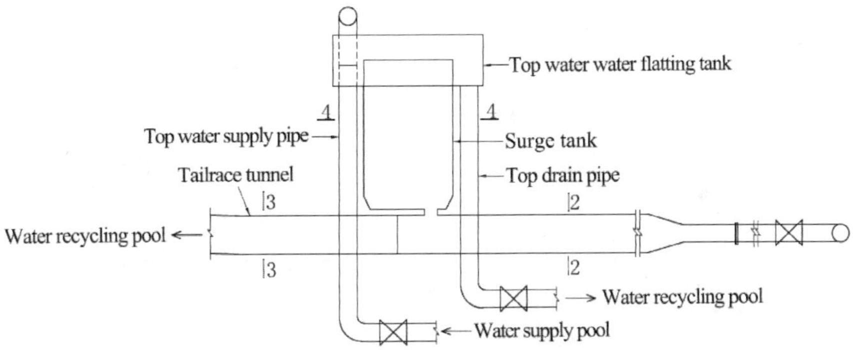

(b) Elevation

Figure 2. Experimental model and layout of measuring section.

Figure 3. Experimental model.

2.2. Experimental scheme

2.2.1. Experimental conditions

According to the specification, the cross-sectional area of the restricted orifice should be 25% - 45% of the section area of the draft tube or tailrace tunnel [3]. In this experiment, three restricted orifice schemes are selected for the study, as shown in table 1. For each restricted orifice scheme, the head loss was measured for conditions of water flowing into and out of the surge tank respectively, and the corresponding head loss coefficient was calculated and analyzed. Water flowing into and out of the surge tank is called the diffluence pattern and the confluence pattern respectively [4-6]. Schematic diagrams of the diffluence pattern and the confluence pattern are as shown in figure 4. In Baihetan hydropower project, one surge tank with double restricted orifices is shared by two units, so each flow pattern includes conditions of single and double unit operation.

Flow pattern simulation and flow control are realized by switching the valves on the draft tube and tailrace tunnel. For each experimental point, the flow and piezometric head are measured when the flow pattern is stable. Steady-state tests with different split ratios are used to simulate the dynamic flow process of the surge tank.

Table 1. Experimental schemes of different restricted orifices.

Scheme	Restricted orifice diameter D/m	Area ratio [a] to surge tank/%	Area ratio to tailrace tunnel/%	Area ratio to draft tube/%
1	7.3	8.7	34.4	19.7
2	7.6	9.4	37.3	21.3
3	8.3	11.2	44.5	25.5

[a] For the area ratio calculation in the table, the area of the restricted orifices refers to the total area of the two holes, and the area of the draft tube refers to the total area of the two branch tunnels.

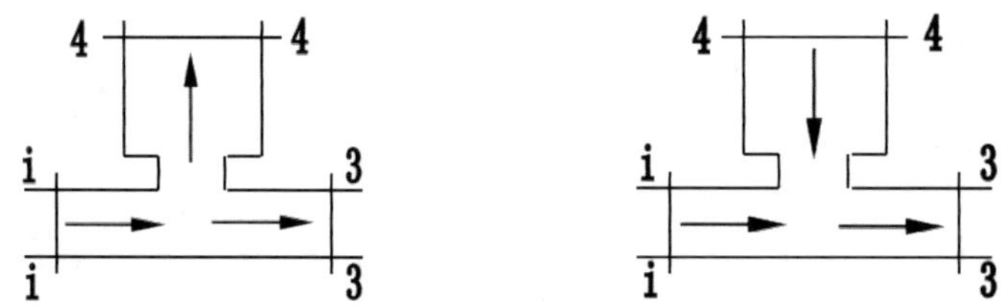

(a) Diffluence pattern (b) Confluence pattern

Figure 4. Schematic diagrams of experimental conditions.

2.2.2. Head loss calculation formula

According to the energy equation, the formula of head loss between adjacent sections is

$$\Delta h_{ij} = (Z_i + \frac{P_i}{\gamma} + \frac{V_i^2}{2g}) - (Z_j + \frac{P_j}{\gamma} + \frac{V_j^2}{2g}) \tag{1}$$

In which, Z is the section elevation, m; P/γ is the pressure head, m and $V/(2g)$ is the velocity head, m. Subscript i and j represent the measuring section.

The calculation formula for head loss coefficient between adjacent sections is

$$\xi_{ij} = \Delta h_{ij} / \frac{V^2}{2g} \tag{2}$$

In which, V is the characteristic velocity, m/s. Under the diffluence condition, the characteristic velocity is the flow velocity in draft tube, and under the confluence condition, that is the flow velocity in tailrace tunnel.

3. Experimental results

3.1. Diffluence condition

Under the diffluence condition, the head loss coefficient variation trend of each restricted orifice scheme is shown in figure 5. The split ratio $q=Q_4 / (Q_1+Q_2)$ (Q represents the flow rate of each section, and the subscript indicates the section position) on the abscissa axis refers to the ratio of the flow into surge tank to the total flow of the two draft tubes. It can be seen that under unsteady flow state, besides the size of the restricted orifice, the head loss coefficient of water flowing into the surge tank is also related to the split ratio and unit operation.

The head loss coefficient of the restricted orifice decreases with the increase of cross-sectional area of the restricted orifice and increases with the increase of split ratio. When the split ratio is 1, all water flows into the surge tank, and the head loss coefficient is the largest, which is 13.32, 11.77 and 8.66 respectively in single unit operation and 53.02, 47.39 and 33.21 respectively in double unit operation. The head loss coefficient of double unit operation is obviously greater than that of single unit operation. The main reason is that the draft tubes are connected at the bottom of the surge tank, when the single unit is running, the water can flow into the surge tank from two restricted orifices, which is equivalent to doubling the area of the resistance area.

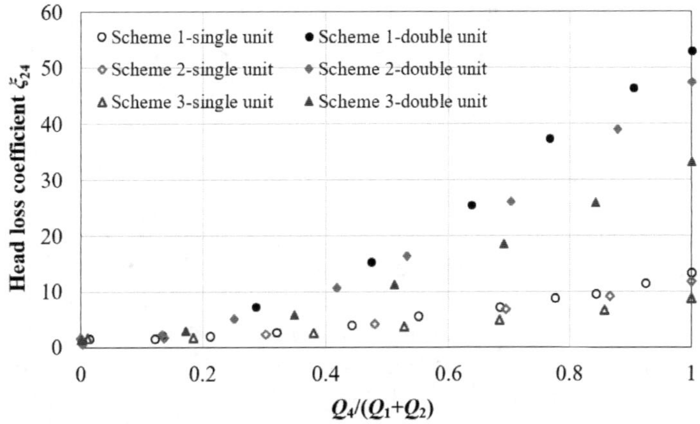

Figure 5. Variation trend of head loss coefficient under diffluence condition.

3.2. Confluence condition

The variation trend of head loss coefficient of each restricted orifice scheme under the confluence condition is shown in figure 6. The split ratio $q=Q_4/Q_3$ on the abscissa axis refers to the ratio of the flow out of surge tank to the flow of the tailrace tunnel. It can be seen that different from the condition of flowing into the surge tank, the variation trend of head loss coefficient has no relevance to the unit operation.

The head loss coefficient of the restricted orifice also decreases with the increase of cross-sectional area of the restricted orifice and increases with the increase of split ratio. When the split ratio is 1, all water flows into the surge tank, and the head loss coefficient is the largest, which is 16.05, 13.22 and 9.42 respectively in single unit operation and 16.35, 13.40 and 9.42 respectively in double unit operation. With the same restricted orifice scheme, the resistance coefficients are nearly the same for both single and double unit operation. The reason is that water flows out of the surge tank from both two restricted orifices whether in single or double unit operation.

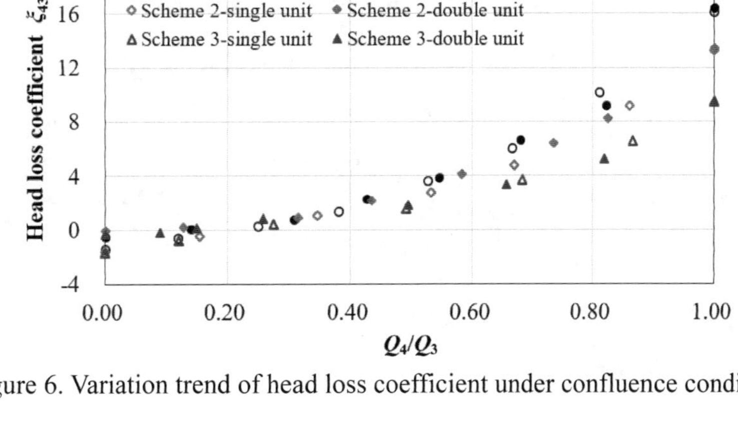

Figure 6. Variation trend of head loss coefficient under confluence condition.

4. Analysis of head loss coefficient

4.1. Flow coefficient analysis for restricted orifice

According to the Design Specification for Surge tank of Hydropower Station (NB T35021-2014), the head loss of restricted orifice h_c can be approximately calculated by the formula $h_c = (1/2g) \cdot (Q/\varphi S)^2$, where φ represents the flow coefficient of resistance, which can be taken as 0.6 - 0.8, S represents the section area of the restricted orifice and Q represents the flow into and out of the surge tank. The flow coefficient calculated from the head loss coefficient obtained from the above formula and experiment, it can be seen that when the split ratio is small, the recommended flow coefficient is inconsistent with the measured value. While when the split ratio is 1, the flow coefficient of water flowing into the surge tank is between 0.67 - 0.68, and that of water flowing out of the surge tank is between 0.73 – 0.74, which are within the range of the recommended value in the specification, as shown in table 2. The above analysis shows that the flow coefficient recommended in the specification is the maximum resistance coefficient, and the head loss characteristics of the surge tank of Baihetan hydropower station conforms to the general law.

In the calculation and analysis of hydraulic transient process, the flow coefficient of restricted orifice obtained by experiment is more in line with the actual situation of Baihetan hydropower station than that recommended in the specification. The data comparison in table 2 shows that the difference of head loss coefficient between each scheme is large, but that of the flow coefficient is small, so the deviation of the recommended method in the specification will be large. Considering from the perspective of economy and safety, model experiment is recommended for large-scale projects.

Furthermore, according to the head loss coefficient and flow coefficient obtained by the experiment, we can know that the head loss of water flowing into the surge tank is greater than that flowing out of the surge tank, showing the general rule of "easy to flow out, hard to flow in". The reason is that the diffusion degree of the flow passage of water flowing from the tailrace tunnel into the surge tank is greater than that of the reverse situation.

Table 2. Table of corresponding flow coefficients of head loss in different conditions.

Unit operating	Scheme	Diffluence condition		Confluence condition	
		Head loss coefficient ξ	Flow coefficient φ	Head loss coefficient ξ	Flow coefficient φ
Single unit	1	13.32	0.68	16.05	0.73
	2	11.77	0.68	13.22	0.74
	3	8.66	0.67	9.42	0.74
Double unit	1	53.02	0.68	16.35	0.73
	2	47.39	0.68	13.4	0.74
	3	33.21	0.68	9.42	0.74

4.2. Fitting formula of dynamic head loss coefficient

For ease of application of computer calculation, a quadratic regression analysis is conducted with the principle of least square method on the relationship between the head loss coefficient and the split ratio obtained by the experiment [7-8], obtaining the formula of the head loss coefficient as shown in table 2. It's hard to judge which scheme is better by the experimental results of the three restricted orifice schemes. However, after the calculation and analysis of transient process of water conveyance and power generation system with the head loss coefficient of the restricted orifice obtained by the experiment, Scheme 2 is selected for Baihetan hydropower station.

Table 3. The formula of the head loss coefficient.

Unit operating	Scheme	Head loss coefficient under diffluence condition	Head loss coefficient under confluence condition
Singles unit	1	$\xi_{14}=10.51q^2+1.3182q+1.4447$	$\xi_{43}=16.713q^2+0.2803q-1.0118$
	2	$\xi_{14}=9.0511q^2+1.1782q+1.5579$	$\xi_{43}=13.48q^2+0.6463q-1.134$
	3	$\xi_{14}=6.3853q^2+0.5733q+1.6621$	$\xi_{43}=8.7059q^2+1.689q-1.1437$
Double unit	1	$\xi_{14}=36.567q^2+17.777q+0.2322$	$\xi_{43}=19.034q^2-3.1703q+0.0273$
	2	$\xi_{14}=36.569q^2+10.462q+1.0409$	$\xi_{43}=15.178q^2-2.2055q+0.2022$
	3	$\xi_{14}=21.798q^2+10.86q+0.8381$	$\xi_{43}=10.575q^2-1.5535q+0.0916$

5. Conclusion

In this paper, the head loss coefficient of throttled tailrace surge tank in Baihetan hydropower station is obtained by model experiment. The results show that the head loss coefficient decreases with the increase of cross-sectional area of the restricted orifice and increases with the increase of split ratio. Under diffluence condition, the head loss coefficient in double unit operation is obviously greater than that in single unit operation. Under confluence condition, the head loss coefficients are the same in both single and double unit operation.

Analysis of flow coefficient of the restricted orifice shows that when the split ratio is 1, the restricted orifice flow coefficient of the surge tank in Baihetan hydropower station is within the recommended value range, proving that the head loss characteristics of the surge tank in Baihetan hydropower station conform to the general law. However, the deviation of the value recommended in the specification is relatively large. From the perspective of economy and safety, it is recommended to use the model experiment to determine the value for large-scale projects.

References

[1] Yongguang C and Jiandong Y 2005 Hydraulic resistance coefficient determination of throttled surge tanks by means of computational fluid dynamics. *Journal of hydraulic engineering*, **2005(07)** 0787-92

[2] Qizhao L and Shouzhuo P 1995 *Design code for surge chamber of hydropower stations* (Beijing: China Water & Power Press) p56-57

[3] China Renewable Energy Engineering Institute 2014 *Design specification for surge tank of hydropower station* (Beijing: China Electric Power Press) p48-52

[4] Xiang L, Bo Y, Shubin Z, Jinheng H and Junling H 2017 Experimental study on damping coefficient of tailrace surge tank shared by two units. *Yangtze River* **48(001)** 80-83

[5] Fulin C, Ming H and Qing C 2001 Coefficient of head loss of throttled surge tanks with long linking pipe. *International Journal Hydroelectric Energy*, **19(004)** 40-42

[6] Lin Y and Xu L 2005 Research on the resistance coefficient of throttled surge chamber in combination with numerical computation and model test. *China Rural Water and Hydropower*, **2005(05)** 109-111

[7] Qizhao L, Wenqi L and Zili Q 1990 Minor head loss factors at bottoms of surge tanks. *Journal of Hohai University (Natural Sciences)*, **000(003)** 47-55

[8] Changfu S and Xiaojuan X 2005 Head loss coefficients of throttled surge tanks with the linking pipe. *Water Conservancy Science and Technology and Economy*, **11(010)** 604-606

Journal of Physics: Conference Series **2215** (2022) 012017 doi:10.1088/1742-6596/2215/1/012017

Design and Simulation of LLC Resonant Converter for Hydrogen Production Based on Proton Exchange Membrane Electrolysis

Haifeng Zhang[a], Zerui Fan[*], Jinghua Zhou[b], Ze Shen[c]

School of Electrical and Control Engineering, North China University of Technology, Beijing, China

[a]ZhanghaiFeng@nuct.edu.cn

*15900326126@163.com

[b]zjh@nuct.edu.cn

[c]1156413679@qq.com

Abstract—This paper studies the relationship between the parameters in LLC resonant circuit based on the fundamental wave analysis method, analyzes the gain curve and the soft-switching conditions of the power tube, gives the selection method of the parameters k and Q and calculates the resonant component parameters. In this paper, a closed-loop frequency modulation control method is used to control the DC output voltage of the resonant converter. At the same time, mathematical modeling and analysis are carried out for the electrochemical model and thermal model of the proton exchange membrane (PEM) hydrogen production electrolyzer, and the Simulink simulation tool is used to complete the joint simulation of the PEM electrolysis model and the LLC resonant converter. The simulation results fully reflect the actual control process.

1. INTRODUCTION

With the launch of the national "dual carbon" plan, hydrogen as clean energy has been gradually valued by society [1-2]. At present, the main method of producing hydrogen is the electrolysis of water to produce hydrogen [3-5]. The hydrogen production power supply generally uses traditional thyristor rectification, which is bulky and low in inefficiency. With power electronic devices, new topologies, and new control strategies, the introduction of hydrogen production power converters has made considerable progress. To pursue higher power density and smaller conversion volume, the switching frequency of hydrogen production power converters has become higher and higher. However, with the increase of switching frequency, switching loss has become one of the bottlenecks in the development of high power density converters [6]. For this reason, domestic and foreign scholars have proposed many new converter topologies and key control technologies to solve this practical problem. Among them, the LLC resonant DC-DC converter suitable for wide voltage or wide load range has natural soft switching characteristics, which can achieve high power density and high conversion efficiency at the same time, and is highly valued by industry and academia [7]. The LLC resonant circuit, as a kind of high-efficiency DC/DC converter, achieves the purpose of reducing switching loss through the opening and closing of the resonant cavity voltage and current. This paper takes the full-bridge LLC resonant converter as the research object, designs the resonance parameters[8-11], and realizes closed-loop control by collecting the output voltage signal feedback [12-15]. With the help of MATLAB software, the control simulation of the LLC resonant converter was completed and combined with the mathematical modeling and analysis of the PEM electrolytic cell[16-18], the joint simulation between the LLC resonant converter and the PEM electrolytic cell was realized. The main factors affecting the PEM electrolyzer are given, which has a good guiding significance for practical applications.

2. THE WORKING PRINCIPLE OF LLC RESONANT FULL-BRIDGE CONVERTER

Figure 1 shows the LLC DC converter and electrolytic cell topology. The primary is an H-bridge structure, and the resonant inductance L_r, the resonant capacitor C_r and the magnetizing inductance L_m form a resonant circuit. The secondary is composed of rectifier diodes VD_1, VD_2, and output filter capacitor C_o.

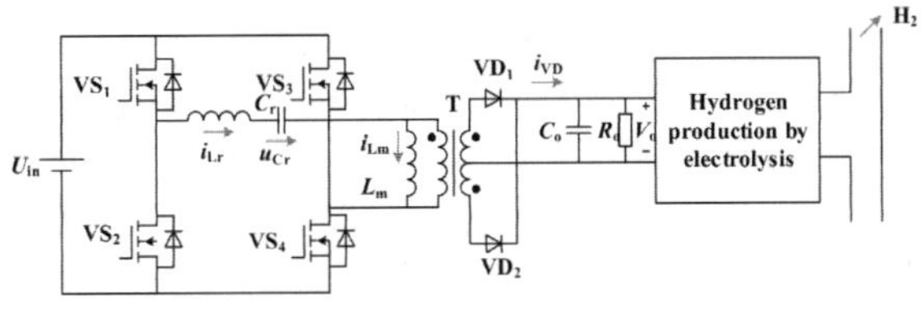

Fig 1. LLC DC converter and electrolytic cell topology

The converter has two resonance points, the resonance frequencies of L_r and C_r are:

$$f_r = 1/(2\pi\sqrt{L_r C_r}) \qquad (1)$$

The resonant network is composed of L_m, L_r, and C_r, and its resonant frequency is:

$$f_m = 1/(2\pi\sqrt{(L_r + L_m)C_r}) \qquad (2)$$

This circuit adopts frequency conversion control. When $f_m < f_s < f_r$ the converter has many advantages. At this time, the main operating waveform of the circuit is shown in Figure 2. One switching cycle can be divided into 8 switching modes for analysis. The working principle of the converter in half a switching cycle is shown in Figure 2.

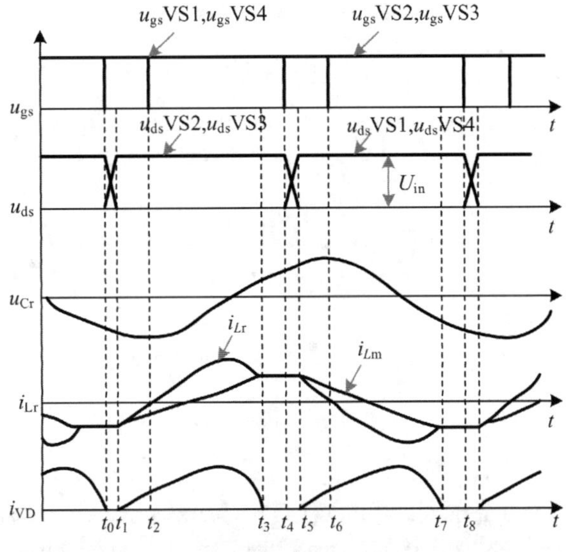

Fig 2. The current and voltage waveforms of each switch tube

(1) Mode 1: t_0-t_1

At time t_0, the switch VS_2 and VS_3 are off, and the secondary current i_{VD}=0. Since L_m is relatively large, it can be considered that the excitation current i_{Lm} remains unchanged. In this interval, i_{Lm} charges the output capacitors of VS_2 and VS_3, and the drain-source voltages u_{ds} (VS_2)and u_{ds}(VS_3) of VS_2 and VS_3 rise linearly. At time t_1, u_{ds}(VS_2) and u_{ds}(VS_3) rise to be equal to E_{in}.

(2)Mode 2: t_1-t_2

Starting from t_1, the primary current i_{Lr} changes the flow path, that is, the parasitic body diode flowing through VS_1 and VS_4, which creates conditions for VS_1 and VS_4 to achieve ZVS. The resonance of i_{Lr} rises, and the secondary diode VD_1 turns on. The voltage across the primary and secondary windings of the transformer is clamped, and i_{Lm} rises linearly.

(3)Mode 3: t_2-t_3

At time t_2, VS_1 and VS_4 are triggered to turn on, and i_{Lr} crosses from negative to zero. L_r and C_r form a resonance network.i_{Lm} continues to rise linearly.

(4)Mode 4: t_4-t_5

As i_{Lr} decreases and i_{Lm} increases, at t_3, i_{Lr}=i_{Lm}, and i_{VD} drops to zero. Therefore, the secondary diode can realize ZCS, and the reverse voltage it bears is clamped to the output voltage, eliminating parasitic oscillation. At t_4, VS_2 and VS_3 are triggered to conduct. The working process of the other half cycle is the same as the first half cycle.

3. Characteristic Analysis and Control Signal Design of LLC Resonant Converter

Because of its frequency-selective characteristics, the resonant cavity is most sensitive to the fundamental wave components near the resonant frequency and has a good filtering effect on the higher harmonic components that deviate from the resonant frequency. It can be considered that only the fundamental wave component can transmit power, so it can be analyzed by the fundamental wave equivalent method. The equivalent circuit diagram is shown in Figure 3.

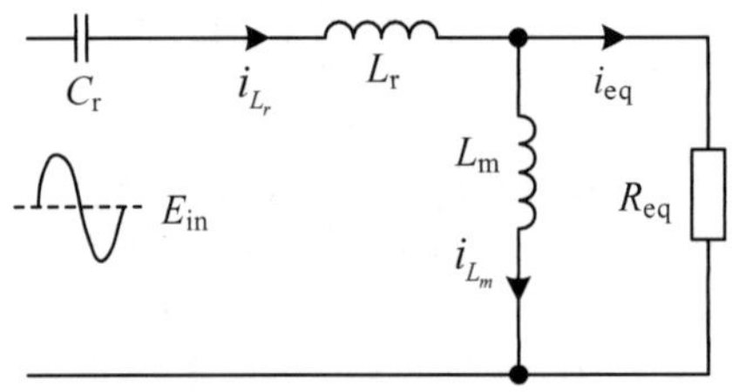

Fig 3. Equivalent circuit diagram

The output voltage of the resonant cavity is E_o, and the input voltage of the resonant cavity is E_{in}. The gain of the resonant cavity is defined as the ratio of the two. The normalized direct current gain formula of the resonant cavity can be obtained:

$$|G(f)| = \frac{Kf_n^2}{\sqrt{\{(1+K)f_n^2-1\}^2+\{QKf_n(f_n^2-1)\}^2}} \tag{3}$$

In the formula: $k=\frac{L_m}{L_r}$, $f_n=\frac{f_s}{f_r}$, $Q=\frac{2\pi f_r L_r}{R_{eq}}$, R_{eq} is the equivalent load resistance, f_s is the switching frequency.

3.1 *Operating characteristics of LLC under no-load condition*

At no load, the operating characteristics of the LLC resonant converter are shown in Figure 4. As the frequency increases, the gain gradually becomes a constant, and the DC output becomes non-adjustable, which causes the ZVS to be not satisfied at light load, and there is a The empty ratio is lost. In the no-load state, it is equivalent to the load impedance approaching infinity, the quality factor Q in formula (3) can be regarded as zero, and the voltage gain can be simply expressed as:

$$Gain = \frac{1}{1+\frac{1}{k}\left(1-\frac{1}{f_n^2}\right)} \tag{4}$$

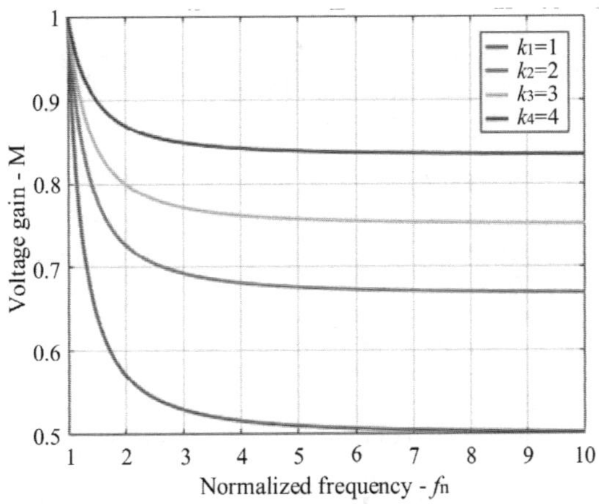

Fig 4. The relationship between DC gain and switching frequency

When the load value tends to infinity, the output current will be small, and the resonance current will also be small. On the other hand, for the parasitic capacitance, the required charging and discharging time will be greatly increased. If it cannot be turned on in the set dead zone time domain, there will be a problem that the switch tube cannot achieve zero voltage turn on. To improve this situation, the peak current i_m of the magnetizing inductance must be completely discharged during the dead time of the junction capacitance at both ends of the switch until the voltage at both ends drops to zero. The switch tube in the off state needs to charge its junction capacitance to the input voltage. The resonant current makes the voltage at both ends of the switch tube body capacitance full within the set time, and at the same time, it also makes the voltage at both ends of the parasitic capacitance commutate. The minimum resonant current that meets the conditions is:

$$I_{min} = 4C_{oss}\frac{E_{in}}{t_{dead}} + 2C_j\frac{nE_o}{t_{dead}} \tag{5}$$

The relationship between the peak value of the inductor current flowing through the transformer and the set dead time value should satisfy:

$$I_{pk} \geq \frac{2E_{ik}C_j}{t_{dead}} \tag{6}$$

Among them, E_{in} is the input bus voltage, C_j is the junction capacitance of the switch MOSFET, C_{oss} is the body capacitance of the selected MOSFET, and t_{dead} is the dead time. The magnetizing inductance L_m should satisfy the relationship:

$$L_m \leq \frac{T \cdot t_{dead}}{C_j} \tag{7}$$

3.2 DC gain analysis of LLC resonant converter

The relationship between the normalized frequency and gain is shown in Figure 5. Given the value of k unchanged, the value of Q is used as a variable. It can be seen from the figure that as the frequency fluctuates within the switching frequency range, the converter gain decreases. When the LLC resonant converter works at the resonant frequency, the gain is not affected by the load parameter settings.

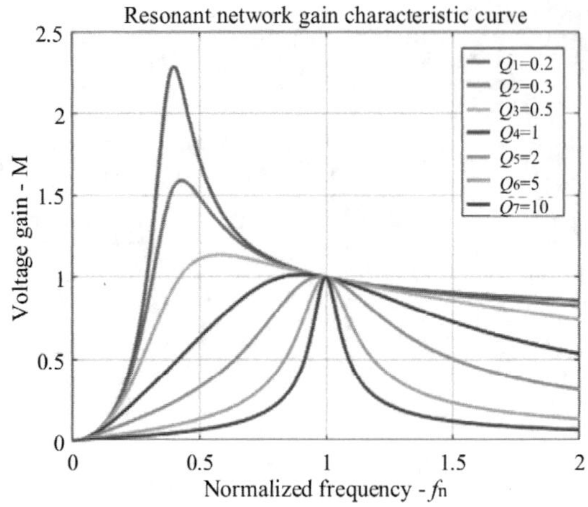

Fig 5. Only change the resonant network characteristic curve of Q

As shown in Figure 6 at the resonance point (f_n> 1)in the area on the right, Q value does not change, change the magnitude of the inductance ratio k, the gain of LLC resonant converter will drop very slowly, showing very hard characteristics. Considering the application of an LLC resonant converter, to achieve the ideal gain, this feature can be used in the region below the resonance point (f_n < 1). When the frequency is in this region, it can be seen that the gain of the LLC resonant converter increases as the value of k decreases. Therefore, when designing the parameters of the resonant converter, the values of Q and k are selected as far as possible to satisfy the region below the resonance point, and the adjustable gain range becomes wider at this time.

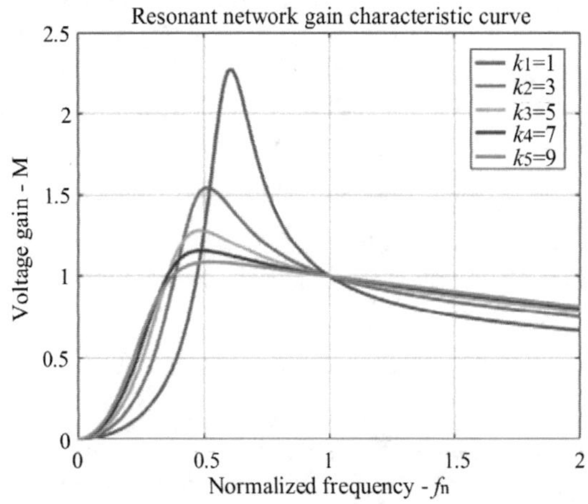

Fig 6. Only change the resonant network characteristic curve of k

LLC resonant converter has the characteristic of natural soft-switching ZVS, and the turn-on loss occupies the main part in the actual process. The parameter selection of inductance and capacitance energy storage components affects the resonant frequency, which is closely related to it. When f_r is constant, the resonant inductance L_r is determined, and the value of k will affect the value of the magnetizing inductance. For the inductor voltage value, the larger the inductance, the smaller the current flowing through the inductor. The inductor loss is proportional to the inductance value and the square of the inductor current value. Within a given frequency range, the greater the difference between the switching frequency and the resonance frequency, the greater the gain of the LLC resonant converter.

Therefore, broadening the gain range of the LLC resonant converter can be achieved by changing the frequency range.

3.3 Design of Driving Signal of LLC Resonant Converter

The pulse control signal module is shown in Figure 7. The full-bridge LLC resonant converter is used as the research object. The difference between the given voltage signal and the DC output voltage signal collected by the load is used as input. After closed-loop PI calculation, the voltage-controlled oscillator is used After adjustment, the switching frequency corresponding to the voltage value is obtained. The trigger signal is generated through the delay module and the dead time is set in the driving signal at the same time to prevent the two switching tubes of the same bridge arm from being turned on at the same time. In this link, proportional integral control can make the dynamic process of the system get a quick response. This process uses a voltage-controlled oscillator to adjust the PI parameters of the error signal, converts the voltage signal into a drive signal with a corresponding response frequency, and uses the output voltage signal feedback to achieve closed-loop control, and the control structure is simplified.

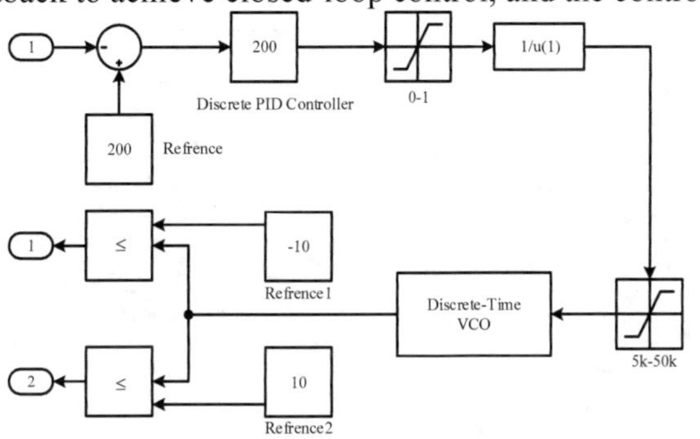

Fig 7. Drive signal generation module

4. MODELING AND ANALYSIS OF ELECTROLYZER

Analyzing the working process of the electrolytic cell, it can be found that in the process of electrolyzing water, there are mainly two parts of work. One part is the electrochemical reaction, which mainly reflects the changes in the voltage of the electrolytic cell during the hydrogen production process, and the other part is the heat exchange. Reflect on the heat transfer and heat exchange between the system and the internal and external environment during work.

4.1 Electrochemical submodel

The electrochemical sub-model defines the working voltage U_{cell}, which reflects the working status of the entire system. Its value is equal to the open-circuit voltage U_{ocv}, the activation overvoltage U_{act}, the ohmic overvoltage U_{ohm} and the concentration overvoltage U_{con}.

$$U_{cell} = U_{ocv} + U_{act} + U_{ohm} + U_{con} \tag{8}$$

4.2 Thermal submodel

To better study the heat transfer and heat exchange between the internal and external environments of the entire electrolytic cell during the working period, a control volume C is defined, and the expression of C is as follows

$$C\frac{dT}{dt} = W_1 - W_2 - Q_{cool} - Q_{loss} - Q_n \tag{9}$$

In the formula: W_1 is the sum of heat and electric energy input during the electrolysis of water; W_2 is work input by the water pump due to hydraulic, mechanical, and total heat dissipation; Q_{cool} is the heat discharged by the system in the heat exchange; Q_{loss} is the heat loss of the system to the environment;

Journal of Physics: Conference Series **2215** (2022) 012017 doi:10.1088/1742-6596/2215/1/012017

Q_n is the sum of the enthalpy leaving the control volume C and the enthalpy of oxygen and hydrogen.

5. Analysis of simulation verification results

In this paper, MATLAB/Simulink software is used to conduct system joint simulation of LLC resonant converter and electrolyzer under PID control of closed-loop frequency conversion. During the simulation, the resonance frequency of the LLC resonant converter is set to 20kHz, and the rated input works at the resonance frequency. According to the above analysis, other parameter settings are shown in Table 1.

The parameters of the LLC converter simulation model are shown in the table below.

TABLE I. LLC Resonant Circuit Parameter Table

Parameters	**Values**
Input voltage	750V
Input voltage	200V
Power	7.5kW
Inductance ratio	6
Resonant frequency	20kHz
Resonance inductance	$0.3*10^{-3}$H
Resonant capacitor	$2*10^{-7}$H
Magnetizing inductance	$1.8*10^{-3}$H
Transformer ratio	4

Figure 8 is a joint simulation diagram of LLC resonant converter and electrolysis hydrogen production. The output voltage of the LLC resonance circuit will link the two simulations into a system simulation. The LLC resonance circuit includes a main circuit module, a resonance circuit module, and a drive signal generation module. The electrolyzer model mainly includes an anode module, cathode module, hydrogen storage module, and output voltage module.

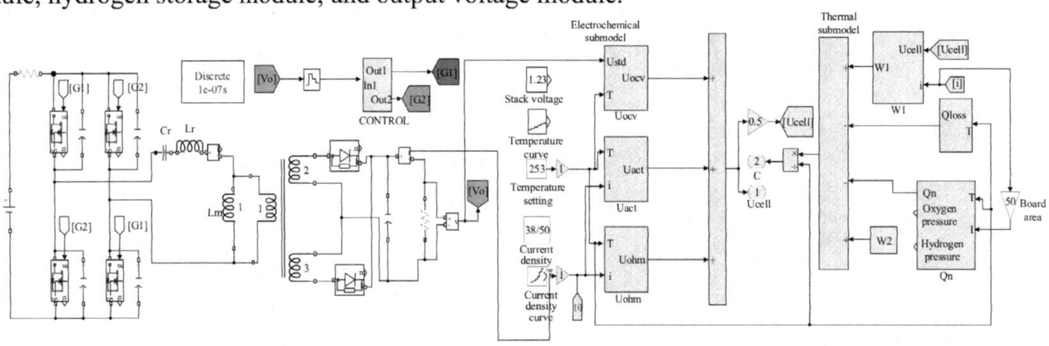

Fig 8. System joint simulation diagram

Figure 9 shows the waveform of the output voltage. It can be seen from the figure that the waveform of the output voltage is relatively smooth and can be well stabilized at 200V, meeting the design goal of ripple less than 1%.

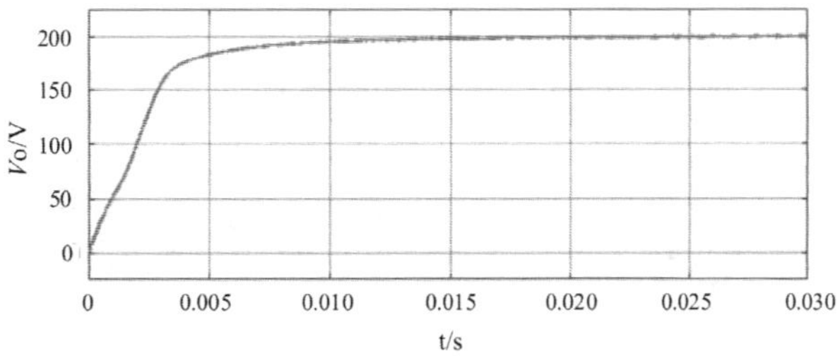

Fig 9. Output voltage waveform diagram

Figure 10 shows the waveforms of $u_{ds}(VS_1)$ and i_{VS1}. It can be seen from the figure that when i_{VS1} is high, S_1 turns on and $u_{ds}(VS_1)$ drops to 0; when i_{VS1} is low, S_1 turns off, and $u_{ds}(VS_1)$ bears a 750V DC voltage. When i_{S1} is still low, $u_{ds}(VS_1)$ quickly drop from 750 to 0, which shows that the charge of C is released to 0 during this period, and then D_1 is naturally turned on. After a while, i_{VS1} rises to high. In other words, $u_{ds}(VS_1)$ drop to 0 first before S_1 is opened, that is, S_1 realizes ZVS. S_2-S_4 achieves similar ZVS waveforms, so I won't repeat them here.

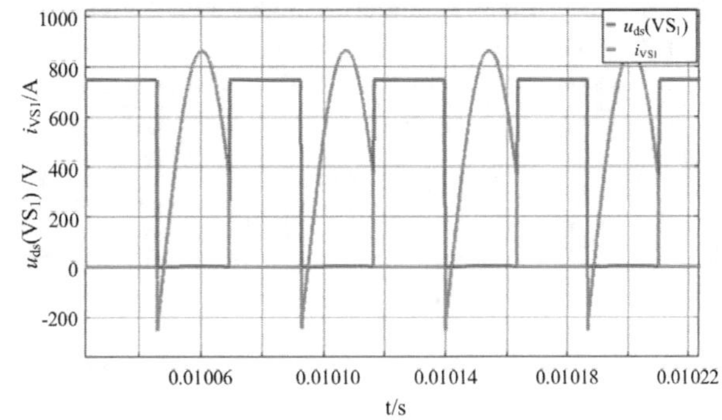

Fig 10. Comparison chart of terminal voltage and current under zero-voltage turn-on(ZVS)

Figure 11 shows the waveforms of i_{VD1} and U_{VD1}. When the rectifier diode is disconnected, the voltage at both ends should be high, and the current flowing through the diode has dropped to 0 at this time, and a zero-current shutdown is reached in this process.

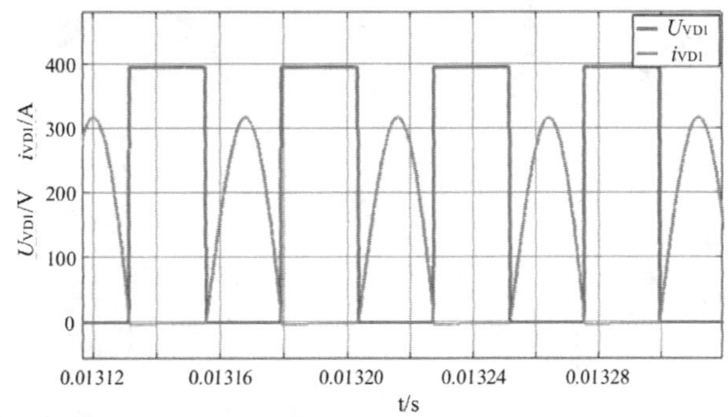

Fig 11. Comparison of voltage and current at the lower end of zero current shutdowns (ZCS)

As shown in Figure 12, it is the voltage waveform of the resonant capacitor voltage in the steady-state. It can be seen that the working process of the circuit is stable, and the output waveform is a sine wave.

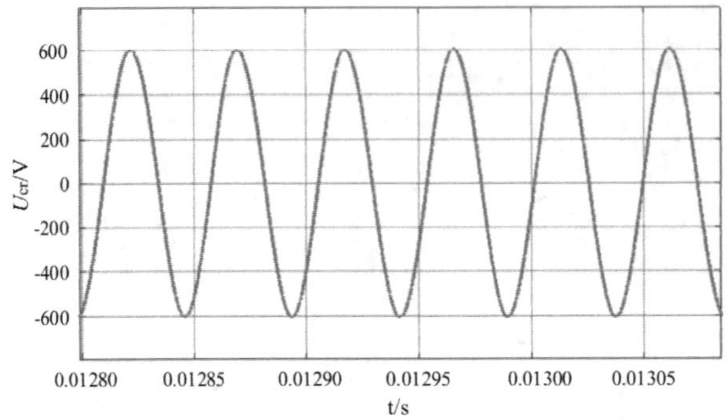

Fig 12. The resonant capacitor voltage waveform

Figure 13 is a static simultion experiment diagram of the system. Figure 13 depicts the stack polarization curve under the conditions of the working pressure of the electrolytic cell stack at 0.3MPa and the stack temperature from 320K to 390K. The curve shows that the stack voltage decreases as the operating temperature of the stack increases. Figure 14 describes the relationship between the voltage of the electrolytic cell and the output pressure under different current densities. The curve shows that when the pressure is constant, the greater the current density, the higher the voltage.

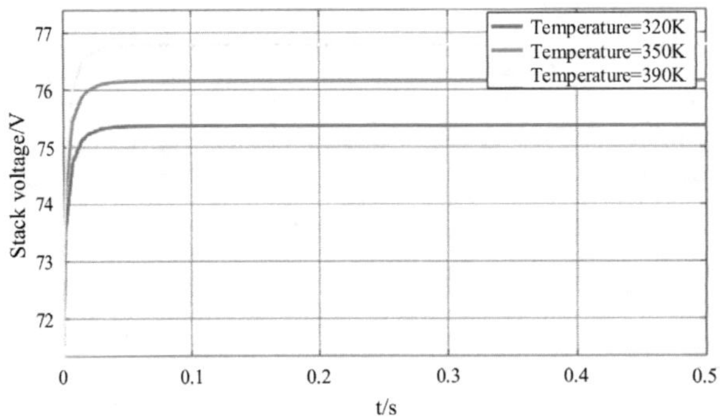

Fig 13. Stack polarization curve (When the temperature changes)

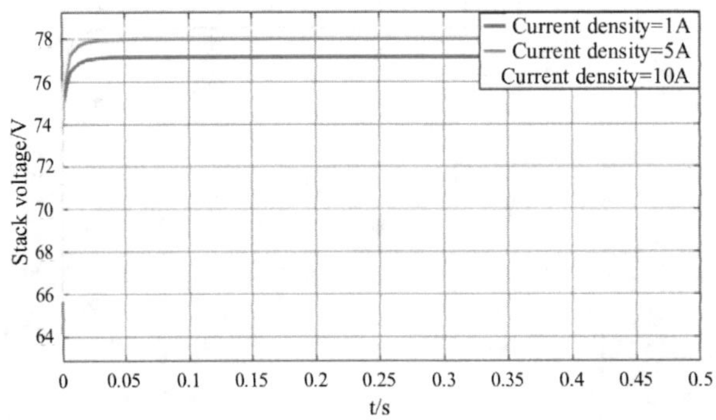

Fig 14. Stack polarization curve (When the current changes)

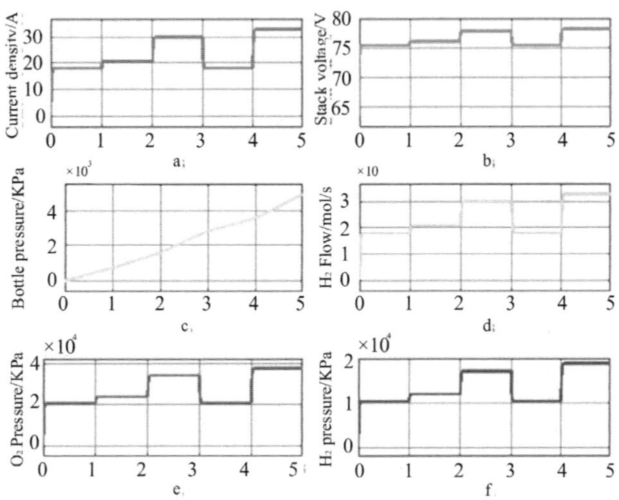

Fig 15. Dynamic simulation experiment results of the electrolytic cell stack described by the model

Figure 15 is a dynamic simulation experiment diagram of the system. An oxidation reaction occurs at the anode of the electrolytic cell, and electrons are lost. After electrons enter the cathode, a reduction reaction occurs. To carry out the dynamic response experiment of the electrolytic cell stack, the following structure and operating parameters are given. The electrolyzer is composed of 50 single cells in series, each cell has an activation area of 50cm^2. Assume that the working temperature is 353K and the working pressure is 100KPa. Figure 15(a)-(f) are the current, voltage, hydrogen flow, storage tank pressure, anode pressure, and cathode pressure of the electrolytic cell stack, respectively.

The simulation results are described as follows:

(1) When t=1s, t=2s, and t=4s, the current of the electrolyzer increases step by step, and the voltage, anode pressure, and cathode pressure all increase rapidly. This is because the purpose of the electrolyzer is to electrolyze water to produce hydrogen and oxygen. The amount of hydrogen and oxygen produced is directly proportional to the current density. The higher the current, the faster the hydrogen and oxygen produced, and the corresponding pressure increases. Similarly, at t=3s, the current decreases, and the voltage, anode pressure, and cathode pressure all decrease accordingly. The relationship between current and voltage conforms to the polarization characteristics of the electrolytic cell battery.

(2) The hydrogen flow rate and the pressure change of the hydrogen storage tank are caused by the current change. When t=1s, t=2s, and t=4s, the current of the electrolytic cell suddenly increases, and the hydrogen flow rate and the pressure of the storage tank also suddenly increase. This is because the increase in current accelerates the production rate of hydrogen. Since the volume of the storage tube remains unchanged, according to the ideal gas law, the gas flow increases, and the pressure in the storage tank increases, which is in line with the actual situation. The flow rate of hydrogen and the pressure of the hydrogen storage tank are changed due to the current change. At t=1s, t=2s, and t=4s, the current of the electrolytic cell suddenly increases, and the flow rate of hydrogen and the pressure of the storage tank also suddenly increase. This is because the increase in current accelerates the production rate of hydrogen. Since the volume of the storage tube remains unchanged, according to the ideal gas law, the gas flow increases, and the pressure in the storage tank increases, which is in line with the actual situation. When t=3s, the current of the electrolytic cell suddenly decreases, and the pressure of the storage tank does not decrease suddenly but slowly increases. This is because the gas in the storage tank does not flow out of the storage tank. Although the current decreases, the electrolyzer electrolyzes the water. The process is still going on, so the gas in the tank continues to increase, but the rate of increase is relatively slow. It is also in line with actual working conditions. However, since the PEM electrolytic cell is composed of proton exchange membranes, catalysts, etc., if the input current changes frequently, the working pressure of the electrolytic cell changes, which also has a great impact on the life of the

electrolytic cell. Therefore, buffer energy storage devices and certain energy management amplification should be used to maintain the stability of the input power of the electrolytic cell. Because the supercapacitor is directly connected to the DC bus, it can quickly output or absorb the energy needed due to power or load fluctuations, alleviate frequent changes in the pressure of the electrolytic cell, and extend the working life.

6. Conclusion

By studying the working principle of the LLC resonant circuit, the fundamental wave analysis method is used to perform equivalent analysis on the circuit, and the relationship between the various resonant parameters is given. Because it is difficult to select the resonant parameter k value and Q value, this paper analyzes the DC gain curve and the conditions for realizing the switch tube ZVS, obtains the selection method of k value and Q value, and calculates the parameters of the corresponding resonant components. For hydrogen production by electrolysis, this article analyzes the various voltage modules in the electrochemical reaction, the energy modules in the heat exchange process, and models them. For hydrogen production by electrolysis, this article analyzes the various voltage modules in the electrochemical reaction, the energy modules in the heat exchange process, and models them. Finally, MATLAB simulation software was used to carry out the system joint simulation: The LLC resonant circuit realizes output voltage stabilization through closed-loop control of the DC converter output voltage signal feedback, which simplifies the design of the control loop, the output response is fast and stable, and the switching tube realizes zero When the voltage is turned on, the rectifier diode completes zero-current turn-off; the electrolysis hydrogen production system builds a Simulink model of electrolysis hydrogen production, and gives the models of each sub-module. In the static experiment, the simulation adjusts the temperature, air pressure, and current density. Then analyze the value of each output module, and explore the factors that have a greater impact on the working state of electrolysis hydrogen production, mainly current density and temperature. When the temperature rises, the value of each voltage module will drop, and the rate of hydrogen and oxygen produced will be slower. In the dynamic experiment, when the electrolyzer current step increases, the hydrogen and oxygen production rate increases accordingly, and the corresponding storage tank pressure increases. When the current of the electrolytic cell decreases step by step, the production rate of hydrogen and oxygen decreases, and the pressure of the storage tank increases slowly, which is in line with the actual situation.

Acknowledgment

This study was supported by the Beijing High-Level Innovation Team Construction Plan (No. IDHT20180502).

References

[1] E. Wikramanayake, P. V. Acharya, M. Kapner and V. Bahadur, "Green hydrogen-based energy storage in Texas for decarbonization of the electric grid," 2021 IEEE Green Technologies Conference (GreenTech), 2021, pp. 409-415.

[2] Y. -S. Huang and S. -J. Liu, "Chinese Green Hydrogen Production Potential Development: A Provincial Case Study," in IEEE Access, vol. 8, pp. 171968-171976, 2020.

[3] N. Naseri, S. El Hani, A. Aghmadi, H. Mediouni, I. Aboudrar and M. Benbouzid, "Solar Photovoltaic Energy Storage as Hydrogen via PEM Fuel Cell for Later Conversion Back to Electricity," IECON 2019 - 45th Annual Conference of the IEEE Industrial Electronics Society, 2019, pp. 4549-4554.

[4] A. Maizelis and A. Pilipenko, "Electrode Materials for Hydrogen Production by Alkaline-Water Electrolysis Powered by Renewable Energy Sources," 2021 IEEE 2nd KhPI Week on Advanced Technology (KhPIWeek), 2021, pp. 313-316,

[5] F. E. Chakik, M. Kaddami and M. Mikou, "Optimization of physico-chemical parameters of hydrogen production by electrolysis of water," 2018 Renewable Energies, Power Systems & Green Inclusive Economy (REPS-GIE), 2018, pp. 1-6.

[6] D. Zhou, X. Zhang, X. Liu, Y. Wang, D. Xu and H. Tian, "Design of LLC Converter Parameters Based on ZVS Characteristic Analysis," 2018 21st International Conference on Electrical Machines and Systems (ICEMS), 2018, pp. 2251-2255.
[7] B. Li, X. Zhang, D. Zhou, S. Sun, Y. Wang and D. Xu, "Soft-switching Characteristics Analysis Based on LLC Resonant Converter," 2018 IEEE Transportation Electrification Conference and Expo, Asia-Pacific (ITEC Asia-Pacific), 2018, pp. 1-6.
[8] J. Xu, Y. Wei, X. Cheng, H. Yang, H. Xue and X. Wang, "Univesal Design of LLC Resonant Converter with Fixed Switching Frequency," 2020 IEEE International Conference on Information Technology,Big Data and Artificial Intelligence (ICIBA), 2020, pp. 1323-1328.
[9] Y. Yang, Y. Li, Z. Cheng and T. Gao, "Optimal Design of Transformer Windings for Full-Bridge LLC Resonant Converters," 2019 14th IEEE Conference on Industrial Electronics and Applications (ICIEA), 2019, pp. 816-820.
[10] V. Suel and S. Kizir, "A Comprehensive Loss Analysis of Half-Bridge LLC Resonant Converter used in LED Street Lights," 2019 1st Global Power, Energy and Communication Conference (GPECOM), 2019, pp. 146-151.
[11] X. He, Y. Chen, C. Chen, Z. Zhao and P. Han, "Resonance Parameter of Variable Mode LLC Converter for Auxiliary Converter," 2019 10th International Conference on Power Electronics and ECCE Asia (ICPE 2019 - ECCE Asia), 2019, pp. 1-6.
[12] T. Zhu, Y. Ji, J. Wang, Y. Liu and C. Gao, "LLC Converter's Design Method Based on Simulation Analysis," 2018 2nd IEEE Advanced Information Management,Communicates,Electronic and Automation Control Conference (IMCEC), 2018, pp. 1996-2000.
[13] D. Indira., V. Sivachidambaranathan. and S. S. Dash, "Closed loop control of hybrid switching scheme for LLC series-resonant half-bridge DC-DC converter," International Conference on Sustainable Energy and Intelligent Systems (SEISCON 2011), 2011, pp. 295-298.
[14] D. Cittanti, M. Gregorio, E. Armando and R. Bojoi, "Digital Multi-Loop Control of an LLC Resonant Converter for Electric Vehicle DC Fast Charging," 2020 IEEE Energy Conversion Congress and Exposition (ECCE), 2020, pp. 4423-4430.
[15] G. S. Nelson and A. Susmitha, "Hybrid control for high efficiency LLC resonant converter," 2016 IEEE 7th Power India International Conference (PIICON), 2016, pp. 1-6.
[16] M. KOUNDI and H. EL FADIL, "Mathematical modeling of PEM electrolyzer and design of a voltage controller by the SMPWM approach," 2019 International Conference on Power Generation Systems and Renewable Energy Technologies (PGSRET), 2019, pp. 1-6.
[17] M. Albarghot and L. Rolland, "MATLAB/Simulink modelling and experimental results of a PEM electrolyzer powered by a solar panel," 2016 IEEE Electrical Power and Energy Conference (EPEC), 2016, pp. 1-6.
[18] A. Beainy, N. Karami and N. Moubayed, "Simulink model for a PEM electrolyzer based on an equivalent electrical circuit," International Conference on Renewable Energies for Developing Countries 2014, 2014, pp. 145-149.

Identification Methods for Transformer Turn-to-Turn Faults and Inrush Current Based on Electrical Information

Xinyi Gong[1,a*], Yuping Zheng[1,2,b], Shuyan Pan[1,c], Chonghao Wu[1,d], Jinzhao Deng[1,e]

[1] NARI Research Institute, NARI Group Corporation (State Grid Electric Power Research Institute), Nanjing, China

[2]State Key Laboratory of Smart Grid Protection and Control

*[a]gongxiny1@126.com, [b]zhengyuping@sgepri.sgcc.com.cn, [c]panshuyan@sgepri.sgcc.com.cn, [d]wuchonghao@sgepri.sgcc.com.cn, [e]dengjinzhao@sgepri.sgcc.com.cn

Abstract—Power transformer is the most imperative and crucial equipment in electrical power system. Accurately monitoring the occurrence of transformer faults and rapidly isolating the faults are of great significance to ensure the reliability of power supply and avoid serious damage to the transformer. There are many methods used in the transformer to monitor transformer running state. This paper provides a comprehensive overview of transformer turn-to-turn fault and inrush current identification methods based on electrical information, and expounds the systematic research content that needs to be carried out for transformer turn-to-turn fault identification methods based on magnetic flux leakage information.

1. INTRODUCTION

Power transformer, which plays a key role in the power generation, transmission, and distribution, is a kind of significant power apparatus in power system. The reliability of power supply and system operation will be affected by transformer failure, which may cause serious economic losses [1].

As the end of 2017, the statistical number of transformers with voltage levels of 220kV and above in China is 15350, an increase of 3237 units from 2013, and the average annual growth rate in the past five years has reached 4.99% [2]. According to the statistics of China Electricity Council, the average annual failure rate of transformers of 220kV and above during the "Twelfth Five-Year Plan" period is 0.1766 times/ 100 units [3].

In order to prevent the expansion of transformer faults from causing more serious accidents, it is necessary to install various relay protection devices that meet the technical requirements according to the possible faults and abnormal operating conditions of the transformer.

Differential protection is usually necessary for transformer to be equipped with differential protection in view of phase-to-phase short-circuit of winding and lead-out wires, single-phase grounding short-circuit of winding and lead-out wires of large current grounding system and short-circuit between turns of winding. The differential protection principle is simple, and it also has rapid and sensitive performance, so it is widely used as the main protection of various equipment. Differential protection is based on KCL and is particularly suitable for generator, motor and line protection [4]. However, differential protection will produce a series of problems under the existing conditions.

Practical operation experience shows that the performance of transformer protection is still not comparable to that of line or bus protection [5].

A brief review of modern transformer fault records shows that transformer internal faults are the most common situation, and 70%-80% of transformer faults are turn-to-turn faults (TTFs). If the faults can be detected in time and corresponding measures are taken rapidly to isolate the faults, the greater losses can be avoided [6].In addition, when the transformer is input in electric network or troubleshooting, large inrush current will be generated, which may result in misoperation of differential protection [7]. Therefore, the transformer protection needs to be able to quickly and accurately identify currents, which is the key to improving the protection performance, reducing transformer protection misoperation rate, and ensuring reliable operation of the power system. At present, many methods have been applied to the offline and online detection of transformer faults and the identification of inrush current. A summary of their content, advantages and disadvantages will help to selectively apply these methods to improve transformer protection performance.

This paper describes the main methods of transformer TTF identification based on electrical information, and then summarizes the research status of transformer inrush current identification. These methods help to find any faults in the transformer and indicate the severity to avoid any unprecedented faults. It also elaborates the identification method based on magnetic flux leakage information and the systematic research content that needs to be carried out, which will provide a fast, reliable and sensitive solution for the identification of TTF and inrush current.

2. OFF-LINE DETECTIOON METHOD FOR TTFS

For a long time, the detection of inter-turn insulation faults in engineering has mainly used traditional test offline methods including DC resistance test, short impedance test, turns ratio test, no-load test, induced voltage withstand test, frequency response test, insulation resistance measurement, absorption ratio measurement, polarization index measurement, dielectric loss factor and capacitance measurement. These techniques have achieved some results in practical application, but they are not effective or sensitive in diagnosis [8].

The winding DC resistance measurement method is to apply DC voltage on the transformer winding, measure the DC resistance of each phase, and judge whether TTF occurs inside by comparing the three-phase resistance value [9]. Due to the transformer DC resistance changes caused by many reasons, the technical operation error in the test process or transformer itself welding and poor contact will make transformer DC resistance changes, so this test method for TTF detection reliability is not high [10]. By applying AC voltage and measuring short impedance, low voltage short impedance test can reflect the change of transformer impedance value before and after winding deformation, and avoid transformer short circuit damage accident. With these two methods, when the transformer has a minor TTF, since the short-circuited turns occupy a small proportion in the winding, the DC resistance test and the short impedance test alone usually cannot find the fault effectively [11].

The transformation ratio test is a traditional test item for transformer products. In the process of manufacturing, on-site installation and operation and maintenance, it can be checked whether the number of winding turns is consistent with the design, whether the tap lead assembly is correct, and whether the voltage ratio of each tap is within the range allowed by the standard. The transformation ratio test is more sensitive when detecting TTF of the transformer, but this method can only make accurate judgments when the transformation ratio changes significantly [12,13].

The no-load test of the transformer also examines the loss of the windings and cores of the transformer, which increases the difficulty of diagnosing the fault. The induced voltage withstand test is to directly apply the voltage amplitude to the winding twice the rated voltage [14]. Although this can directly investigate the inter-turn insulation performance, the magnetization characteristic curve of the transformer core has been saturated at this time, so the no-load current is greater than the maximum current allowed by the winding which will cause irreversible damage to the transformer [15].

The transformer coil may deform due to the heat and electromotive force generated by the short-circuit current or the impact during transportation and installation which may lead to TTFs of transformer. The application of frequency response analysis method to detect transformer winding deformation was first proposed by Dick in Canada. In the early 1990s, my country began to study frequency response analysis to detect transformer winding deformation. A large number of tests and on-site measurement results show that the frequency response of transformer winding in the range of 1 kHz to 1MHz can reflect faults such as short circuit between turns, twist and bulge of winding. However, for a long time, the use of frequency response analysis is still based on the comparison of graphs. There is no clear quantitative standard and lack of in-depth diagnostic methods. With the wide application of frequency response analysis in winding deformation detection, the accumulation of a large amount of frequency response data provides favorable conditions for the development of intelligent diagnosis research on winding deformation such as artificial neural network, data mining and pattern recognition [16,17].

However, these methods can only detect severe TTFs, and the detection sensitivity needs to be improved. At the same time, most of them require equipment to be powered off. In recent years, the requirements for the production efficiency and economic benefits of power grid operation have also been continuously improved. In view of the problems exposed by the traditional regular maintenance system and offline tests, on the one hand, the majority of intact equipment is blindly repaired regularly, resulting in waste of manpower and material resources [18]. And this kind of over-maintenance may also introduce new hidden troubles; on the other hand, there is the possibility that some product performance defects, including insulation defects, cannot be detected and repaired in time and may develop into major failures. Therefore, people begin to pay attention to the research and application of transformer online monitoring [19].

3. ON-LINE MONITORING METHOD FOR TTFS

3.1 Leakage Inductance Method

When the transformer is in normal operation, external faults and inrush currents, the number of turns of the transformer windings and the magnetic circuit which the flux leakage passes do not change, and the leakage inductance of the transformer windings will not change; however, when a fault occurs between partial turns, the number of turns which the winding current passes will change, and the leakage inductance will definitely change. Based on the characteristics, the leakage inductance and resistance of the transformer windings can be used as the criteria to distinguish the normal operation, internal faults, external faults, and inrush current conditions [20,21].

Many methods have been applied to calculate leakage inductance. In order to improve the sensitivity of the algorithm, the least square algorithm is used to estimate the transformer leakage inductance parameters [22,23]. In addition, using the dynamic vector model based on Fourier theory, the winding resistance parameters are identified when the transformer is closed at no load, and the identified resistance parameters are used to modify the parameter identification model when the transformer is operating in a steady state, and the leakage inductance parameters are identified [24]. Based on the finite element method, the large-scale finite element software ANSYS is used to establish the transformer core and the winding model; the field-circuit coupling method is used to simulate the leakage magnetic field of the transformer, and the energy method is used to calculate the leakage inductance parameters of the transformer [25,26].

However, the transformer protection with leakage inductance method has the following problems in the identification of winding faults: 1) the accuracy of the identification value is not high, only when the transformer fault occurs in the area, the parameters change greatly, so that it can have enough sensitivity to detect the fault. 2) The transformer model established is relatively simple, and usually only a single model is established to detect transformer faults.

3.2 Power Loss Method

When the transformer is operating in a non-fault state, the active power loss of the transformer is relatively minor. When the transformer winding has a TTF, not only may it consume a lot of energy due to the increase of the current, but also the additional loss of the transformer will increase due to the change of the leakage magnetic field caused by the short-circuit. It is feasible to use the power loss as the characteristic quantity of transformer TTF diagnosis, which is the reason why this method has been maturely applied in engineering practice. Generally, only the electrical quantities on both sides of the transformer need to be collected, and the relevant formula can be used to calculate the power loss of the transformer at this time, and this value can be compared with the reference value to determine whether a TTF has occurred and the severity of the fault. In addition, the criterion is not affected by inrush current and external faults [27].

When an TTF occurs in the transformer windings, due to magnetic coupling, in addition to the large change in the power loss of the faulty phase, the power loss of the non-faulty phase also increases correspondingly, but it is not obvious compared to the faulty phase, so it does not affect the power loss method for faulty phase diagnosis of transformer TTFs. However, a significant shortcoming of this method is that it cannot be used for faulty turns diagnosis. The traveling wave analysis technique is combined with the power loss method to diagnosis fault occurrence and location with high accuracy. The positive sequence component and the negative sequence component are used to calculate the corresponding power loss, eliminating the influence of the zero sequence component, improving the diagnostic accuracy, and achieving good results. Since the power loss is not sensitive enough to identify minor TTFs, integrating the power and taking the changes in the electrical energy loss before and after the fault as the monitoring object can improve the sensitivity but reduce the rapidity [28].

3.3 Magnetic Flux Leakage Method

The distribution of magnetic flux leakage of the transformer is very complicated, especially when the transformer has a TTF, the internal magnetic field will change greatly. In addition, since the magnetic flux leakage changes immediately after the fault, a protection scheme with both rapidity and sensitivity can be formulated according to the distribution law of magnetic flux leakage before and after the fault.

To formulate a protection scheme, it is first necessary to clearly analyze the distribution law of the transformer magnetic field. The theoretical law of the axial symmetry of the leakage flux distribution is obtained by simplified analysis [29,30]. And simple sensors, based on air core coils, are used to measure the leakage flux of the transformer, which provides a low cost scheme for TTF identification based on flux leakage information [31,32]. Based on the symmetry principle of magnetic flux leakage distribution, a novel approach to detect transformer TTF by the search coil induced voltage analysis is proposed [33,34]. The distorted leakage flux linking with the search coil induces voltage which can detect presence of fault and also predict its severity. A two-dimensional simulation model was established to verify that the axial symmetry obtained by the approximate calculation is correct. Against the traditional differential protection schemes, tap changer operation, inrush current and over-fluxing cannot impact the proposed technique performance and the results show that the proposed algorithm is a success to detect the minor TTFs as small as ten turns (lower than 0.25% of the winding) [35,36].

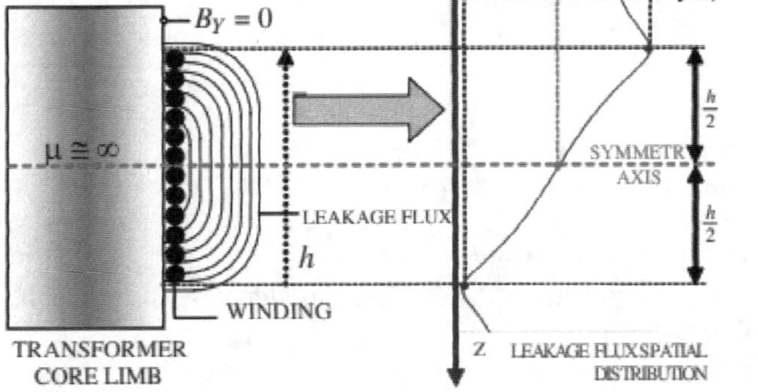

Figure 1 Theoretical calculation of axial magnetic density symmetry

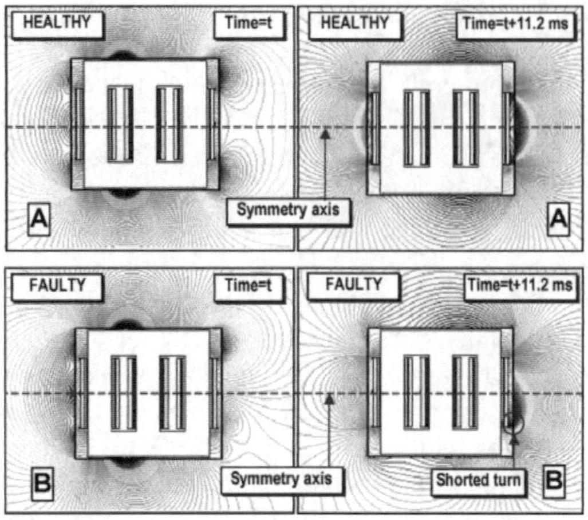

Figure 2 Leakage flux lines at two time instances of a steady-state operation cycle of the machine. A: healthy machine. B: transformer with one shorted turn.

4. IDENTIFICATION METHOD OF INRUSH CURRENT

4.1 Criterion of Inrush Current Based on Current Information

The method of identifying inrush current based on the information of current mainly includes the second harmonic method, the principle of dead angle, and the principle of waveform symmetry.

The second harmonic method uses the characteristics of a large number of second harmonic components in the transformer magnetizing inrush current waveform. When the second harmonic component of the differential current flowing into the differential relay is detected, the dynamic protection is blocked to avoid the misoperation of the differential protection due to the inrush current [37].

The dead angle principle mainly uses the characteristics of the dead angle in the inrush current waveform. When it is detected that the discontinuous angle of the differential current flowing into the differential relay is greater than the set value, the differential protection is blocked.

This method mainly faces the problem of discontinuous angular deformation caused by non-periodic components. In addition, current transformer saturation increases the discontinuous angle of the differential current when an internal fault occurs, and may also cause the protection to refuse to operate [38,39].

The principle of waveform symmetry is to use the front/post half-wave symmetry comparison of the derivative of the difference current (to reduce the influence of aperiodic components), and to identify the inrush current based on the comparison result. The principle of waveform symmetrical inrush current recognition is not affected by the offset of the DC component, and is an extension of the discontinuous angle inrush current criterion, and is easier to implement than the discontinuous angle inrush current criterion [40,41].

4.2 Criterion of Inrush Current Based on Voltage Information

The basic idea of using the transformer voltage to identify the inrush current is that when the transformer is severely saturated due to the inrush current, the terminal voltage will be severely distorted. Compared with current harmonic braking, the principle of harmonic voltage braking is relatively insensitive to LC oscillations, and some deficiencies of second harmonic braking can be improved to some extent [42].

The application of the principle is closely related to the size of the system impedance. Therefore, the protection using this principle inevitably requires a more accurate understanding of the system impedance, and there is complexity in setting. When the system impedance is minor, the action characteristics of the protection of this principle may deteriorate.

4.3 *Criterion of Inrush Current Based on Current and Voltage Information*

Based on the inrush current identification method of the transformer admittance type equivalent circuit, the internal and external faults of the transformer are identified by detecting changes in the admittance against the ground. The test results of the criterion used to distinguish between internal faults and inrush current on a simulated transformer prove the effectiveness of the method. The advantage of this algorithm is that it can quickly identify internal faults and inrush currents. Even if the internal fault situation is superimposed on the inrush current situation, the correct judgment result can generally be given within half a cycle [43].

The active power loss of the transformer can be used to identify the inrush current. During normal operation, the active power loss of the transformer is minor; but when the transformer generates the inrush current, the active power loss at the initial stage of the inrush current may be large, but it will soon be reduced to a small value. When the transformer has an internal fault, the active power loss greatly increases, and its value far exceeds the loss during normal operation and magnetizing inrush current. However, it is difficult to obtain the copper loss, which makes parameter setting difficult, which limits its practicability [44].

4.4 *Criterion of Inrush Current Based on Magnetic Flux Leakage Information*

Since when any fault occurs in the transformer winding, the symmetrical form of the magnetic flux leakage distribution will be disturbed, but the magnetic flux leakage distribution remains symmetrical under the inrush current condition, so it can be regarded as an appropriate standard to implement a suitable protection algorithm.

By setting the action value of the induced voltage of the search coil for detecting the magnetic field to be higher than the value in the case of the excitation inrush current, the protection can be ensured that the protection does not operate in the excitation inrush current state, but the higher action value will reduce the sensitivity to turn-to-turn fault identification [33,34].

A new technology based on flux leakage is used to diagnose the inrush current of transformers. The proposed technology uses a pre-installed flux leakage optical fiber sensor as a magnetic flux sensor near the transformer yoke to measure the flux leakage passing through the yoke at various positions. Since the magnetic flux density when the no-load transformer is energized is very different from the magnetic flux density detected in the normal state, the inrush current of the transformer can be identified. By comparing the different closing angles of the transformer under normal conditions and the difference of the flux leakage near the yoke, the change of the magnetic flux density can effectively identify the transformer magnetizing inrush current. This identification method has been initially verified in Maxwell's simulation [45].

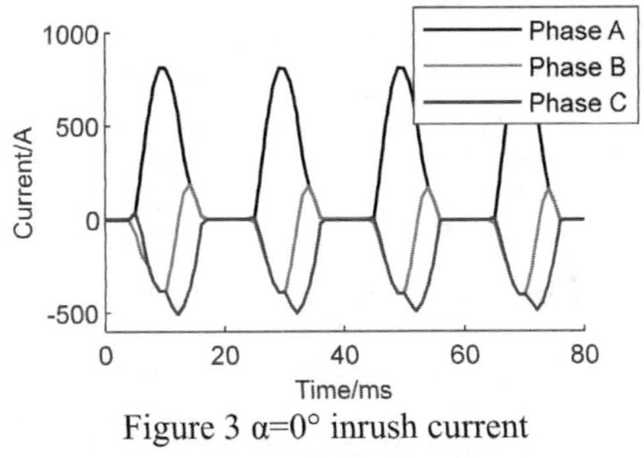

Figure 3 α=0° inrush current

5. CONCLUSION

Safe and reliable power supply is the guarantee and prerequisite for promoting better and faster economic and social development. Excellent operating conditions of transformers play a very significant role in the security and stability of the power grid and the quality of power.

The main methods of transformer TTF and inrush current identification based on electrical information are detailed reviewed in this paper and their advantages and disadvantages are introduced, which helps to selectively apply these methods to online or offline detection of faults and indication of the severity of transformers faults.

In summary, when magnetic leakage information is used as the basis for transformer turn-to-turn fault identification, it has the advantages of sensitivity, rapidity, and ability of accurately distinguish inrush current conditions. However, the current research on the magnetic flux leakage of transformers mostly focuses on the influence of magnetic flux leakage during the operation of the transformer, and there is a lack of systematic research on the use of magnetic flux leakage for turn-to-turn fault identification:

- Establish a mathematical model of transformer windings for theoretical analysis of magnetic flux leakage distribution;
- Establish a high-precision transformer simulation model to complete the simulation calculation;
- Develop an installation plan for magnetic flux leakage sensors based on theoretical calculations and simulation results;
- Apply the sensor installation plan in the true transformer for test verification.

Therefore, further research on the distribution and change rule of magnetic flux leakage during transformer operation has important theoretical significance and application value for promoting the development of transformer protection theory.

ACKNOWLEDGMENT

This project supported by Joint Funds integration project of the National Natural Science Foundation of China (U1866603): Fundamental Research on Multivariable-Based Adaptive Protection and Safe Operation for Power Transformers

REFERENCES

[1] Y. Xu, J. Ma, and Z. Wang, "Research on novel power transformer protection based on self-correction function," in TENCON 2005 - 2005 IEEE Region 10 Conference, November 21, 2005 - November 24, 2005, Melbourne, Australia, 2007, vol. 2007: Institute of Electrical and Electronics Engineers Inc.

[2] Reliability Management Center of China Electricity Council. 2017 National Electric Power Reliability Index. http://news.bjx.com.cn/html/20180607/903893.shtml

[3] Reliability Management Center of China Electricity Council. 2015 National Electric Power Reliability Index. http://chinaer.cec.org.cn/zhibiaofabu/linianzhibiao/2017-12-15/176136.html

[4] E. Ali, A. Helal, H. Desouki, K. Shebl, S. Abdelkader, and O. P. Malik, "Power transformer differential protection using current and voltage ratios," Electric Power Systems Research, vol. 154, pp. 140-150, 2018.

[5] S. Mageshwari and K. Sandeep, "Design and simulation of self adaptive differential protection for a power transformer," in 7th International Conference on Electrical Energy Systems, ICEES 2021, February 11, 2021 - February 13, 2021, Virtual, Chennai, India, 2021: Institute of Electrical and Electronics Engineers Inc., pp. 454-459.

[6] K. Antunes Tavares and K. Melo Silva, "Evaluation of Power Transformer Differential Protection Using the ATP Software," in IEEE Latin America Transactions, vol. 12, no. 2, pp. 161-168, March 2014, doi: 10.1109/TLA.2014.6749533.

[7] L.-C. Wu, C.-W. Liu, S.-E. Chien, and C.-S. Chen, "The effect of inrush current on transformer protection," in 2006 38th Annual North American Power Symposium, NAPS-2006, September 17, 2006 - September 19, 2006, Carbondale, IL, United states, 2006: IEEE Computer Society, pp. 449-456.

[8] 1 000 kV preventive test procedures for AC electrical equipment, GB/T 24846-2018, 2018.

[9] H. P. Correa and F. H. T. Vieira, "An approach to steady-state power transformer modeling considering direct current resistance test measurements," Sensors, vol. 21, no. 18, 2021.

[10] A. N. Jahromi, M. Hosseinkhanloo, and L. Lamare, "Under Load Tap Changer Diagnostics Based on Transformer DGA and DC Resistance Tests," in 2018 IEEE Electrical Insulation Conference, EIC 2018, June 17, 2018 - June 20, 2018, San Antonio, TX, United states, 2018: Institute of Electrical and Electronics Engineers Inc., pp. 323-326.
[11] H. Zheng, B. Pu, Y. Shao, Y. Li, W. Wang, and W. Xue, "Study on transformer winding deformation using low-voltage impedance diagnostic method," in 2015 IEEE International Conference on Information and Automation, ICIA 2015 - In conjunction with 2015 IEEE International Conference on Automation and Logistics, August 8, 2015 - August 10, 2015, Yunnan, China, 2015: Institute of Electrical and Electronics Engineers Inc., pp. 1947-1952.
[12] A. Prasetyo and E. Supriyanto, "Failure Analysis of Power Transformer Based on Transformer Turn Ratio Test and SFRA," in 1st International Conference on Computational Intelligence and Energy Advancements, ICCIEA 2020, December 12, 2020 - December 13, 2020, Telangana, Virtual, India, 2021, vol. 1817: IOP Publishing Ltd.
[13] S. Zhai, L. Wei, W. Chen, T. Deng, and G. Ji, "Research on automatic calibration technology of transformation ratio tester," in 2nd International Conference on Wireless Communications and Smart Grid, ICWCSG 2020, June 12, 2020 - June 14, 2020, Qingdao, China, 2020: Institute of Electrical and Electronics Engineers Inc., pp. 221-224.
[14] Y.-F. Wu, H.-R. Hu, W. Luo, T. Wang, and L. Ruan, "Research on no-load test of 1000kV ultra-high voltage transformer," in 2011 Asia-Pacific Power and Energy Engineering Conference, APPEEC 2011 - Proceedings, 2011: IEEE Computer Society.
[15] F. Hu, G. Yang, R. Guo, and N. Zhang, "Field No-load Test and the Uncertainty Calculation of No-load Loss Measurement Uncertainty for UHV Disintegration Transformer," in 3rd IEEE Advanced Information Management, Communicates, Electronic and Automation Control Conference, IMCEC 2019, October 11, 2019 - October 13, 2019, Chongqing, China, 2019: Institute of Electrical and Electronics Engineers Inc., pp. 1844-1849.
[16] N. Shanmugam, S. Gopal, B. Madanmohan, S. P. Balaji, and R. Rajamani, "Diagnosis of Inter-Turn Shorts of Loaded Transformer under Various Load Currents and Power Factors; Impulse Voltage-Based Frequency Response Approach," IEEE Access, vol. 9, pp. 40811-40822, 2021.
[17] R. Talukdar, "Transformer winding movement differentiated using frequency response analysis," in 1st IEEE International Conference on Advances in Electrical, Computing, Communications and Sustainable Technologies, ICAECT 2021, February 19, 2021 - February 20, 2021, Bhilai, India, 2021: Institute of Electrical and Electronics Engineers Inc.
[18] B. Chen and G. Liu, "Research Progress in On-line Monitoring Methods of Micro-water Content in Transformer Oil," Gaodianya Jishu/High Voltage Engineering, vol. 46, no. 4, pp. 1405-1416, 2020.
[19] H. Fan, Y. Zhu, W. Zheng, X. Qin, D. Wu, and D. Zhong, "Research on On-line Monitoring and Regulation Technology of Pressure of High Voltage Transformer," in 2019 5th International Conference on Advances in Energy Resources and Environment Engineering, ICAESEE 2019, December 6, 2019 - December 8, 2019, Chongqing, China, 2020, vol. 446: IOP Publishing Ltd.
[20] L. M. R. Oliveira and A. J. M. Cardoso, "Leakage inductances calculation for power transformers interturn fault studies," IEEE Transactions on Power Delivery, vol. 30, no. 3, pp. 1213-1220, 2015.
[21] T. Zheng, Y. Shu, S. Dong, and Y. Cao, "Interturn fault protection for transformer type controllable shunt reactor based on leakage inductance change," Dianli Xitong Zidonghua/Automation of Electric Power Systems, vol. 35, no. 12, pp. 65-69, 2011.
[22] H. Yang, X. Zeng, and J. Luo, "A three-phase transformer leakage inductances parameter identification algorithm based on the constrained least squares method," Dianli Xitong Zidonghua/Automation of Electric Power Systems, vol. 33, no. 13, pp. 68-72, 2009.

[23] Z. Zhao et al., "Leakage inductances matrix identification of new converter transformer based on Levenberg-Marquadt nonlinear least squares algorithm," Diangong Jishu Xuebao/Transactions of China Electrotechnical Society, vol. 28, no. 4, pp. 212-220, 2013.
[24] X. Deng, X. Xiong, L. Gao, Y. Fu, and Y. Chen, "Method of on-line monitoring of transformer winding deformation based on parameter identification," Zhongguo Dianji Gongcheng Xuebao/Proceedings of the Chinese Society of Electrical Engineering, vol. 34, no. 28, pp. 4950-4958, 2014.
[25] A. Ehsanifar, M. Dehghani, and M. Allahbakhshi, "Calculating the leakage inductance for transformer inter-turn fault detection using finite element method," in 25th Iranian Conference on Electrical Engineering, ICEE 2017, May 2, 2017 - May 4, 2017, Tehran, Iran, 2017: Institute of Electrical and Electronics Engineers Inc., pp. 1372-1377.
[26] N. Wang, K. Xu, J. Yang, and J. Hu, "Leakage inductance characteristics of power transformer winding fault based on ANSOFT," Journal of Power Technologies, vol. 99, no. 3, pp. 231-238, 2019.
[27] Y. Zheng, X. Liu, B. Yu, Z. Zhang, S. Pan, and F. Long, "A method of adaptive transformer turn-turn fault protection based on power loss," Dianli Xitong Zidonghua/Automation of Electric Power Systems, vol. 37, no. 10, pp. 104-107, 2013.
[28] X. Zhu, "Design of on-line monitoring system for transformer turn-to-turn short circuit Fault," Master, Shandong University, 2017.
[29] M. F. Cabanas et al., "Detection of insulation faults on disc-type winding transformers by means of leakage flux analysis," in 2009 IEEE International Symposium on Diagnostics for Electric Machines, Power Electronics and Drives, SDEMPED 2009, August 31, 2009 - September 3, 2009, Cargese, France, 2009: IEEE Computer Society.
[30] M. F. Cabanas et al., "Insulation fault diagnosis in high voltage power transformers by means of leakage flux analysis," Progress in Electromagnetics Research, vol. 114, pp. 211-234, 2011.
[31] P. A. Venikar, M. S. Ballal, B. S. Umre, and H. M. Suryawanshi, "Search coil based transformer inter-turn fault detection," in 2014 18th National Power Systems Conference, NPSC 2014, December 18, 2014 - December 20, 2014, Guwahati, India, 2014: Institute of Electrical and Electronics Engineers Inc.
[32] M. Arivamudhan, N. Vanamadevi, and S. Santhi, "Measurement of axial leakage flux for the detection of winding deformation in transformer," in 2008 IEEE Region 10 Conference, TENCON 2008, November 19, 2008 - November 21, 2008, Hyderabad, India, 2008: Institute of Electrical and Electronics Engineers Inc.
[33] F. Haghjoo, M. Mostafaei, and H. Mohammadi, "A New Leakage Flux-Based Technique for Turn-to-Turn Fault Protection and Faulty Region Identification in Transformers," IEEE Transactions on Power Delivery, vol. 33, no. 2, pp. 671-679, 2018.
[34] F. Haghjoo, M. Mostafaei, and M. Mohammadzadeh, "A novel flux-based protection scheme for power transformers," in 9th Power Systems Protection and Control Conference, PSPC 2015, January 14, 2015 - January 15, 2015, No. 424, Hafez Ave., Tehran, Iran, 2015: Institute of Electrical and Electronics Engineers Inc., pp. 25-30.
[35] F. Haghjoo and H. Mohammadi, "Planar Sensors for Online Detection and Region Identification of Turn-to-Turn Faults in Transformers," IEEE Sensors Journal, vol. 17, no. 17, pp. 5450-5459, 2017.
[36] M. Mostafaei and F. Haghjoo, "Flux-based turn-to-turn fault protection for power transformers," IET Generation, Transmission and Distribution, vol. 10, no. 5, pp. 1154-1163, 2016.
[37] J. Song et al., "Magnetizing Inrush Current Identification Method Based on Second-Order Taylor Derivative," Zhongguo Dianji Gongcheng Xuebao/Proceedings of the Chinese Society of Electrical Engineering, vol. 40, no. 3, pp. 1020-1029, 2020.
[38] Y. Quan, H. Li, and Z. Yan, "New method for measuring dead angle with the wavelet transform," Dianli Xitong Zidonghua/Automation of Electric Power Systems, vol. 22, no. 1, pp. 33-35, 1998.

[39] M. Tripathy and N. Nirala, "Power transformer differential protection algorithm based on dead angle of Wavelet Energy Waveform," in 35th IEEE Region 10 Conference, TENCON 2015, November 1, 2015 - November 4, 2015, Macau, China, 2015, vol. 2016-January: Institute of Electrical and Electronics Engineers Inc.

[40] Y. Li, J. Chen, and B. Yao, "A novel method to identify the inrush current based on waveform longitudinal symmetry coefficient for transformer protection," in IEEE TENCON 2004 - 2004 IEEE Region 10 Conference: Analog and Digital Techniques in Electrical Engineering, November 21, 2004 - November 24, 2004, Chiang Mai, Thailand, 2004, vol. C: Institute of Electrical and Electronics Engineers Inc., pp. C393-C396.

[41] L. Zhang, M. Li, T. Ji, and J. Zeng, "Identifying magnetizing inrush in power transformers based on symmetry of current waveforms," IEEJ Transactions on Electrical and Electronic Engineering, vol. 12, no. 6, pp. 959-960, 2017.

[42] J. S. Thorp and A. G. Phadke, "A Microprocessor Based Three-Phase Transformer Differential Relay," in IEEE Transactions on Power Apparatus and Systems, vol. PAS-101, no. 2, pp. 426-432, Feb. 1982, doi: 10.1109/TPAS.1982.317124.

[43] K. Inagaki et al., "Digital protection method for power transformers based on an equivalent circuit composed of inverse inductance," in IEEE Transactions on Power Delivery, vol. 3, no. 4, pp. 1501-1510, Oct. 1988, doi: 10.1109/61.193949.

[44] K. Yabe, "Power differential method for discrimination between fault and magnetizing inrush current in transformers," in IEEE Transactions on Power Delivery, vol. 12, no. 3, pp. 1109-1118, July 1997, doi: 10.1109/61.636909.

[45] Y.-P. Zheng, Z.-Y. Zhang, C.-H. Wu, S.-Y. Pan, and Z. Li, "Leakage flux-based method to identify the inrush current in transformers," in 12th International Conference on Measuring Technology and Mechatronics Automation, ICMTMA 2020, February 28, 2020 - February 29, 2020, Phuket, Thailand, 2020: Institute of Electrical and Electronics Engineers Inc., pp. 1083-1088.

Analysis for Gas in Transformer Oil based on improved GM (1,N) Grey

QiFei Zhu[1,a] , ZhaoPeng Lv[2,b] , LiPeng Liu[2,c] , Jing Zhang[2,d] , Gao Lou[2,e] , XingJun Tian[2,f]

[1]Shuohuang Railway Development Co. LTD, National Energy Group, Cangzhou, Hebei, China

[2]School of Electrical and Electronic Engineering, Shijiazhuang Tiedao University, Shijiazhuang, Hebei, China

[a]e-mail: 331679904@qq.com; [b]e-mail: 244700614@qq.com; *Corresponding author:[c]1332947646@qq.com; [d]e-mail: 1648271853@qq.com; [e]e-mail: 754765643@qq.com; [f]e-mail: 358939042@qq.com

Abstract—Based on the traditional gray GM (1, N) model, with seven kinds of gas fusion in transformer oil predicted, the actual detection times for the gas is less, the sequence accuracy is greatly reduced with an large error, and it will not conform to the standard judgment for the smooth curve. Therefore, an adaptive regression algorithm is proposed to revise each calculation result of its GM model to modify the model error. Through simulation, the sequence curves of the N data by error correction are obviously closer with a greater correlation degree. The improved gray prediction model can effectively improve the accuracy of traditional GM (1, N).

1. Introduction

In traction power systems, power transformers work as important core equipment, which are responsible for converting, distributing and transmitting electrical energy. With a fault occurring, the power transformer will directly affect the safe and stable operation in power systems[5]. Being not interfered by external electromagnetic field factors, dissolved gas analysis can not only reflect the types of transformer faults effectively, but also be made an effective means to discover transformer defects and potential faults[6]. Moreover, the transformer failure will lead to a change of its internal gas components[7]. By diagnosing and predicting the mass concentration of the dissolved gas in the transformer oil, it can realize the online real-time monitoring, find the fault immediately, eliminate the danger in order to guarantee safety in power systems[8].

Currently, the commonly used prediction models include time series model, fuzzy model, gray model, artificial neural network, support vector machine (SVM), Kalman filter, interval prediction[1]. Among them, the regression analysis prediction model is simple and easy to implement, but the prediction accuracy is relatively low; The neural network prediction model requires a large number of samples for training, which is difficult to implement[13]. Considering that the transformer itself can be regarded as a gray system containing known components and unknown connections, the gray prediction model is widely used due to its simple principle and low data requirements[3]. The traditional GM(1,1) model considers the change of a single variable over time, and there is a certain relationship between the dissolved gas components in transformer oil, so the GM(1,N) gray multivariate model can be used to extract Multiple related feature variables are used for comprehensive prediction[14], and the prediction accuracy is higher. Coupled with the adaptive regression algorithm to correct the error, improve the accuracy of prediction.

2. Grey multivariate GM (1, N) prediction model

Grey correlation degree is one of the gray theory analysis, represents the system factors or factors on the uncertainty of a connection between the main behavior. The variation in content of dissolved gas in transformer oil, analysis of the correlation with the content of a gas as the main body, to predict factor analysis of other gas content the extent of the impact on the gas content changes. Essence is the history curve shape of gas content in the sequence of similar level as the standard, namely judging other gases and closer proximity to predict gas curve, the closer the shows the greater the degree of correlation between the corresponding sequence, the opposite correlation degree is smaller.

The characteristic data sequence with a high correlation is $x_i^{(0)} = (x_i^{(0)}(1), x_i^{(0)}(2), \cdots, x_i^{(0)}(n))$ $i = 2,3,\cdots,n$, n is a factor sequence[2]. The matrix can be obtained separately from $x_1^{(0)}, x_2^{(0)}, \cdots, x_n^{(0)}$.

$$\begin{cases} X_j^{(1)} = (x_j^{(1)}(1), x_j^{(1)}(2) \ldots x_j^{(1)}(n)) \\ x_j^{(1)}(i) = \sum_{k=1}^{i} x_j^{(0)}(k) \end{cases} \tag{1}$$

The matrix form of GM (1, N) model is

$$dX^{(1)}(t)/dt = AX^{(1)}(t) + B \tag{2}$$

The response formula of equation (2) is

$$X^{(1)}(t) = \mathrm{e}^{A(t-1)}(X^{(1)}(1) + A^{-1}B) - A^{-1}B \tag{3}$$

Discretize equation (2) to obtain

$$x_j^{(0)}(k) = \sum_{l=1}^{m} a_{jl} z_l(k) + b_j \tag{4}$$

In equation (4): $z_l(k) = 0.5(x_l^{(1)}(k-1) + x_l^{(1)}(k))$
Obtained by the least square method

$$\hat{a}_j = (\hat{a}_{j1}, \hat{a}_{j2}, \cdots, \hat{a}_{jm}, \hat{a}_j)^T = (P^T P)^{-1} P^T Y_j \tag{5}$$

In equation (5):

$$P^T = \begin{pmatrix} z_1^{(1)}(2) & z_1^{(1)}(3) & z_1^{(1)}(n) \\ z_1^{(1)}(2) & z_1^{(1)}(3) & z_1^{(1)}(n) \\ \vdots & \ddots & \vdots \\ z_1^{(1)}(2) & z_1^{(1)}(3) & z_1^{(1)}(n) \\ 1 & 1 & 1 \end{pmatrix} \quad Y_j = (x_j^{(0)}(2), x_j^{(0)}(3), \cdots, x_j^{(0)}(n))^T$$

It can be obtained that the identifications of A and B are $\hat{A} = (\hat{a}_{ij})_{m \times m}, \hat{B} = (\hat{b}_1, \hat{b}_2 \cdots, \hat{b}_m)$.
The response formula for GM (1, N) is equation (6).

$$\hat{X}^{(1)}(k) = e^{\hat{A}(k-1)}(X^{(1)}(1) + \hat{A}^{-1}\hat{B}) - \hat{A}^{-1}\hat{B} \tag{6}$$

Model data restored to equation (7).

$$\hat{X}^{(0)}(k) = \hat{X}^{(1)}(k) - \hat{X}^{(1)}(k-1) \tag{7}$$

In equation (7): $k = 2,3,\cdots,n$.

2.1 Deformation to approximate method

From the above, the grey differential equation can be obtained as equation (8).

$$x_i^{(0)}(k)+az^{(1)}(k)=b_2x_2^{(1)}(k)+b_3x_3^{(1)}(k) +\cdots b_Nx_N^{(1)}(k) \quad (8)$$

From the generated sequence, the whitening equation can be obtained as equation (9).

$$\frac{dx_1^{(1)}(t)}{dt}+ax_1^{(1)}(t)=b_2x_2^{(1)}(t)+b_3x_3^{(1)}(t)+\cdots+b_Nx_N^{(1)}(t) \quad (9)$$

In equation (8): If the change of $x_i(i=2,3,\cdots,N)$ is small, Take the gray constant as $\sum_{i=2}^{N}b_ix_i^{(1)}(k)$. The approximate response equation can be obtained from the grey differential equation (8) and transformed into equation (10)

$$\hat{x}_1^{(1)}(k+1)=x_1^{(1)}(0)-\frac{1}{a}\sum_{i=2}^{N}b_ix_1^{(1)}(k+1)e^{-ak} +\frac{1}{a}\sum_{i=2}^{N}b_ix_i^{(1)}(k+1) \quad (10)$$

This method of treating variables as constants is called approximation method.

2.2 Trapezoidal method

By using the ordinary differential method, the solution of the whitening equation in any interval can be obtained as equation (11).

$$x_1^{(1)}(t)=e^{-a(t-t_k)}x_1^{(1)}(t_k)+e^{-at}\int_{t_k}^{t}e^{-at}\sum_{i=2}^{N}b_ix_i^{(1)}(t)dt \quad (11)$$

In equation (11): $t=t_{k+1}$, According to the characteristics of time series, When $t_k=k$.

$$x_1^{(1)}(k+1)=e^{-a}x_1^{(1)}(k)+e^{-a(k+1)}\int_{k}^{k+1}e^{-at}\sum_{i=2}^{N}b_ix_1^{(1)}(t)dt \quad (12)$$

In equation (12): use trapezoidal formula, we use computer programs to simplify the calculation process. the corresponding time expression of the grey differential equation can be obtained.

$$\hat{x}_1^{(1)}(k+1)=e^{-a}\hat{x}_1^{(1)}(k)+ \frac{1}{2}\sum_{i=2}^{N}b_i(x_1^{(1)}(k)e^{-a}+x_i^{(1)}(k+1)) \quad (13)$$

This method is called trapezoidal method .

3. Adaptive regression residual correction GM (1, N) model

When establishing GM (1, N) model for prediction, each number has a predicted value at a stage[11]. The residual value is obtained by comparing the fitting results with the actual results, regression analysis of a series of residual values can correct the prediction results of GM ((1, N) on the test set, avoid the continuous accumulation of residuals and increasing the prediction error, so as to improve the accuracy[10].

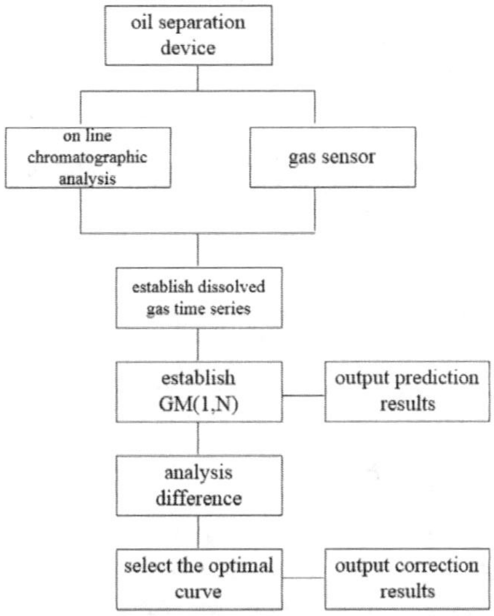

Figure.1 Adaptive regression residual correction process

The residual between the fitted value and the actual value is

$$\sigma^{(0)}{}_i(k) = x^{(0)}{}_i(k) - \hat{x}^{(0)}{}_i(k) \tag{14}$$

Since the fitting result may be greater than or less than the actual result, that is, the residual value may have a negative value, in order to facilitate the subsequent cumulative generation, a number a can be appropriately added to make all the residual values non negative[4]. Then, the one-time accumulation generation method of grey system theory is adopted to obtain:

$$\delta_i^{(1)}(k) = \sum_{j=1}^{k} \delta_i^{(0)}(j) = \begin{bmatrix} \delta_1^{(1)}(1) & \delta_2^{(1)}(1) & \cdots & \delta_n^{(1)}(1) \\ \delta_1^{(1)}(2) & \delta_2^{(1)}(2) & \cdots & \delta_n^{(1)}(2) \\ \vdots & \vdots & \ddots & \vdots \\ \delta_1^{(1)}(m) & \delta_2^{(1)}(m) & \cdots & \delta_n^{(1)}(m) \end{bmatrix} \tag{15}$$

The change trend of $\delta_i^{(1)}(k)$ discrete point diagram, combined with several commonly used curve regression equations, The fitting results are compared by using the determination coefficient, root mean square difference and standard deviation of the equation[15], select the best one as the regression equation of the $\delta_i^{(1)}(k)$, Then the residual correction value of the original data is obtained according to the progressive reduction processing:

$$\sigma_i^{(0)}(k) = \sigma_i^{(1)}(k) - \sigma_i^{(1)}(k-1) - a \tag{16}$$

The revised forecast results are:

$$\bar{x}^{(0)}(k) = \hat{x}^{(0)}(k) + \hat{x}_i^{(0)}(k) \tag{17}$$

4. Grey correlation analysis of GM (1, N) model variables

When using GM (1, N) model for fault prediction, it is necessary to clarify the relationship between various variables[12]. After processing n original data sequences, the relationship between these sequences is judged according to the geometric similarity of the sequence curves[4]. The closer the curves are, the greater the correlation degree between the corresponding sequences, and vice versa. This is the principle of grey correlation analysis[9].

The parent sequence of grey correlation analysis is made as $X_1^{(0)}, X_2^{(0)}, \cdots X_m^{(0)}$, The remaining sequences are subsequences $x_i(k)$.

$$x_0^{'}(k) = x_0(k) / x_0(1) \tag{18}$$

$$\xi_i^{'}(k) = \frac{\left(\min_i \min_k |x_0'(k) - x_i'(k)| + \rho \max_i \max_k |x_0'(k) - x_i'(k)|\right)}{\left(|x_0'(k) - x_i'(k)| + \rho \max_i \max_k |x_0'(k) - x_i'(k)|\right)} \tag{19}$$

In equation (19): $\xi_i(k)$ is the grey correlation coefficient of sequence x_i and x_i at point *K*. $\rho = 0.5$.

The grey correlation degree between subsequence and parent sequence is

$$r_i = \frac{1}{m}\sum_{i=1}^{m} \xi_i(k) \tag{20}$$
$$i = 1, 2 \cdots m-1;$$

M is the number of sequences. If $r_i \geq 0.5$, it is considered that there is correlation between child and parent sequences.

5. Accuracy test of prediction model

There is always a prediction error when predicting the state parameters of power transformer with grey prediction model, The reason is that there is always a certain difference between the prediction model and the objective reality it describes. In order to measure whether the prediction model is reasonable and whether the prediction results are credible, it is necessary to test its accuracy.

5.1 Mean residual

Residual test is a accuracy test method in grey theory. Set original data sequence $X^{(0)} = (x^{(0)}(1), x^{(0)}(2)\cdots, x^{(0)}(n))$.

The corresponding model simulation sequence is $\hat{X}^{(0)} = (\hat{X}^{(0)}(1), \hat{X}^{(0)}(2), \cdots \hat{X}^{(0)}(n))$. The residual of the grey model is

$$\xi^{(0)}(k) = \frac{x^{(0)}(k) - \hat{x}^{(0)}(k)}{x^{(0)}(k)} \times 100\% \tag{21}$$

Sequence is $\xi^{(0)}(k) = (\xi^{(0)}(1), \xi^{(0)}(2)\cdots, \xi^{(0)}(n))$, The average residual is equation (22).

$$\varepsilon(\text{avg}) = \frac{1}{n-1}\sum_{k=2}^{n} |\varepsilon(k)| \tag{22}$$

5.2Then verify the relative error

Verifying the relative error is the relative error between the model prediction data and the actual data. That is

$$\delta(k) = \frac{1}{p}\sum_{i=1}^{p} \left| \frac{y_i(k) - \hat{y}_i(k)}{y_i(k)} \right| \times 100\% \tag{23}$$

In equation (23): Verifying the actual value used to $y_i(k)$;the predicted value of the model is $y_i(k)$; the number of data inspected is *p*.

6. Case analysis

6.1 Data prediction

Taking the chromatographic data of the measured mass concentration of dissolved gas in oil of a 220kV transformer as an example, five groups of data were obtained for error analysis.

Table.1 gas composition and concentration in transformer oil

date	H_2 (μL/L)	CH_4 (μL/L)	C_2H_6 (μL/L)	C_2H_4 (μL/L)	C_2H_2 (μL/L)
2020-05-14	28.5	80.2	40.2	135.1	1.0
2020-06-07	41.4	95.4	54.1	162.7	1.1
2020-07-18	70.1	114.5	71.1	200.4	1.8
2020-08-14	80.4	121.3	85.1	220.4	1.9
2020-09-20	101.4	150.2	100.5	270.2	1.9
2020-10-05	114.2	190.2	130.4	350.1	2.1
2020-11-15	150.4	220.1	131.1	415.2	2.5

The improved GM (1, 5) prediction model is used to analyze the oil chromatographic data in the oil of multiple power transformers to verify that the improved GM (1, 5) prediction model is better than GM (1, 1) and traditional GM (1, 5).) Superiority. Before the experiment, the original data sequence was normalized to reduce the difference between the various gas levels and the prediction results.

In order to verify the effectiveness of the improved GM (1, 5) model prediction, the first 3 data in the original data sequence are used to establish an equal-dimensional and new-information model, that is, $m = 2$ and the remaining 2 data are used to verify the effectiveness of the prediction. The results obtained are shown in Table 2.

Table 2 predicted gas composition and concentration

date	H_2 (μL/L)	CH_4 (μL/L)	C_2H_6 (μL/L)	C_2H_4 (μL/L)	C_2H_2 (μL/L)
2020-05-14	28.5	80.2	40.2	135.1	1.0
2020-06-07	41.4	95.4	54.1	162.7	1.1
2020-07-18	69.2	100.5	80.1	210.1	1.8
2020-08-14	78.2	120.1	84.2	218.2	1.8
2020-09-20	100.2	148.2	90.2	259.7	2.0
2020-10-05	100.2	180.5	130.1	360.4	2.1
2020-11-15	148.5	221.8	130.8	400.8	2.5

Table 3 Residual adaptive regression correction results

date	Actual value	Predictive value	Deviation rate(/%)	State interval
2020-05-14	1.0	1.03	3	S3
2020-06-07	1.3	1.33	2.03	S3
2020-07-18	1.9	1.92	1	S3
2020-08-14	1.9	1.96	3.1	S3
2020-09-20	2.4	2.35	-2	S2
2020-10-05	2.9	2.94	1.3	S3
2020-11-15	2.9	2.86	-1.3	S2

Through the state interval, can directly see the adjusted error is very small. All state in the smallest range, and is distributed evenly on both sides. The correction results accord with theoretical analysis, practical engineering requirements.

6.2 Result analysis

Based on GM (1, N) grey multivariable model prediction and analysis method, multiple characteristic variables can be used for comprehensive prediction, but the error is large and the accuracy is low. In view of this problem, this paper uses adaptive regression method to correct the formula and data results from the perspective of residual correction. The limitation of grey prediction caused by fluctuation of original data is improved.

The example shows that the adaptive regression algorithm can significantly improve the traditional GM (1, N) grey prediction model and greatly improve the prediction accuracy.

7. Conclusions

In view of the rich data of dissolved gas content in transformer oil, and the characteristics of a certain trend, the transformer interior is regarded as one. The GM(1, N) grey multivariable model was established for prediction analysis.

To improve the model prediction effect, Based on the adaptive regression algorithm provides a new algorithm to predict. Using residual error correction forecast results. As a case study of dissolved gas in transformer oil are analyzed.

The predictive analysis method is simple based on GM (1, N) grey multivariable model in dealing with various data, on the other hand, can extract related multiple characteristic variables for comprehensive prediction, but the accuracy of the series will still be reduced with a long time span.

The adaptive regression method is used from the Angle of residual correction, which improves the limitation of grey prediction caused by the fluctuation of original data, and determines the difference of data with the help, which is different grey model formula. Correcting the residual result. The algorithm makes the prediction of gas composition more accurate and expands the application range of grey model. More in line with the transformer oil interaction and influence of multidimensional gas component in the reality.

References

[1] BORUCKI S. Diagnosis of technical condition of power transformers based on the analysis of vibroacoustic signals measured in transient operating conditions [J]. IEEE Transactions on Power Deliver, 2012, 27(2): 670-676.
[2] SUURMOND C. Design and implementation of vibration on-line monitoring system of power transformer [C]. Power and Energy Society General Meeting IEEE, 2013: 1-5.
[3] Luo Y.B, (2003) Prediction of dissolved gas content in transformer oil by grey model J Chinese Journal of Electrical Engineering, 20(3): 65-69.
[4] Sun C.X, (2003) A new model of grey prediction parameter model and its application in electrical insulation fault prediction [J]. Control theory and Application, 20(5): 797-801.
[5] SERGEJS Z, POISS G. Power transformer mechanical condition assessment with a vibration-based diagnostic method [C]. International Scientific Conference on Power and Electrical Engineering of Riga Technical University IEEE, 2015: 1-4.
[6] SICA F C, GUIMARAES F G, DUARTE R D O, et al. A cognitive system for fault prognosis in power transformers [J]. Electric Power Systems Research, 2015(127): 109-117.
[7] SIMA L P,SHU N P,ZUO J, et al. Prediction of dissolved gas concentration in transformer oil based on grey correlation and fuzzy support vector machine [J]. Power System Protection and Control, 2012, 40(19): 41-46.
[8] WAN F, YANG M L, HE P, et al. Gas Raman spectrum detection and signal processing method in transformer oil [J]. Chinese Journal of Scientific Instrument, 2016, 37(11): 2482-2488.
[9] LI Y, CHENG H B, XIN J B, et al. GM (1， n) prediction of dissolved gas concentration in transformer oil [J]. Journal of East China Jiaotong University, 2017, 34(3): 131-136.
[10] TANG Y B, GUI W H, PENG T. Multinuclear principal component regression prediction model for gas in transformer oil [J]. Electric Machines and Control, 2012, 16(11): 92-98.
[11] HE B, MENG Q. Research on extension method of grey prediction model[J]. System Engineering Theory and Practice, 2012, 22(9):4.
[12] LIU Q H, LUO L M, ZHANG Z W. Combined weighting method for grey correlation fault diagnosis model of power transformer [J]. High Voltage Appliance, 2012, 48(1):5.
[13] LIU W. Forecasting the semiconductor industry cycles by bootstrap prediction intervals [J].Applied Economics, 2007, 39(13):1731-1742.

[14] TRUONG D G, AHK K K. Wave prediction based on a modified grey model MGM(1,1) for real-time control of wave energy converters in irregular waves [J]. Renewable Energy, 2012, 43(6):242-255.
[15] WANG K, LI J ZH, ZHANG SH Q, et al. New characteristic parameters of dissolved gas in transform fault diagnosis oil [J].Proceedings of the CSEE, 2016, 36(23):6570-6578.

Journal of Physics: Conference Series **2215** (2022) 012020 doi:10.1088/1742-6596/2215/1/012020

Research on Fault Location for Direct Traction Network

Defang Li[1,a] , Hao Zhang[2,b] , Hui Yan[2,c*] , Xingjun Tian[2,d] , Nanqing Wang[2,e] , Zexuan Yuan[2,f]

[1]Shuohuang Railway Development Co. LTD, National Energy Group, Cangzhou, Hebei, China

[2]School of Electrical and Electronic Engineering, Shijiazhuang Tiedao University, Shijiazhuang, Hebei, China

[a]e-mail: 348555964@qq.com; [b]e-mail: 616397085@qq.com; *Corresponding author: [c]1102429881@qq.com; [d]e-mail: 358939042@qq.com; [e]e-mail: 1399728263@qq.com; [f]e-mail: 1712471242@qq.com

Abstract—In view of the low accuracy of existing fault location methods for traction networks with longer power supply lines, the equivalent model of the traction network for double-track electrified railways is calculated, and the power supply lines and contacts are introduced on the basis of traditional traction network fault location methods. The parameter of grid equivalent impedance ratio proposes a modified uplink and downlink current ratio ranging method, which effectively solves the problem of accurate fault location of the double-track electrified railway direct-supply traction network. The RT-plus smart grid real-time digital simulation instrument is used to establish the traction power supply dynamics. The simulation test system is simulated and tested, and the test results show that the proposed algorithm has good feasibility and accuracy, and fully meets the engineering requirements.

1. INTRODUCTION

The electrified railway currently in operation in China mainly adopts two forms of direct power supply and AT power supply, each of which can be divided into two modes, single-line and double-line[1]. Regardless of the mode, the transmission line of the traction network is composed of two parts: the power supply line and the contact wire (catenary), and the parameters of the two lines are generally different[2]. Since the length of the contact line is much longer than the length of the power supply line, the power supply line can be equivalent to a contact net to approximate the distance to the fault when the fault location accuracy requirements are not high. With the rapid development of electrified railways in China, the accuracy requirements for fault location of traction networks are becoming higher and higher. This has also become one of the urgent problems in the field of traction power supply systems.

At present, there are many solutions to the problem of fault location of electrified railway traction power supply system at home and abroad[3~5]. In view of the large fault location error of direct supply traction network, the method of calculating fault distance by using two sets of traction network reactance setting tables is proposed in document [6], but the fault location device needs to automatically identify the operation mode of traction network. Literature [7] adds the reactance setting table of the districts on the basis of the original reactance setting table. Cross-link current ratio ranging method, AT suction current ratio ranging method, and sub-section uplink and downlink current ratio ranging method [8] are suitable for AT power supply mode. In the theoretical derivation of the above methods, the influence of power supply line parameters on the distance measurement accuracy of the traction network is not considered. In order to reduce the error, some parameters need to be measured in advance through certain field tests to correct the distance measurement results in actual application.

Based on the above analysis, this article focuses on the problem of accurate fault location of the traction network of the double-track electrified railway direct power supply mode taking into account the parameters of the power supply line. After rigorous theoretical derivation, the corrected uplink and downlink current ratio ranging method is obtained. Finally, the traction network dynamic simulation test platform based on RT-plus real-time power grid intelligent simulation instrument verifies the effectiveness of the ranging algorithm.

2. Traditional ranging method

The direct power supply mode is a power supply mode in which the traction substation directly provides electric power to the electric locomotive via the catenary. It is mostly used for freight dedicated lines and early passenger dedicated lines.

2.1 Single-line direct supply ranging method

Theoretically, the equivalent unit impedance of the single-line direct-supply traction network is constant and proportional to the fault distance, which satisfies formula (1).

$$Z_d = Z_0 l \tag{1}$$

In the formula: Z_d is the impedance measured at the time of failure; Z_0 is the equivalent impedance per unit length of the traction network; l is the distance from the exit of the traction substation to the point of failure.

If there is a transition resistance during a short circuit, it is necessary to replace the impedance in formula (1) with reactance. This method is called impedance or reactance ranging method.

2.2 Double-track direct supply ranging method

In the double-line direct supply mode, the upstream and downstream catenaries are connected in parallel in the districts. Due to the mutual inductance between the two catenaries and the rails, the equivalent impedance per unit length of the traction network is not constant, so from the substation to the short-circuit point The measured impedance and the fault distance are no longer a linear relationship, that is, equation (1) no longer holds.

In order to use the measured impedance to find the distance, the actual engineering adopts the idea of piecewise linearization to divide the impedance curve of the double-track traction network into several sections, and uses the formula (1) to carry out the fault location in each section. The impedance characteristic curve of single-line and double-line direct-supply traction network is shown in Figure 1.

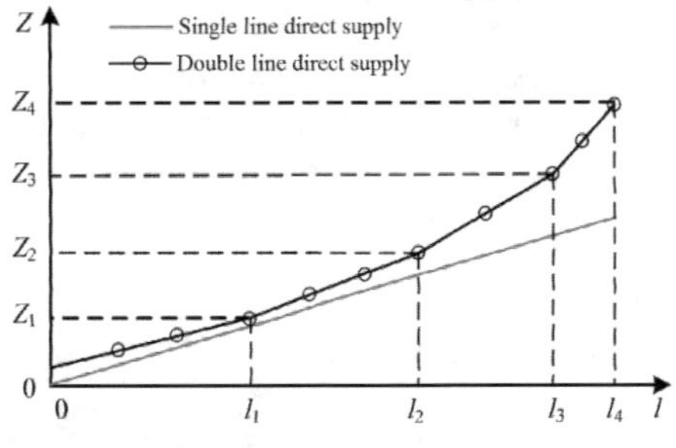

Figure 1 Impedance characteristic curve of single-line and double-line direct-supply traction network

In addition to the impedance ranging method, in the double-line direct supply mode, the relationship between the upstream and downstream currents shown in equation (2) can also be used to calculate the fault distance. This method is called the upstream and downstream current ratio ranging method [7].

$$\left|\frac{I_{t1}}{I_{t1}+I_{t1}}\right| = \frac{l}{2L} \tag{2}$$

In the formula: I_{t1} and I_{t2} are the non-fault side current and the fault side current measured at the time of the fault, respectively, L is the length of the power supply arm from the substation to the substation.

2.3 Shortcomings of traditional ranging methods

Through the discussion and analysis of the main power supply modes and ranging methods of the above traction network, it can be seen that each traction power supply mode has a corresponding ranging method and ranging formula, but these formulas ignore the length of the power supply line in the actual project. When the length of the power supply line is long, the accuracy of the distance measurement will inevitably be greatly affected. This article mainly analyzes the problem of accurate fault location when considering power supply line parameters in the form of double-line direct supply.

3. ACCURATE FAULT LOCATION OF DOUBLE-TRACK DIRECT-SUPPLY TRACTION NETWORK

3.1 Equivalent of double-track traction network model

The layout of the single-chain suspension double-track traction net is shown in Figure 2(a). C_1, J_1 and C_2, J_2 represent the load-bearing cables and contact wires of lines I and II, and R_{11}, R_{12} and R_{21}, R_{22} represent the rails of lines I and II, respectively. The load-bearing cable and the contact wire are collectively referred to as the contact net.

In the double-track electrification section, the upper and lower line catenary is generally connected in parallel at the end, and the rails of the two lines must be connected in parallel at intervals. Therefore, the two catenaries should be equivalent to two circuits, and the two rails should be equal to each other. The effect is a loop, and the equivalent circuit is shown in Figure 2(b).

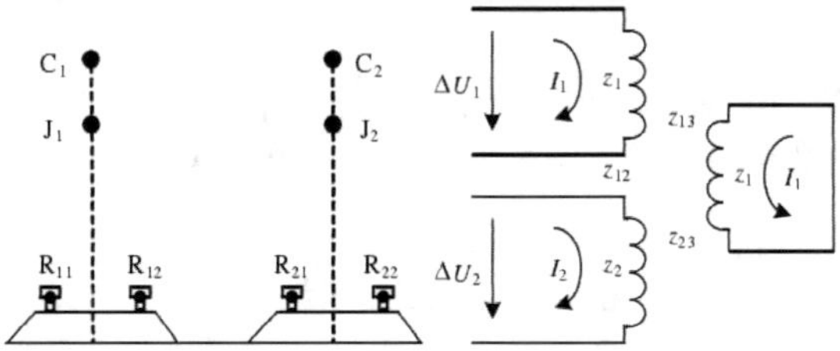

(a) Arrangement of double-track (b) Equivalent circuit of double-track traction network traction network

Figure 2 Double-track traction network structure

3.2 Modified uplink and downlink current ratio ranging method

From the analysis in the previous section, we can see that when the parameters of the uplink and downlink are the same, $z_1 = z_2$ in Figure 2(b), the equivalent model of the double-line direct-supply traction network can be obtained as shown in Figure 3.

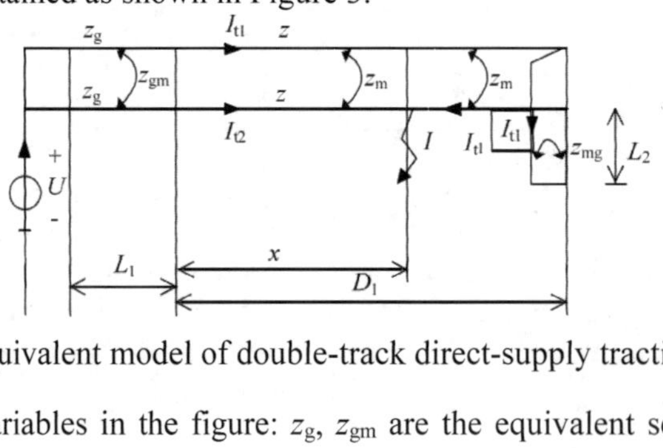

Figure 3 Equivalent model of double-track direct-supply traction network

The meaning of the variables in the figure: z_g, z_{gm} are the equivalent self-impedance and mutual impedance of the power supply line respectively; z, z_m are the equivalent self-impedance and mutual impedance of the catenary respectively; L_1, L_2 are the length of the power supply line of the substation and the district respectively ; D_1 is the length of the traction network from the substation to the district; U is the power supply voltage; I_{t1} and I_{t2} are the currents of the downlink and uplink respectively; x is the distance from the fault point to the grid point of the substation.

The short circuit voltage equation in the figure is

$$\left(z_{gm}L_1 + z_m x\right)I_{t1} + \left(z_g L_1 + zx\right)I_{t2} = U \tag{3}$$

The long loop voltage equation is

$$\begin{aligned}&\left[z_g L_1 + z\left(2D_1 - x\right) - 2z_m\left(D_1 - x\right)\right]I_{t1}\\&+2L_2\left(z_g - z_{gm}\right)I_{t1} + \left(z_{gm}L_1 + z_m x\right)I_{t2} = U\end{aligned} \tag{4}$$

Current relationship is satisfied

$$I_{t1} + I_{t2} - I = 0 \tag{5}$$

It can be solved that the ratios of the up and down currents to the total current are:

$$\begin{cases}\dfrac{I_{t1}}{I} = \dfrac{x + kL_1}{2\left(D_1 + k\left(L_1 + L_2\right)\right)}\\[2ex]\dfrac{I_{t2}}{I} = \dfrac{-x + 2D_1 + k\left(L_1 + 2L_2\right)}{2\left(D_1 + k\left(L_1 + L_2\right)\right)}\end{cases} \tag{6}$$

In the formula, $k = \dfrac{z_g - z_{gm}}{z - z_m}$ is called the equivalent impedance ratio of the power supply line and the catenary.

The relationship between the ratio coefficient of the upstream and downstream currents and the distance to the fault is shown in Figure 4.

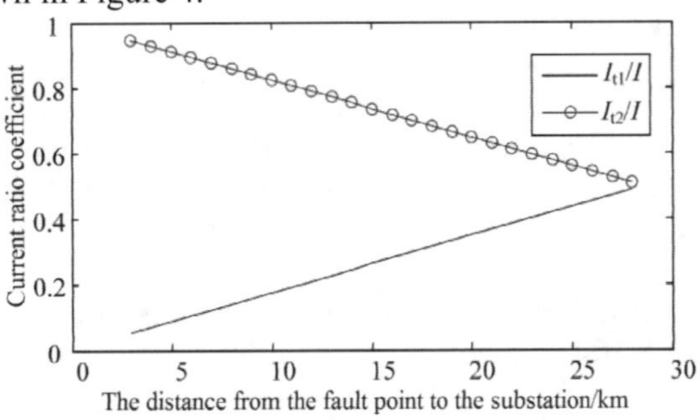

Figure 4 Relationship between current ratio coefficient and the distance from the fault point to the substation

From equation (6) and Figure 4, it can be seen that the current ratio coefficient of the faulty line and the fault distance satisfy a linear decreasing relationship, and the non-faulty line is the opposite. This equation can be used to accurately calculate the fault distance. This method It is called the uplink and downlink current ratio ranging method.

4. DYNAMIC SIMULATION TEST SYSTEM SIMULATION

The traction power supply dynamic simulation test system based on the RT-plus smart grid real-time digital simulation instrument can simulate various operating conditions and failure scenarios of the traction power supply system, comprehensively assess the functions and performance of the relay protection and safety automatic devices. The structure of the test system is shown in Figure 5. The main hardware equipment includes RT-plus real-time digital simulator, fault location protection device prototype (KF6571), clock synchronization device, background system, switch and Ethernet.

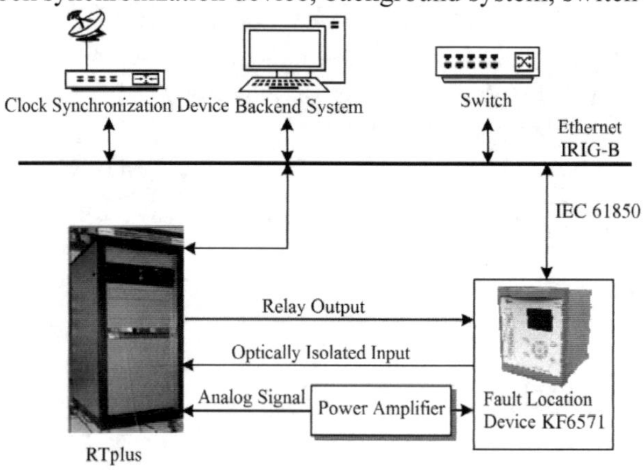

Figure 5 Traction power supply dynamic simulation test system

The RT-plus traction power supply real-time digital simulator is used to simulate the real traction power supply system, and the background system control is used to control the simulation model. The electrical quantity of the traction network can be transmitted to the fault location device through direct transmission or Ethernet transmission.

Build a full-parallel AT power supply traction network model based on RT-plus, and control the position of the circuit breaker of the AT and the district to make the system run in the double-line direct supply mode.

The calculation process of the relevant parameters of the traction power supply system can be found in literature [9,10] and so on. The main parameters in the model: the length of the power supply line of the substation L_1=3km, the length of the power supply line of the substation L_2=0.5km, the length of the catenary from the substation to the substation D_1=25km, the reactance ratio coefficient of the power supply line and the catenary k is 1.4.

A metallic instantaneous short-circuit grounding fault is set 8km away from the substation. The fault waveform recorded by the prototype KF6571 of the fault location device is shown in Figure 6. The effective value of voltage U=14791V, the effective value of current I_{t2}=3114A, I_{t1}=566A. When considering the influence of the power supply line parameters, the calculated fault distance is 8.0004km, and the error is only 0.4m, which meets the accuracy requirements of the standard requirement of ranging error not greater than 500m. When ignoring the influence of power supply line parameters, the calculated fault distance is 8.7696km with an error of 770m, which cannot meet the accuracy requirements.

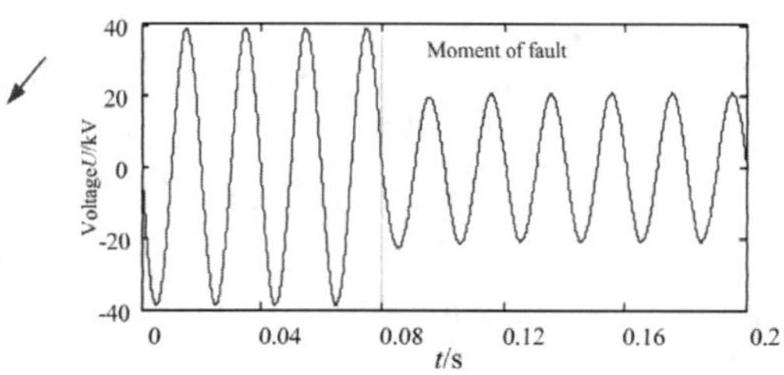

(a) Voltage waveform of substation

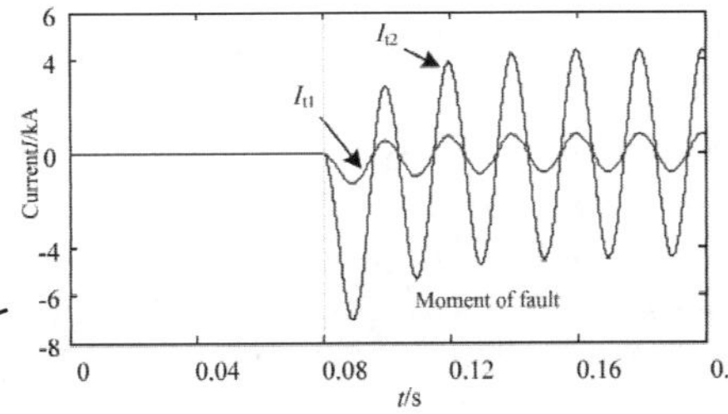

(b) Waveform of upstream and downstream current of substation

Figure 6 Voltage and current waveforms before and after the fault in the substation

Perform simulation analysis in other locations of the line, and summarize some results as shown in Table 1.

Table 1 Fault location results of uplink and downlink current ratio

U/V	I_{t2}/A	I_{t1}/A	Fault Distance /km			Absolute Error /m	
			Actual Distance	k=1.4	k=1	k=1.4	k=1
12347	3984	379	4	4.0008	4.95703	0.8	957
14791	3114	566	8	8.0004	8.7696	0.4	770
16364	2521	714	12	12.0000	12.5819	0.0	582
17428	2090	844	16	15.9995	16.3942	0.5	394
18164	1757	965	20	19.9991	20.2065	0.9	206
18673	1491	1086	24	23.9986	24.0187	1.4	19
19011	1268	1210	28	27.9981	27.8310	1.9	169

It can be obtained by analyzing the data in the table that when the influence of power supply line parameters is considered, the ranging result of the modified uplink and downlink current ratio ranging method is very close to the actual value. Its ranging error is almost zero, and all are within the allowable range of error specified by the national standard. However, the traditional uplink and downlink current is very unstable compared with the ranging method, that is, when k=1, a large error may occur.

5. Conclusion

After discussing and analyzing the main operation modes and ranging methods of traction power supply, this article concludes that the traditional traction network fault location method does not consider the influence of power supply line parameters. When the power supply line is long, the ranging error is large and the accuracy is large. Not high; A modified uplink and downlink current ratio ranging method is proposed for the accurate fault location of the double-line direct-supply traction network with power supply line parameters. The simulation test results of the traction power supply dynamic simulation test system show that after introducing the equivalent impedance ratio coefficient of the power supply line and the catenary, the method proposed in this paper greatly improves the accuracy of the fault location of the traction network in the double-line direct supply mode.

References

[1] Tan X.B. AC electrified railway traction power supply system [M]. Chengdu: Southwest Jiaotong University Press, 2009.
[2] Li Q.Z., Lian J.S., Gao S.B. High-speed railway electrification engineering [M]. Chengdu: Southwest Jiaotong University Press, 2006.

[3] Park J D. Ground Fault Detection and Location for Ungrounded DC Traction Power Systems[J]. Vehicular Technology, IEEE Transactions on Vehicular Technology,2015, 64(12):5667-5676.
[4] Wang J.F. Research on Fault Location of All Parallel AT Power Supply Traction Network [D]. Chengdu: Southwest Jiaotong University, 2006.
[5] Jiang G.L. Research on fault location of AT power supply traction network based on support vector machine [J]. Electrified Railway, 2009 (2): 22-24.
[6] Fang H.L. Technical improvement of catenary fault location device [J]. Shanghai Railway Science and Technology, 2008, 18 (3): 89-91.
[7] Lin G.S., Ma J.F. Electrified railway fault location scheme based on the reactance of sub-zones[J]. Journal of Railway Engineering Society, 2018, 35(3): 79-83.
[8] Lin G.S. Research on New Protection and Ranging Principles of Traction Power Supply System[D]. Chengdu: Southwest Jiaotong University,2010
[9] B. Mohamed, P. Arboleya, I. El-Sayed, et al. High-speed 2×25 kV Traction System Model and Solver for Extensive Network Simulations[J]. IEEE Transactions on Power Systems, 2019：34(5)：3837-3847.
[10] Dou X.W., Lang Bing, Chen X.T., et al. Simulation and calculation of short-circuit faults of high-speed railway AT traction power supply system based on PSCAD[J]. Railway Standard Design, 2018, 62(2): 154-159.

Research and Demonstration of Intelligent Multi-Physical Quantity Integration Sensor for Transmission Lines

Liang WEI[1], Yongcheng LI[1], Zhimei CUI[1], Baoshi WANG[1], Guanlang HUANG[1], Zhiming WANG[2], Jie WEI[2] *

[1]Nanning Power Supply Bureau, Guangxi Power Grid Co., Ltd., Nanning, China

[2]Digital Grid Research Institute (Joint Laboratory of Digital Grid), China Southern Power Grid, Guangzhou, China

* Corresponding author: weijie1@csg.cn

Abstract—The digital grid requires sensors to be widely deployed, safe, and reliable. At present, the single functional sensors suffered from redundant sensing system, repeated construction and low utilization rate. Therefore, this paper researched the multi-physical quantity integration sensor for a quad-bundle spacer on 500kV transmission lines. The sensor was composed of self-supplying rings and the monitoring host. Based on the non-contact current measurement technology, energy supply technology, intelligent sensing technology, low power consumption technology, reliability design technology, and other advanced technologies, this paper designed the hardware, software, and protective structure of the sensor, which can monitor the current and temperature of bubbles, channel side image and infrared image on transmission lines, ambient temperature, humidity, and air pressure. Finally, the multi-physical quantity integration sensors have been used on transmission lines in China Southern Power Grid, and the functions were verified. The sensors promote the automation, information, and digitization of monitoring transmission lines.

1. INTRODUCTION

In China, transmission lines have the characteristics of large dispersion, long-distance, crossing many areas with complex terrain and climate due to unbalanced distribution of energy resources and power demands. Monitoring operation conditions of transmission lines are important in the construction of a smart grid. At present, the transmission line monitoring technology has transitioned from the traditional 'Human Patrol' to 'Human Patrol + Machine Patrol". However, unmanned aerial vehicle (UAV) Patrol has some shortcomings, such as being vulnerable to weather, a short distance of continuous line patrol, only image monitoring volume, and so on. With the development of the Internet of things, artificial intelligent technology, and the needs of power grid digital transformation, it is necessary to establish the IoT monitoring system of perception layer, transmission layer, and platform layer for transmission line operation status monitoring [1]. Besides, the intelligent sensing of multi-parameters is the core and basic technology of the transmission line sensing layer.

Many scholars have done a lot of research on monitoring conditions of transmission lines. In Ref.[2-4], the equivalent icing thickness of the conductor was calculated by insulator tension for monitoring transmission line icing conditions. Ref. [5] predicted transmission line mountain fires based on satellite images. Ref. [6-7] studied the image recognition algorithm of deep learning based on the images of transmission line channels and conductors and realized the fault early warning of the transmission line.

Ref. [8-9] realized conductor galloping monitoring based on Beidou technology and image technology. The above Reference monitored conditions of the transmission line by configuring different sensors, however, the single sensing terminal only has a single state monitoring function and lacks integrated sensing equipment. To sense transmission lines in the overall view, many different types of single-function sensors need to be installed, which face the problems of the redundant sensing systems, repeated construction, and low utilization rate.

In this paper, multi-physical quantity integrated sensors were proposed for the needs of power grid digital transformation. The sensors overcome the difficulties of sensor energy acquisition, integration, and structural protection technology, which realized the acquisition and transmission of multiple state parameters by a single integrated sensor terminal. This technology improved the efficiency of data acquisition and promotes the goal of automation, informatization, and digitization of intelligent monitoring of transmission lines.

2. Transmission Line Operation Condition and Sensing Demand Analysis

There are various abnormal conditions in transmission line operation. Generally, the operation conditions of equipment gradually deteriorate with time. The monitoring of the operation state is the basis of line operation and maintenance. Monitoring equipment on transmission lines is usually divided into real-time monitoring and non-real-time monitoring. Common monitoring technologies are shown in TABLE 1.

The non-real-time monitoring equipment needs to be operated by maintenance personnel. Its advantages are high patrol accuracy, Low misjudgment rate of human observation, however, its disadvantages are high labor cost and difficult operation in complex terrain areas. At present, non-real-time monitoring gradually adopts human-machine collaborative patrol. Based on the cycle and operation cooperation of human patrol and machine patrol, the patrol information interaction is opened up, which improves the efficiency of transmission line patrol. However, the technology still fails to quickly warn the fault.

TABLE 1 Common transmission line monitoring technology

Classification	Monitoring technology	Function
real-time	UAV patrol	Channel patrol inspection
	Human Patrol	Fine patrol inspection
non-real-time	Monitoring device	Daily monitoring
	Distributed fault location	Fault location
	Galloping monitoring	Abnormal galloping alarm
	Icing monitoring	Icing thickness alarm
	Sag monitoring	Sag alarm
	Insulator monitoring	Monitoring tension and zero value of insulator
	Tower inclination monitoring	Inclination and settlement monitoring

The monitoring equipment senses operation conditions by the sensing device, and the data information is transmitted to the background terminal by the power communication network. The advantages are real-time online and rapid fault early warning. However, the disadvantages are a large amount of data, unstable power supply, and so on. At present, many separate and single sensors are used to monitor the operating conditions of the power system, which cannot realize effective integration and information sharing at the sensing level. Therefore, the development of multi-parameter information fusion and diagnosis technology is slow, which cannot meet the development requirements of the digital power grid.

Therefore, to fully coverage monitor location information of transmission lines, it is necessary to establish a perfect and systematic monitoring system through the environment, conductor, channel, and tower monitoring. The specific monitoring data are as follows:

(a) Environmental monitoring. The volume of micrometeorological monitoring devices (i.e. wind speed and rainfall) is large, which is mostly installed on the tower. While, temperature, humidity, wind speed, and air pressure sensors can be installed on various positions of the transmission line by integrated design;

(b) Wire monitoring. Its monitoring quantity is the electrical quantity and mechanical quantity of the conductor. By measuring the current and temperature of the conductor, the energy transmission state of the conductor can be monitored. By monitoring the galloping, sag, and vibration on the conductor, the mechanical characteristics can be evaluated, and the operation risks (e.g., wire breakage and strand breakage) can be prevented;

(c) Channel monitoring. The channel monitoring network of the transmission line can be constructed by visible and infrared images. The mountain fire, external damage, tree barrier, and bird's nest are quickly diagnosed based on AI recognition algorithm;

(d) Tower monitoring. By installing displacement, tilt, and other sensors on the tower, the stress state and the stability of the tower are monitored for preventing tower tilt and other faults.

According to the above monitoring requirements of transmission lines, multi-physical quantity integrated sensors are designed, which make the best of the sensor, video monitoring, communication, and image recognition technology. The integrated sensor will carry out real-time monitoring and control on the site for replacing human inspection.

3. INTEGRATED SENSOR DESIGN

3.1.Architecture Design

In this paper, the multi physical quantity integrated sensor were designed based on 500kV four bundle conductor structure from hardware development, software development and application scenario.

The multi-physical quantity integrated sensor was proposed to collect the current of each split conductor, which supports the subsequent development of fault location and other functions. In addition, the sensor integrates various sensing functions and wireless long-distance communication modules, which causes large system power consumption. Therefore, it is necessary to design the measuring power claw (ring) to realize the self-power supply. Therefore, the independent measuring power ring was designed, which can be installed on each conductor and measure conductor current and temperature. The monitoring host was installed in the middle of the spacer, which did not affect the mechanical performance of spacer fittings. The host was connected with the measuring power rings by flexible connecting lines, as shown in Figure 1(a).

During operation, the measuring power ring was in contact with the conductor for measuring the current and temperature. Besides, the monitoring host was responsible for the data acquisition, processing and, energy management of the system.

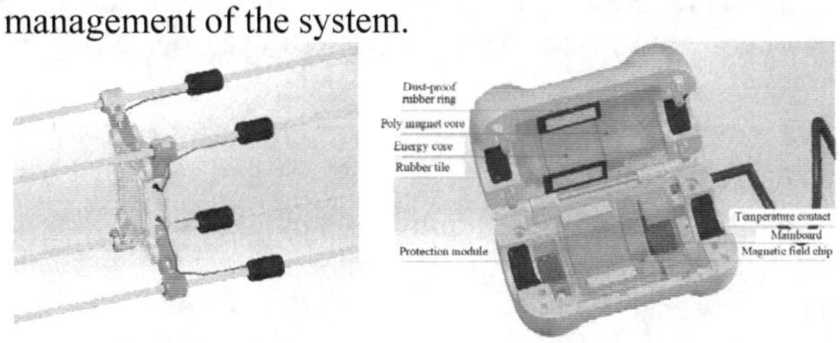

(a) Architecture of sensor (b) Measuring power ring

Figure 1. Architecture of sensor and Measuring power ring

3.2.Sensor Integration

3.2.1.Measuring power ring design

The measuring power ring is divided into three functional modules, i.e., coil energy acquisition module, current acquisition module, and temperature acquisition module.

(a) Coil energy acquisition module. The sensor obtains electric energy from the conductor by coil power acquisition and processes electric energy for supplying to the acquisition module. Specifically, the output voltage of the coil energy acquisition module is 12V, with a maximum output of 7.2W. The output voltage is larger than the power consumption of 1.9W, so the monitoring host is supplied by excess energy by cables.

(b) Current acquisition module. The current on a conductor is measured by a microsensor. Then, the current is processed into an analog signal and transmitted to the host.

(c) Temperature acquisition module. the temperature on the conductor surface is measured and processed into an analog signal for transmitting to the host.

To prevent the failure of the energy extraction module and ensure stable operation, the adaptive work management strategy was designed between the main power supply and standby battery. The battery is a lithium battery with a capacity of 3200 mAh, which can provide the system to work continuously for more than 8h. The working temperature is between -10 and 80°C, which has strong environmental adaptability. The battery is charged under high current, and the standby battery works under low current, which improves the stability and reliability of the sensing system. The structure of the measuring power ring is shown in Figure 1(b).

3.2.2.Hardware Design

The monitoring host is divided into five functional modules, i.e., power management module, sensing module, data acquisition module, main control module, and data encryption communication module. The hardware circuit architecture is shown in Figure 2.

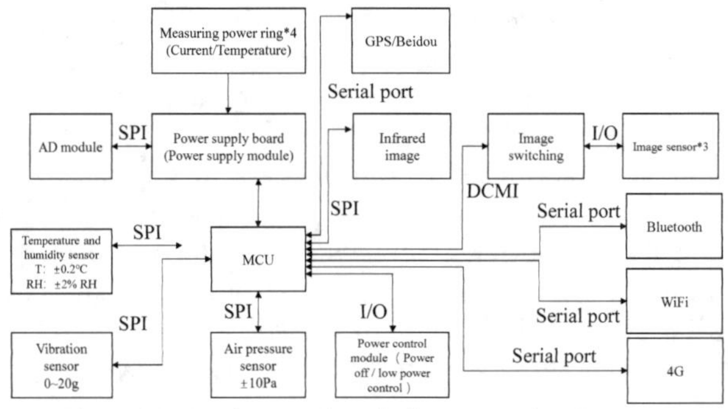

Figure 2 Hardware circuit framework of sensor

In the design of sensor hardware, each data interface is reserved, and the standardized interfaces (e.g., SPI, I2C, or serial port) are used to transmit data. Therefore, the sensor module can be integrated standardized, which can replace the sensor module according to the application scenario. The standardized interface includes power terminal, enable terminal, function control terminal, data terminal, etc. Besides, the sensor hardware adopts a pin and cable layout design with a locking function.

The design can freely combine the sensor modules of the system. When monitoring functions are less, some sensor modules can be removed to reduce the sensor cost; When the monitoring function needs to be added, the number of sensors (such as cameras) can be increased to improve the monitoring accuracy.

To ensure stable and reliable data transmission, three communication interfaces are designed, i.e., Bluetooth, WiFi, and 4G. Bluetooth communication has the function of a communication relay, which can receive data from other monitoring devices near the sensor. The WiFi module is used for sensor debugging and configuration, which can support Ad Hoc network transmission. Besides, the 4G module is the main channel for sensor data transmission. It uses the special IOT network card of CSG for communication. The data is directly connected to the IoT platform, which is safe and reliable.

3.2.3.Software Design

The software system was developed in layers according to business, function, and hardware. As shown in Figure 3.

Software application is divided into three parts, i.e., configuration system, operation system, and upgrading system, as shown in Figure 4. Specifically, the configuration system is used for calibration, installation, and commissioning in the field. The operation system is the logic for the normal operation of the sensor. The upgrading system provides system function expansion with the remote wireless upgrade function. According to the adjustable working interval, the sensor can maintain a low-power standby state for the rest of the time.

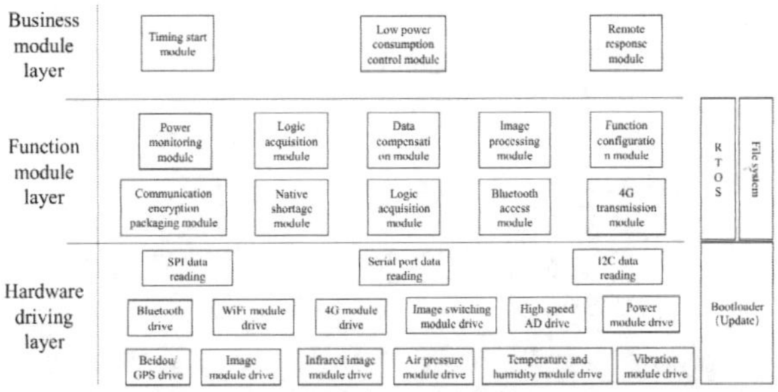

Figure 3 Software architecture

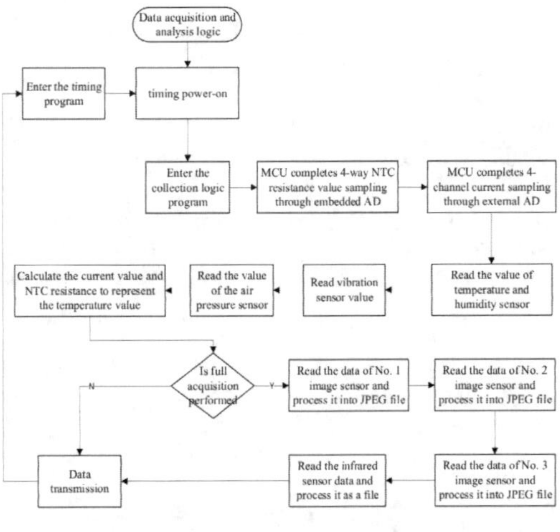

Figure 4 Normal workflow

3.2.4.Protective Structure Design

The multi-physical quantity integrated sensor often operates outdoor with high temperature, high humidity, icing, and other harsh environments for a long time. Therefore, the device must meet the high IP protection design, and heat dissipation, and insulation functions.

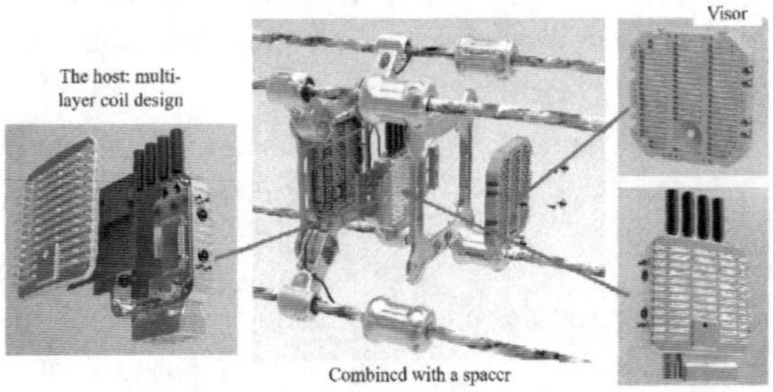

Figure 5 Structure of sensor

The host is protected by a multi-layer coil for IP67 protection grade. Because the internal heating component is mainly MCU, however, the total calorific value is not high, the overall heating of the device is mainly caused by external sunlight. Therefore, referring to the design of louvers, the mesh sunshade and ventilation hood were added to the shell on the host for dissipating heat. The structure is shown in Figure 5.

4. POWER GRID OPERATION TEST AND RESULTS

At present, the multi-physical quantity integrated sensors have been applied in the five provinces of the China Southern Power Grid. The device can be installed on energized or de-energized. Besides, special installation tools were developed for quick installation. Multiple key parameters can be measured at one installation which can reduce the total cost. The sensor was installed on the field as shown in Figure 6.

Figure 6 Sensor installed on the line

The images from the sensor installed in Diqing transmission line are shown in Figure 7.

(a) In front of the channel (b) Behind the channel

(c) Under the channel (d) Infrared image below channel

Figure 7 Images monitores by sensor

The measurement images and data from the integrated sensor verify the pre-designed functions, which realized the image acquisition in front, behind, and below the transmission line channel. Besides, the current and temperature of the conductor can be measured, and the parameters (e.g., conductor galloping and sag) can be calculated. This technology established the foundation for the automation, informatization, and digitization of intelligent monitoring of transmission line.

5. CONCLUSIONS

For supporting the construction of digital power grid and build the new power system, a large number of micro intelligent sensors must be deployed. Therefore, an intelligent multi-physical quantity sensor was proposed in this paper. Relying on non-contact current measurement, energy supply, intelligent sensing, low power consumption, and reliability design technology, different types of sensing modules were integrated into one sensor for monitoring the condition of the transmission line conductor side and channel side,. Finally, the pilot application and functions of the sensor were carried out in the field, which established the technical foundation for intelligent monitoring of transmission lines.

Nevertheless, the proposed intelligent sensors were performed on a few transmission lines. According to the plans, the intelligent sensors will widely apply in CSG for intelligent and digital monitoring transmission lines. In addition, there is a need for mathematical relationships between sensing data and operation conditions of transmission lines.

References

[1] H. Bai, C. Zhou, Z. Yuan, and J. Lei, "Prospect and Thinking of Digital Power Grid Based on Digital Twin," Southern Power System Technology, vol. 14, pp. 18-24+40, 2020.

[2] Y. Gan, Z. Du, W. Zhou, and et al., "Equivalent ice thickness calculation model of strain tower line based on icing tension monitoring system," Electrical Measurement and Instrumentation, vol. 58, pp. 39-45, 2021.

[3] J. Li, Y. Tang, and Q. Wu, "On-line Monitoring Technology of High Voltage Transmission Cable Based on Temperature and Humidity Sensor," Instrumentation and Measurment, vol. 40, pp. 108-111, 2021.

[4] W. Wang, L. Li, H. Li, and et al., "System for Monitoring Ice Thickness on Transmission Lines Based on Wavelet Transform and Morphological Processing," Automatic Control Systems and Equipments, vol. 42, pp. 63-65, 2020.

[5] M. Sun, Q. Wang, Y. Sun, and Y. Chen, "Mountain Fire Monitoring and Positioning of Transmission Line Satellite Based on Image Recognition Technology," Identification Modeling and Simulation, vol. 38, pp. 65-69, 2019.

[6] Y. Wang, H. Liu, L. Li, and et al., "Application of Image Recognition Technology Based on Edge Intelligence Analysis in Transmission Line Online Monitoring," Electric Power ICT, vol. 17, pp. 35-40, 2019.

[7] J. Zhao, X. Zhang, J. Zhao, and et al., "Research on Icing Monitoring Method of Transmission Lines Based on Image and Stress," Automation and Instrumentation, vol. 33, pp. 54-58, 2018.

[8] F. Guo, Z. Fan, W. Yu, "Research on 110kV Transmission Line Galloping Monitoring Technology Based on Power Internet of Things," Electrical technology, vol. 22, pp. 113-115, 2020.

[9] Z. Zhang, L. Lang, J. Yan, and et al., "Transmission Line Galloping Monitoring System Based on Beidou Ground-based Augment System," Electric Power ICT, vol. 18, pp. 42-47, 2020.

the Technology of Space Electric Field Detection of Deteriorated Insulator Strings Based on Multi Rotor UAV

JinPeng HAO[1]*, ShiTao LIU[2], Kai YANG[1], Hong WU[1], BoYu ZHANG[3], ZhiYong XIAO[3]

[1]Electric Power Research Institute, State Grid Ningxia electric power company, Yinchuan, China

[2]State Grid Ningxia electric power company, Yinchuan, China

[3]Hebei Key Laboratory of power transmission and transformation equipment security defense

North China Electric Power University, Baoding, China

* 690197939@qq.com

Abstract**—**Porcelain insulator with good insulation performance is the key to the stable operation of power system. In order to characterize its insulation performance safely and accurately, a 1:1 three-dimensional electric field simulation model of 220 kV porcelain tension insulator string is established by using COMSOL finite element software, and a UAV live detection platform for degraded insulator string is built according to the optical electric field sensor. The results show that when the insulator deteriorates, the distribution electric field curve will appear obvious distortion; comparing the simulation results with the live measurement results, it is found that the electric field waveform curve obtained by the live detection platform of degraded insulators based on UAV can be accurate and can effectively judge the insulators with deterioration.

1. INTRODUCTION

Insulator plays the role of electrical insulation and support connection in transmission line. It is a key insulating component of power system to maintain the safe operation of power grid[1-3]. Transmission lines work in complex external environment all year round. With the increase of operation time year by year, the insulator will inevitably deteriorate to varying degrees due to the influence of electromagnetic interference, mechanical stress, temperature and humidity, resulting in flashover phenomenon, which greatly threatens the safety of power grid[4,5]. Therefore, it is very important to carry out regular maintenance on insulators and accurately judge their insulation state.

At present, in the live detection, the detection methods of insulators include: electric quantity detection method (pulse current detection method, electric field measurement method, insulation resistance detection method, etc.) and non-electric quantity detection method (infrared thermal imaging method, ultraviolet imaging method, etc.)[6]. References[7,8] found that the sensitivity of deteriorated insulators based on UV imaging detection technology is low, and misjudgment will occur if the deterioration position occurs in the middle of insulator string. References[9,10] use infrared technology to detect the temperature of insulators in 220 kV substation. It is found that the detection sensitivity of porcelain insulators is low, because the overall temperature difference of porcelain insulators is small, and they are easy to be disturbed by radiation and surrounding environment in the actual detection process. The non electric quantity detection method is greatly disturbed by the environment, which will affect the accuracy and sensitivity of the detection results. Reference[11] takes the pulse discharge current as the starting point and uses the discharge intensity vector to reflect the distribution of back electric pulse, which is an important basis for judging the deteriorated insulator, but it is vulnerable to the external environment and has insufficient accuracy and sensitivity; The insulation resistance detection method must ensure that the insulator surface is clean, otherwise it is impossible to judge whether the decrease of insulator resistance is caused by the damage of insulator integrity or because there are impurities on the insulator surface [12].

The electric field detection method in the electric quantity detection method has a broader application prospect compared with other methods. It tests the insulation performance of Insulators based on the distortion of electric field distribution curve[13-15]. The electric quantity detection method is less affected by external factors and can directly judge the insulation performance of insulators. If there are defects, the voltage drop between different insulator pieces will change, and the electric field intensity will also change[16].

In reference[17], the live detection tests of composite insulators in good condition and core rod conduction are carried out respectively, which proves that the electric field detection can obtain more reasonable detection results; In reference[18], the finite element electric field model of composite insulator umbrella skirt with different damage positions and damage degrees is established. It is concluded that the spatial axial electric field curve has no effect when the umbrella skirt at the high voltage end of insulator string is damaged. The potential and electric field distribution of composite insulators of 330 kV transmission line are simulated by finite element method in document[19]. It shows that the wire will equalize the potential of the insulator, and the maximum field strength on the high voltage side is 60% lower than that without wire.At present, there are many researches on composite insulators in the research results, and porcelain insulators have become the basic equipment in ultra-high voltage and high voltage transmission lines. Its internal structure contains a variety of materials with large differences in thermal expansion coefficient. The material gap is easy to produce heat and be broken down, which seriously endangers the safe operation of lines.

In order to accurately judge the insulation performance of porcelain insulator and find out the deteriorated insulator accurately and efficiently, according to the electric field distribution characteristics of porcelain insulator, a three-dimensional finite element electric field simulation calculation model of U210BP/170 insulator string is established, and the surface electric field curve when the insulator string has good performance and deterioration and the distribution electric field curve on the outermost side of porcelain piece are obtained; Based on the electric field sensor of photoelectric crystal Pockels effect, a live inspection platform for UAV with deteriorated insulator is built; The live measurement of insulator space electric field in a line in Yinchuan is carried out. The accuracy of the live detection device of UAV degraded insulator is verified by comparing the known degraded insulator electric field detection waveform with the simulation waveform.

2. INSULATOR ELECTRIC FIELD MEASUREMENT METHOD

2.1 Measurement Principle of Electric Field Distribution in Live Line

Due to the influence of stray capacitance between wire, tower and insulator, the distributed electric field of insulator presents a U-shaped curve with high at both ends and low in the middle[20]. The equivalent circuit diagram of insulator to tower and wire is shown in Fig.1.

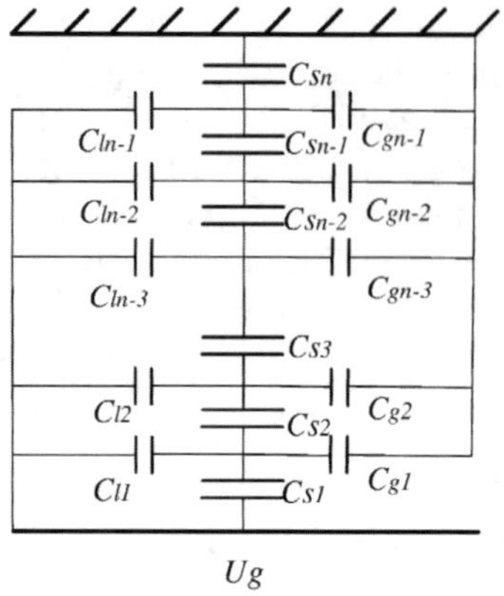

Fig.1 Equivalent circuit diagram of insulator string to tower to conductor

Where C_S is the capacitance of insulator itself, C_g is the equivalent stray capacitance between insulator and tower, and C_l is the stray capacitance between insulator and wire. The stray capacitance C_g of the insulator to the tower increases the voltage drop at the high voltage end of the insulator because its shunting effect is greater than the confluence effect; The shunt effect of the insulator on the stray capacitance C_l of the wire is less than the confluence effect, so that the voltage drop borne by the insulator near the low-voltage end increases accordingly. Therefore, under the combined action of stray capacitance C_g and C_l, the distributed electric field intensity curve of insulator approximately presents a U-shaped curve with high at both ends and low in the middle. When the insulator is deteriorated, its own capacitance C_S will become very small or even become a short-circuit state, and the electric field strength will suddenly decrease. Obvious distortion points can be found on the electric field strength curve of the insulator. Therefore, the insulation performance can be judged by deteriorating the electric field characteristics of insulators.

2.2 Maintaining the Integrity of the Specifications

Since the spatial electric field of porcelain insulator is a three-dimensional field without boundary, the calculation model includes porcelain insulator, cross arm, conductor and connecting hardware. The actual line referenced by the simulation is Yinchuan Nafeng double line three-phase line, and the insulator is tension insulator.

The established COMSOL 3D simulation model is shown in the figure below.

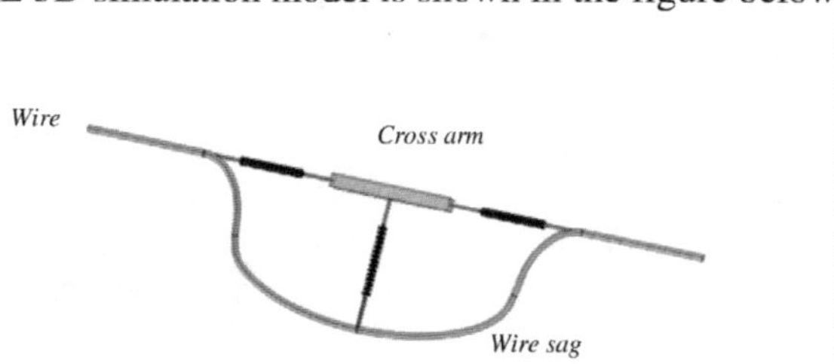

Fig.2 Three dimensional simulation model of transmission line insulator

On the premise of not affecting the calculation results, the calculation model can be simplified: in the actual modeling, only the line on one side of the tower is considered, and the influence of the electric field of other two phases is ignored; In the actual line, due to the existence of split wire and the growth of line length, the stray capacitance of insulator to conductor plays a leading role in the influence of insulator electric field, so the complex network structure of tower is simplified as cross arm; Because the connection fittings have little influence on the insulator electric field, all connections are simplified as cylinders. The insulator string is U210BP/170 double umbrella 15 piece porcelain insulator string, and the simulation parameter settings are shown in Tab.1.

Tab.1 Simulation parameter setting

Parameter	**Relative dielectric constant**	**Conductivity / （S /m）**
Air domain	1	0
Porcelain	6	10-12
Wire	1000	107

2.3 Analysis of Simulation Results

127 kV AC phase voltage is applied to the high voltage side of the insulator string in Figure 2 model to simulate the insulator electric field at 220 kV voltage level, and the potential degree of freedom coupling is carried out between the steel pin and steel cap of the insulator piece to simulate the deteriorated insulator. The surface electric field curve of insulator strings with intact insulation performance and deterioration at different positions is obtained, as shown in Fig. 3.

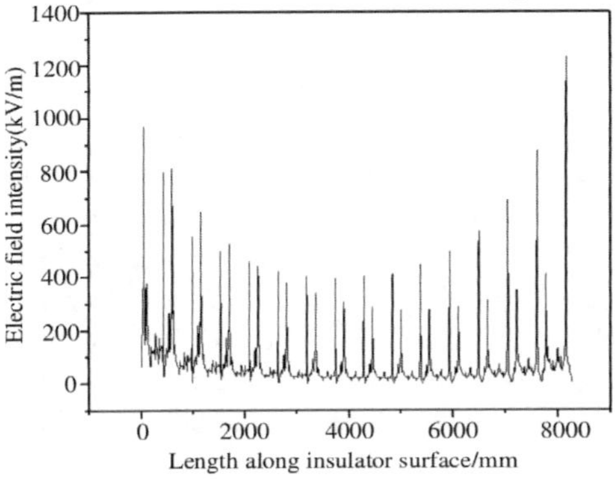

a) Surface electric field of non deteriorated insulator

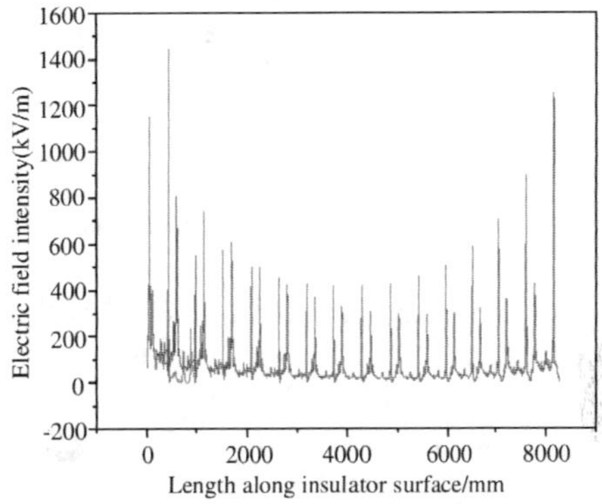

b) Surface electric field of deteriorated insulator at high voltage side

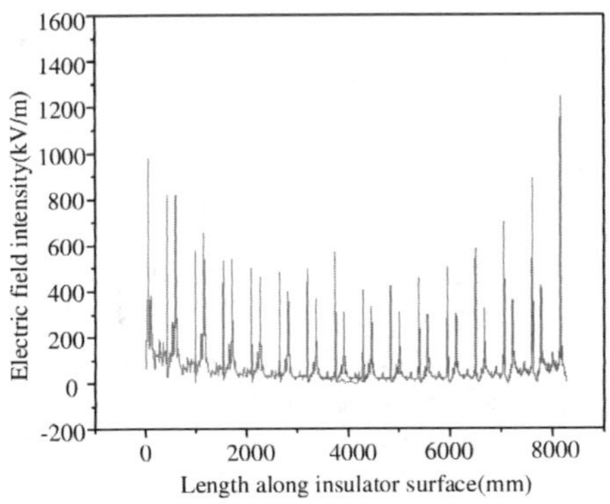

c) Surface electric field of middle deteriorated insulator

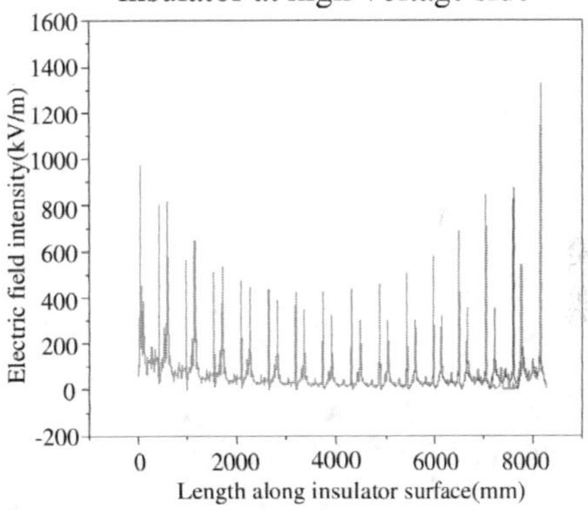

d)Surface electric field of deteriorated insulator at low voltage side

Fig.3 Electric field intensity curve along insulator surface

It can be seen from Figure 3 that for insulator strings with good insulation performance, the surface electric field of insulator pieces near the high-voltage end and low-voltage end is high, and the electric field intensity of insulator pieces near the middle is relatively low. When the deteriorated insulator piece is set at different positions of the insulator, the electric field intensity at the deteriorated position is distorted to varying degrees, and the electric field intensity of the insulator on both sides of the piece of insulator increases abnormally.

The surface electric field curve of insulator accurately reflects the electric field on the insulator surface, but it can not directly judge the deteriorated insulator for engineering practice. Extract the electric field intensity on the outermost side of the insulator tile in Fig. 3 to obtain the distributed electric field curve of the insulator, and the obtained distributed electric field curve of the insulator is shown in Fig. 4.

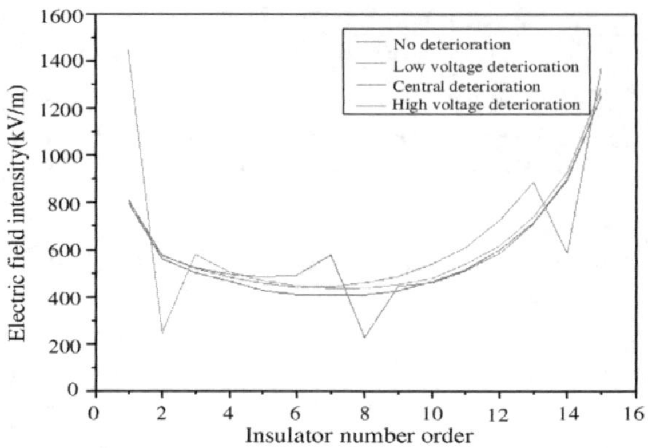

Fig.4 Electric field intensity curve of insulator distribution

When the high voltage end of the insulator is deteriorated, the electric field strength decreases to 55.8% of that without deterioration; When there is deterioration in the middle of the insulator string, the electric field intensity decreases to 44.1% without deterioration; When there is deterioration at the low voltage end of the insulator, the electric field intensity is 35% of that without deterioration. It can be seen that when the insulator deteriorates, its spatial electric field strength will decrease significantly, among which the electric field at the high voltage end decreases most, followed by the middle part, and the electric field at the low voltage end decreases least. Using the distributed electric field characteristics of insulator string, the simulated image can be compared with the measured insulator electric field intensity waveform of UAV to infer whether the insulator in the actual line is deteriorated.

3. Live detection platform of deteriorated insulator based on UAV

Using the spatial electric field distribution characteristics of insulators can effectively judge the insulation performance of insulators. Based on this principle, a live detection platform of degraded insulators based on UAV can be built for non-contact rapid detection of insulator electric field.

3.1 Optical Electric Field Measurement System

3.1.1 Optical electric field sensor

The transient high-voltage electric field sensor based on photoelectric crystal Pockels effect can convert the electric field signal into optical signal for transmission, and then convert the optical signal into voltage signal through photoelectric converter to complete the transient electric field measurement. In the whole measurement process, the field strength measurement circuit is completely isolated from the high-voltage circuit, which has the characteristics of strong anti-interference and good transient follow-up.

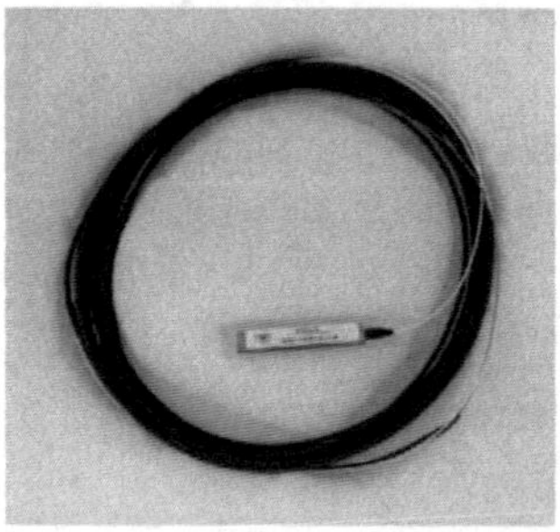

Fig.5 Probe and optical fiber of electric field sensor

Test and study the power frequency response characteristics of the sensor: place the sensor probe in the center of the 0.5m plate to plate gap, and then apply a power frequency voltage of 12 kV ~ 120 kV to the gap on the premise of ensuring no corona, and the corresponding electric field strength is 24 kV / M ~ 240 kV / m. The power frequency electric field waveforms measured by the sensor under different electric field intensities are shown in Figure 6. The electric field range measured by the optical electric field sensor is 24 kV / m ~ 240 kV / m, which can be applied to insulator electric field measurement under most field conditions. The power frequency electric field waveform has no distortion and good response. Compared with the actual electric field strength, it can be seen from Figure 7 that the power frequency amplitude characteristic linearity of the electric field sensor is very good[20].

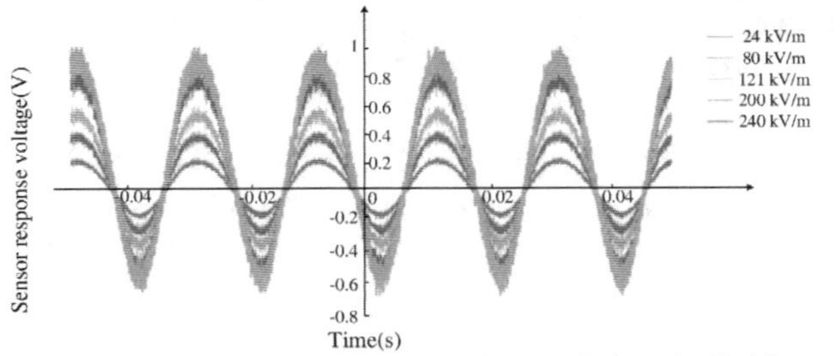

Fig.6 Power frequency response waveform of electric field sensor

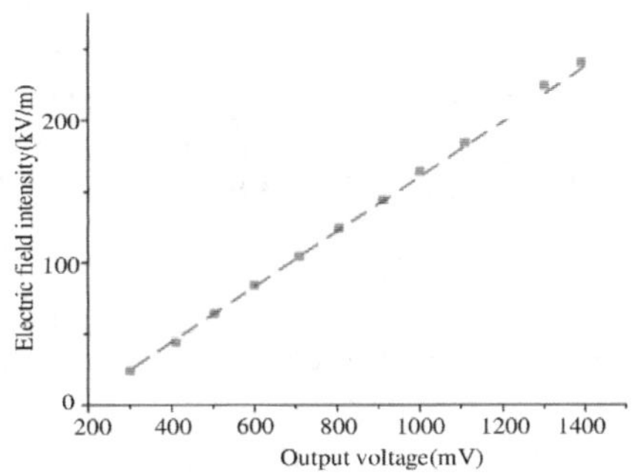

Fig.7 Power frequency response characteristics of electric field sensor

3.1.2 Miniaturization design of electric field transmission module

The electric field data transmission module is composed of power adapter, light source driving circuit, polarization maintaining isolator, polarization beam splitter and detector. In order to carry it on M600 UAV for electric field data transmission of degraded insulator detection, it needs to be miniaturized. Since the UAV system is powered by 12V DC power supply, the power adapter of the electric field data transmission module is changed from AC to DC; At the same time, small integrated components are designed according to the fuselage parameters of UAV to replace large-scale polarization maintaining isolator, polarization beam splitter and detector. On the premise of ensuring the accurate transmission of electric field data, three electric field measurement channels are designed to compact all integrated components. The size of the electric field data transmission module after miniaturization design is 100 * 150 * 30MM, as shown in the figure.

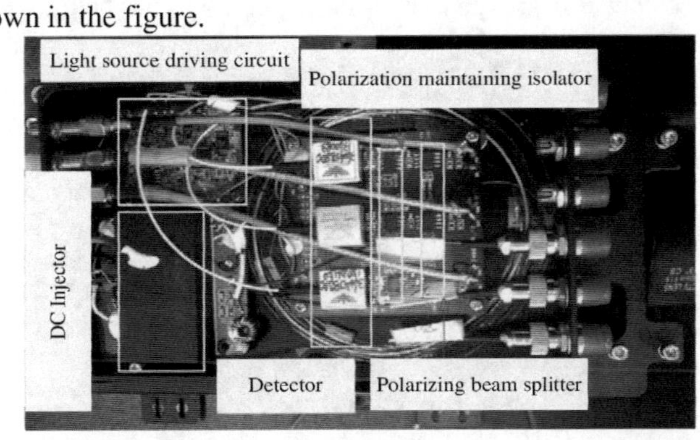

Fig.8 Data conversion model after miniaturization design

3.2 Composition of UAV Degraded Insulator Detection Platform

The airborne terminal of the detection platform is in front of the M600 UAV, equipped with an electric field sensor based on the Pockels effect of photoelectric crystal, and the miniaturized electric field measurement data conversion module is integrated under the UAV. In order to accurately locate and detect the insulator, the UAV is also equipped with laser positioning lens and visible light lens to accurately locate the insulator.

When observing the insulator electric field at the ground terminal, the electric field measurement data receiving module is used to receive the electric field data transmitted by the front end of the UAV, and the insulator electric field waveform is displayed through a micro oscilloscope. The airborne end and ground end of the degraded insulator detection platform are mainly interconnected through the wireless signal module composed of data link. The composition of the detection platform is shown in the figure below.

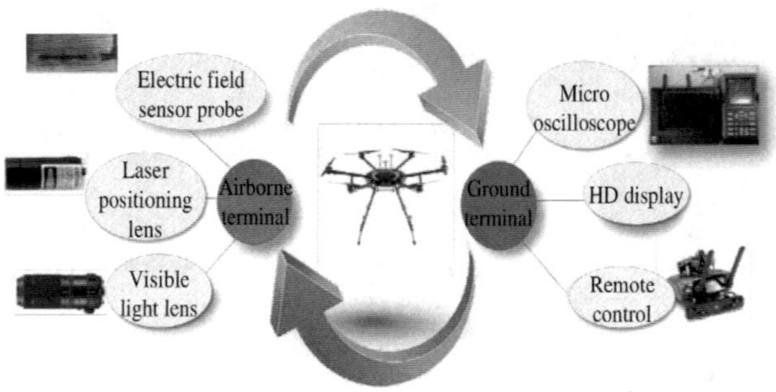

Fig.9 UAV degraded insulator detection platform

4. INSULATOR FAULT MEASUREMENT

4.1 Live Measurement of Insulators with Good Insulation Performance

In order to verify the accuracy of UAV deterioration insulator detection device, the electric field detection of tension insulator string is carried out on Yinchuan 220kV transmission line. The detected insulator is double umbrella porcelain insulator, and the model is consistent with that in the simulation model. The temperature on the day of testing is 15 °C, and the relative humidity is 50%. In the process of live detection, the staff at the airborne end and the ground end shall ensure standardized operation, and keep the UAV at a safe distance of 3M from the line and insulator during flight under the voltage level of 220 kV. The UAV carries out real-time ranging through the laser rangefinder to ensure that the distance between the probe of the electric field sensor and the outside of the insulator umbrella skirt remains unchanged during the measurement process, and the flight speed of the UAV is about 0.12 m / s. The UAV is used to scan the electric field distribution around a single insulator in the double string insulator at a uniform speed. The test site diagram is shown in the figure below.

Fig.10 Test site

For insulators with good insulation performance in 220kV line, the measured waveform of electric field is as follows:

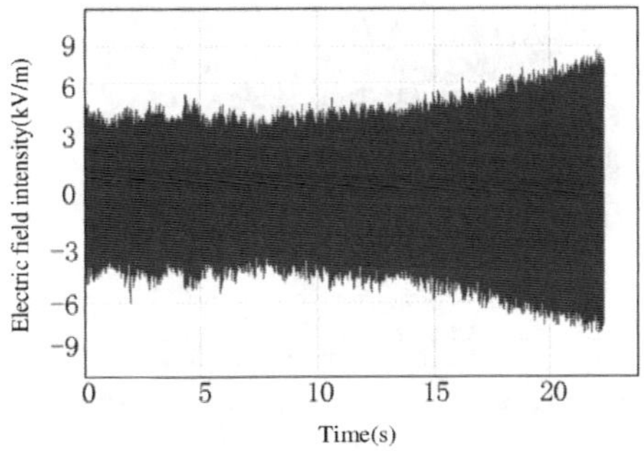

Fig.11 Measured electric field waveform of insulator without deterioration

By comparing with the simulation figure, it can be seen that there is a difference between the measured electric field distribution of insulator string and the simulation calculated value, which is mainly due to the fact that the UAV degraded insulator detection platform has a certain distance from the insulator surface when detecting the degraded insulator string, but in order to realize the non-contact rapid detection of insulator, The insulation performance of insulator string can be judged by observing whether there are distortion points in its own electric field waveform.

The UAV flies at a uniform speed relative to the insulator string, and the time node corresponding to the electric field intensity amplitude of each insulator is calculated according to the measured time. The corresponding position of each time node is taken as the center position of each insulator skirt. Each measured strain insulator string has 15 insulators. During data processing, 15 typical observation points can be selected according to the center position of each insulator umbrella skirt for key analysis, and the curves are as follows:

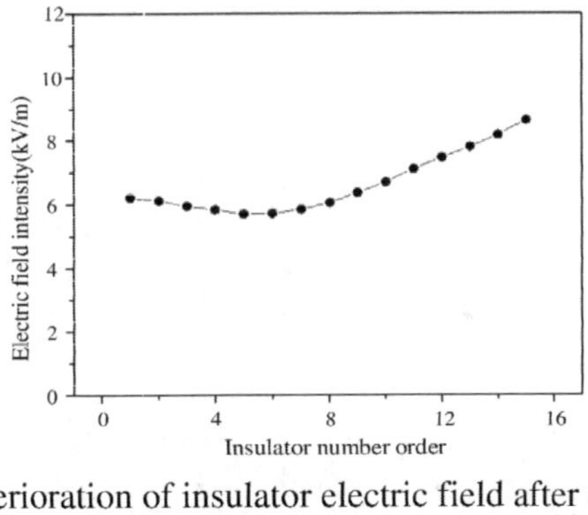

Fig.12 No deterioration of insulator electric field after data processing

It can be seen from the figure that although the measured electric field intensity is lower than that of the simulation model due to the long distance from the insulator, the lowest measured electric field of the insulator is 5.93kV/m, and the electric field amplitudes at both ends of the insulator string are 5.39kV/m and 8.88kV/m respectively. The distributed electric field of insulators is a U-shaped curve with high at both ends and low in the middle. The measured electric field waveform of Yinchuan line is similar to the COMSOL simulation results. The UAV degraded insulator detection device is reasonable.

4.2 Comparison of Measured Electric Field of Deteriorated Insulator

Based on the judgment results of manual detection: there are degraded insulators in the tower of a line and they are running on line. The live measurement results by using UAV degradation detection system are as follows:

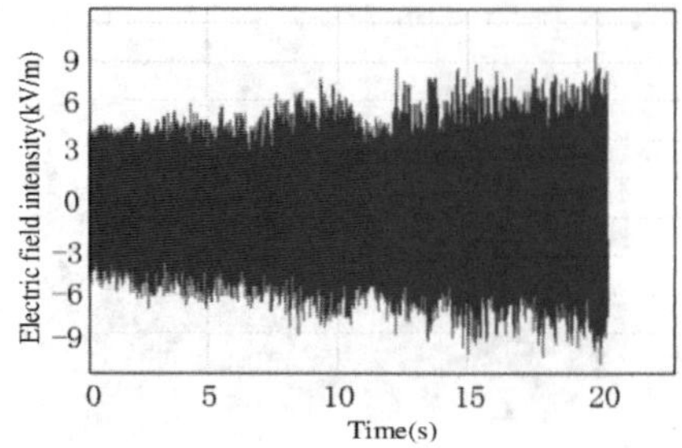

Fig.13 The measured electric field waveform when the insulator is deteriorated

During the field test, there will be an obvious "drop" phenomenon along the scanning electric field amplitude of the insulator, resulting in the depression of the overall measurement waveform in a certain period of time. Considering that the insulation effect of some insulator pieces of deteriorated insulators is weakened and the bearing voltage at both ends is low, resulting in the weakening of radiated electric field, it is reasonable to think that when the amplitude of scanning electric field drops significantly, the insulator pieces will deteriorate at the corresponding measurement time. Therefore, in the measured line, if there is an obvious abnormal decrease in the amplitude of insulator electric field, it can be judged as a deteriorated insulator.

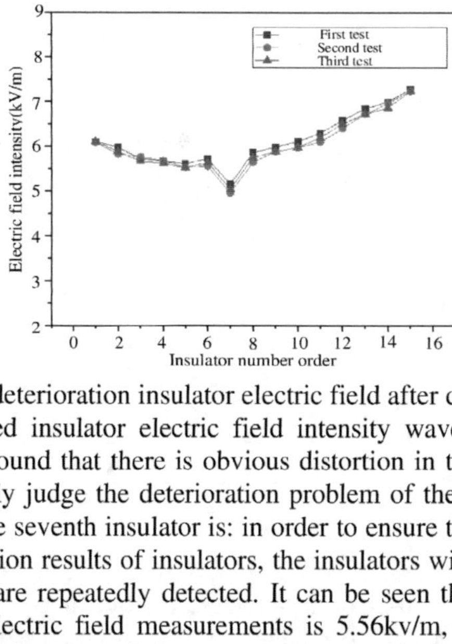

Fig.14 The deterioration insulator electric field after data processing

According to the measured insulator electric field intensity waveform of UAV and COMSOL simulation results, it can be found that there is obvious distortion in the insulator string electric field waveform, which can basically judge the deterioration problem of the seventh insulator. The electric field intensity amplitude of the seventh insulator is: in order to ensure the accuracy and repeatability of the spatial electric field detection results of insulators, the insulators with significant electric field drop during the detection process are repeatedly detected. It can be seen that at the seventh insulator, the average value of the three electric field measurements is 5.56kv/m, and the effective value of the electric field is significantly reduced, It can be considered that it is in a degraded state near the seventh insulator. The test shows that the electric field distortion near the deteriorated insulator measured by the electric field measuring device carried by UAV has high repeatability, and can effectively identify the existence and location of the deteriorated insulator.

5. Conclusions

(1) Under the influence of stray capacitance of lines and towers, the electric field curve of insulators with good insulation performance presents a U-shape of high at both ends and low in the middle. The electric field at the deteriorated insulator has a steep drop, while the electric field on both sides increases abnormally. The deteriorated insulator has the greatest impact on the electric field at the high-voltage end, followed by the middle part, and the electric field at the low-voltage end has the least obvious decline.

(2) The transient high voltage electric field sensor based on photoelectric crystal Pockels effect has good response to the power frequency electric field waveform less than 60 kV / m, good linearity of power frequency amplitude, and no distortion and distortion. After the miniaturization design of the electric field data transmission module, the size is 100 * 150 * 30MM, which can be integrated into the UAV fuselage. The space electric field measurement system can carry out insulator detection based on UAV platform to meet the needs of engineering.

(3) Based on the live detection of the spatial electric field of insulators by the photoelectric sensor detection device of UAV, it can be found that the device can efficiently and accurately collect the distributed electric field of insulators, extract the electric field amplitude at the center of each insulator umbrella skirt after data processing of the measured electric field waveform, and determine whether there are distortion points in the obtained insulator electric field curve, It can accurately judge whether there are deteriorated insulators in the insulator string. The UAV degraded insulator detection platform based on optical electric field sensor can be used for non-contact accurate judgment of insulator insulation performance.

ACKNOWLEDGMENT
National Natural Science Foundation of China (Youth Science Foundation Project) (52007064)

REFERENCES

[1] LI Yonggang, HUANG Chengcai, WANG Fochi, et al.Influences of haze weather on operating performance of composite insulators [J]. Power System Technology, 2014,38(7) : 1792-1797.
[2] JIA Baiyan. Physics and chemistry analysis on composite insulators after operated[J]. High Voltage Engineering, 2012, 38(4): 914-921.
[3] TU Youping, CHEN Jingjing, XU Zhuo, et al. Uneven change characteristic of composite insulators hydrophobic- ity in long time service[J]. High Voltage Engineering, 2013, 39(6) : 1469 -1475.
[4] Guan Zhicheng, Liu Yingyan, Zhou Yuanxiang, et al. External insulation of insulator and transmission line equipment[M]. Beijing: Tsinghua University Press, 2006.
[5] Zeng Qin, Zhou Kai, Zhao Shilin. Theidentification of several flashover factors for the transmission line insulation [J]. Journal of Electric Power Science and Technology, 2020.
[6] WAN Xun. LIU Zhengting. GONG Zhengxiong,et al. Microwave Transmission Method for Detecting Deteriorated Ceramic Insula-tors[J].Insulators and Surge Arresters,2020(04):207-214.
[7] CHEN Heng, HOU Shan-jing. Infrared Fault Diagnosis of Power Equipment[M]. Beijing: China Electric Power Press, 1997.
[8] Zhang Xiaochun, Ouyang Guangze, He Hongying, et al. Zero Insulator Detection Based on Infrared Image Matching[J]. Electrical Measure- ment & Instrument, 2019, 56(6) : 100-105.
[9] CHEN Tao, Study on non-contact detection for faulty insula-tors[D].Chongqing University, 2006.
[10] Lu Ming, Yao Degui, et al. Study on methods of detection of faulty in- sulators[J]. Insulator and Surge Arresters,2006, 5(2) : 13-14.
[11] [11] Z,edan F.M. Analytical Studies of HV Insulators in Saudi Ara-bia[J].IEEETrans. on Elec.Insul., l 993. 28(3): 379-391.
[12] LI Chengrong, CHENG Yangchun, CHEN yu, et al. On-line detection of internal defects of composite insulator by electric field mapping[J]. High Voltage Engineering, 1999, 25(1): 39-41.
[13] XIE Congzhen, LIU Shan, LIU Qin, et al.Internal defects influence of 500 kV AC composite insulator on the electric field distribution along the axis[J].High Voltage Engineering, 2012, 38(4): 922-928

[14] CHEN Yao, ZOUDehua, NIU Jie, et al. The impact of Zero and low resistance insulator on potential and electric field distribution of long insulator strings in the 500 kV transmission line[J]. Insulators and Surge Arresters, 2015(3) : 29-34.

[15] WANG Liming, LIAO Yongli, GUAN Zhicheng, Analysis of Electric Field Distribution Along the Axis of Composite Insulators in Compact Transmission Line[J].High Voltage Engineering,2009, 35(8): 1862-1868.

[16] Zhang Zhimeng, Li Kang, Liu Jie, et al. Live detection technology of deteriorated insulator based on unmanned aerial vehicle inspection platform[J]. Science Technology and Engineering, 2020,20(21): 8616-8621

[17] Jia Boyan, Zhang Zhimeng, Liu Jie, et al. Electric field distribution simulation analysis of composite insulators with sheds damaged[J]. Science Technology and Engineering, 2019, 19(12) : 176-181

[18] YUAN Xiao- xian, CHEN Jun- wu, ZHOU Zhi- cheng,et al.A Finite Element Method for the Calculation of the Electric Field and Potential Distribution along the 330kV Line Composite Insulators[J]. Insulators and Surge Arresters, 2006(04):9-12.

[19] LIU Xiaowei, WANG Linong, WANG Junhua, et al. Research on automatic detection technology for faulty porcelain insulators on AC transmission lines[J]. Electrical Measurement & Instrumentation, 2016, 53(11):110-115.

[20] ZHENG Zhongsheng, DENG Wei, LI Bing, et al. Detection system of deterioration insulators based on optica l transducer[J]. Insulator sand Surge Arresters, 2014(6): 38-44.

Numerical Simulation of RC Beams Subjected to Continuous Explosive Loading

Guanjun Cao[1*], Jinhuo Yu[1]

[1] School of Highway, Chang'an University, Xi'an 710064, China;

Author's brief introduction: Guanjun Cao (1995-), Male, Master, E-mail address: 2019121017@chd.edu.cn;

*Corresponding author's E-mail address: 2019121017@chd.edu.cn; caogunjun95@163.com

Abstract. In order to study the performance of reinforced concrete (RC) beams Subjected to continuous explosive load, the numerical model of RC beam is established by LS-DYNA finite element program. And the numerical model experiences two explosive cases continuously. The reliability of the numerical model is verified by comparing with the field test results of a RC beam under the same explosive cases. It provides a reference for the future numerical simulation and experimental research of Building components under continuous explosive load.

1. Introduction

In the new century, with the rapid development of national economy, people's demand for energy is increasing. This also makes the fire and explosion of various energy transportation vehicles and gas pipelines inevitable. This kind of accidents often lead to continuous explosions, which cause great damage to bridges and other engineering structures. RC beams is a common stressed component in bridge engineering. Studying its performance under continuous explosion load is the basis of anti-explosion design of engineering structures.

Many scholars have carried out a lot of research on the anti-explosion of RC beams. Wang et al.[1]conducted contact explosion tests of RC beams with the same size under different charge quantities. The results show that under the contact explosion load, RC beams have local failure modes such as front cratering, side collapse, back collapse and section punching. The depth of blasting pit, collapse thickness, surface damage area and longitudinal reinforcement deformation in the compression area all increase linearly with the cubic root of charge. Zhang et al.[2] studied the damage state of RC beam under close range explosion through field test. The results show that the macroscopic damage and fracture of the material are similar under all test conditions; Compared with large-size beams, the local damage of small-size beams is reduced. Lin et al.[3] conducted field explosion tests and a series of numerical simulations on 9 scale beam models to study the dynamic response and local damage of the specimens under explosion. The results show that the damage of RC beam decreases gradually with the movement of explosive from the middle of the span to the end of the beam. Ganchai et al.[4] conducted one, two and three independent explosion tests on FRP Reinforced RC slab (single-sided FRP, single-layer FRP and double-layer FRP). The results show that the RC slab with FRP interlayer can withstand the secondary explosion better. Kumar and Petroski [5] established an ideal rigid plastic model to study the plastic response of beams with central cracks under transverse impact load. The results show that the initial defects have a great influence on the failure mode and failure behavior of the beam. Hu et al. [6] studied the mechanical response of RC bridge under explosion by combining simulation and test. The results show that when the car bomb explodes above the bridge deck, the car steel plate can effectively block the propagation of shock wave.

In conclusion, scholars at home and abroad have made rich achievements in the research of RC beams under one explosion, but there is almost no research on RC beams under continuous explosion. In addition, the field explosion test is not only expensive, but also difficult to collect complete test data. So it is necessary to explore the finite element method to study the performance of RC beams under continuous explosion. In this paper, the numerical model of RC beam is established by LS-DYNA. The reliability of the numerical model is verified by comparing the damage and acceleration of a field test and numerical simulation. It lays a foundation for the numerical simulation and experimental research of RC beams under continuous explosion.

2. finite element model

In this paper, the numerical model is established by the nonlinear finite element program LS-DYNA, which is based on explicit integration and has been widely used in the simulation analysis of structural anti-explosion. The numerical model adopts ALE (Arbitrary-Lagrangian-Eulerian) method. The explosive and air are simulated by Euler element, and the structure is simulated by Lagrange element. Although ALE method will spend more calculation time, it can well simulate the formation and propagation of explosion shock wave and the interaction between shock wave and structure.

The finite element model includes five parts: reinforcement, concrete, explosive, air and boundary conditions, as shown in Figure 1. Air, concrete and explosive use SOLID element and reinforcement use BEAM element. The element size of the air model is 4cm, totaling 63100 elements. The element size of RC beam model is 2cm, totaling 52000 elements. These element sizes are determined after convergence analysis. Further reduction of the mesh size will not significantly improve the accuracy of the calculation results, but will cause the number of element to multiply and require higher computer performance and computing time. For a limited air area, in order to avoid the influence of wave reflection at the boundary on the solution area, it is necessary to impose non-reflective boundary conditions on each surface to simulate infinite space. The non-reflective boundary passes through the keyword *BOUNDARY_NON_REFLECTING. The Lagrange element (RC beam) and the Euler element (air and explosive) are via *CONSTRAINED LAGRANGE_IN_SOLID is embedded together. The RC beam in the model is articulated at both ends.

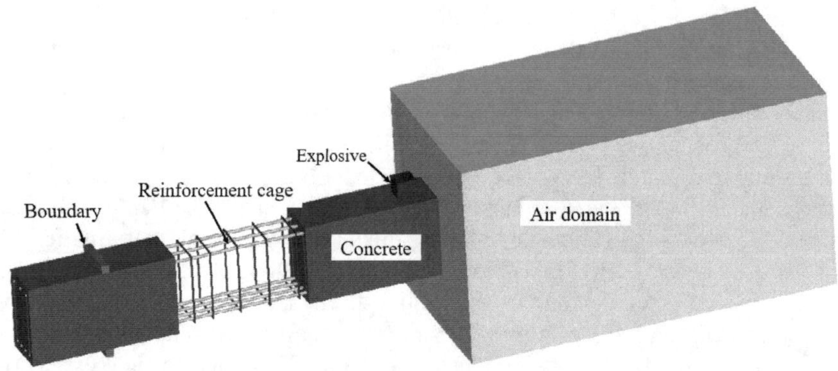

Figure 1. Numerical Model for RC Beams

3. Material Model

Select *MAT_HIGH_EXPLOSIVE_BURN and EOS_JWL simulates the detonation of explosives. The P-V relation of explosive material model and state equation is as follows.

$$p = A\left(1-\frac{\omega}{R_1 V}\right)e^{-R_1 V} + B\left(1-\frac{\omega}{R_2 V}\right)e^{-R_2 V} + \frac{\omega E}{V} \tag{1}$$

In equation: P is the detonation pressure; V is the relative volume; E is the initial unit volume energy; ω、A, B, R1 and R2 are material constants determined by experiments. The values of parameters in the model are given in Table 1.

Table 1. Material Parameters of Explosive Model

Parameter	A (GPa)	B (GPa)	R_1	R_2	ω	V	E ($kJ \cdot m^{-3}$)	ρ ($kg \cdot m^{-3}$)	V_D ($m \cdot s^{-1}$)	P_{CJ} (GPa)
Value	3.74e11	3.747	4.15	0.95	0.35	1.0	7.0e9	1.695e3[a]	8.2e3[b]	21[c]

[a] Notes are explosive density.
[b] explosive velocity.
[c] explosive pressure.

Select *MAT_NULL simulates air and its linear polynomial equation of state is:

$$p = C_0 + C_1\mu + C_2\mu^2 + C_3\mu^3 + \left(C_4 + C_5\mu + C_6\mu^2\right)E \tag{2}$$

$$\mu = \frac{\rho}{\rho_0} - 1 \tag{3}$$

In equation: E is the ratio of gas density to initial gas density.C0, C1, C2, C3 and C6 are all 0, C4 and C5 are 0.4.

Concrete model use *MAT_CONCRETE_DAMAGE_REL3, which can effectively simulate the mechanical behavior of concrete under high strain rate and large deformation, refers to Table 2 for detailed parameters.

Table 2. Material Parameters of Concrete Model

Parameter	Rho ($kg \cdot m^{-3}$)	PR	FT (MPa)	A0 (MPa)	RSIZE	UCF	LCRATE
Value	2400	0.25	2.93e6	-4.076e7	39.37[a]	1.45e-4[b]	--[c]

[a] Notes are the length unit conversion factor.
[b] Stress unit conversion factor.
[c] Relationship curve between strain rate and strength of concrete.

Steel is a strain rate sensitive material, and use *MAT_PLASTIC_KINEMATIC describes the dynamic characteristics of reinforcement, and its detailed parameters are shown in Table 3.

Table 3. Material Parameters of Reinforcement Model

Parameter	Rho ($kg \cdot m^{-3}$)	E (GPa)	NU	SIGY MPa	ETAN MPa	BETA	SRC	SRP	FS
Value	7850	200	0.3	450	2000[a]	1.0[b]	40.0[c]	5.0[d]	0.1[e]

[a] Notes are tangent modulus.
[b] BETA is the hardening parameter.
[c] SRC and [d] SRP are strain rate parameters.
[e] FS is the failure strain.

4. Erosion criteria and restart

The Lagrange element will produce large deformation under explosive load. In order to avoid excessive deformation of the element and accurately simulate crushing and flaking of concrete, erosion algorithm is used for concrete in the model. In this algorithm, there are many judgment that can be used to set material failure, such as pressure, maximum principal stress, equivalent stress, maximum principal strain and maximum shear strain. The model uses the maximum principal strain as the erosion judgment and deletes the corresponding units when the criterion is met. Based on a lot of data, the range of maximum principal strain is determined, and it is found through a lot of trial calculations that the critical value of maximum principal strain is 0.04.

This paper divides the numerical simulation of RC beam under continuous explosion into two stages according to the actual test process by utilizing the complete restart function of LS-DYNA. The first stage is 0.5kg of explosive equivalent and the second stage is 1.0kg of explosive equivalent based on the first stage. By defining the keyword *SRTESS_INITIALIZATION inherits stress and deformation. The calculation time for both stages is 10ms.

5. Results and Verification

The accuracy of the numerical model is verified by the RC beam of an explosion test. The RC beam continues to experience case1 (0.5kg explosive equivalent) and case2 (1.0 explosive equivalent). See Table 4 for detailed parameters. The comparison between the test results and the simulation results of specimen B1 under case1 is shown in Figure 2.

Table 4. Test Cases

Test case	Specimen NO.	TNT equivalent (kg)	TNT location	Blast type
Cases1	B1	0.5	Mid span of RC beam	Contact explosion
Cases2	B1	1.0	Mid span of RC beam	Contact explosion

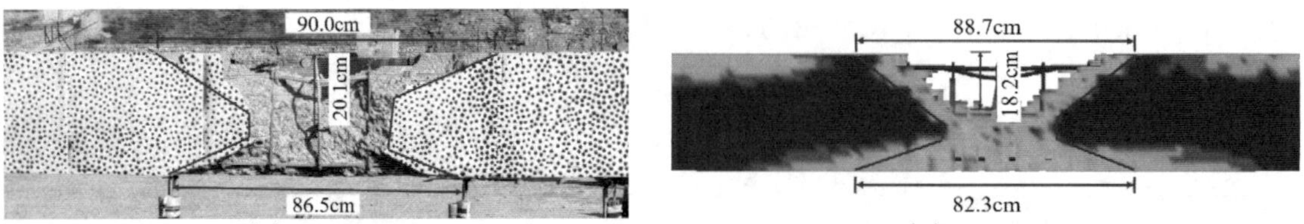

(a) Field Test (b) Simulation

Figure 2. Damage Comparison between Test and Simulation (Case 1)

The comparison of test results and simulation results of specimen B1 under case2 is shown in Figure 3.

(a) Field Test (b) Simulation

Figure 3. Damage Comparison between Test and Simulation (Case 2)

Test acceleration time history of specimen B1 at 0.5m mid-span under case1 compared with simulation results as shown in Figure 4. The simulated peak acceleration and vibration trend are in agreement with the test results. The maximum acceleration of the test and simulation is reached at 1.27ms, the maximum of the test is 51000m·s^{-2}, the maximum of the simulation is 45800 m·s^{-2}, and the error is 10.2%. The main fluctuation data are concentrated between 1.1m s and 3ms. In the field test, the acceleration sensor under case2 was completely destroyed, so its acceleration was not analyzed.

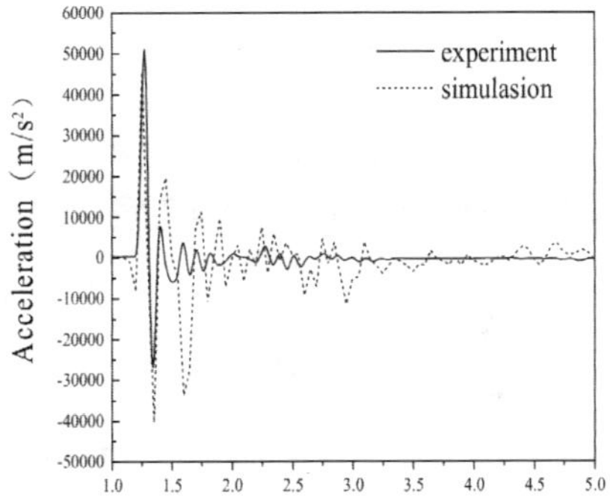

Figure 4. Acceleration Time-history Comparison between Test and Simulation (Case 1)

6. Conclusion

The numerical model of reinforced concrete beam under continuous explosive load is established by using finite element program, and the damage state and acceleration time history of numerical simulation and field test are compared. The results show that the numerical simulation results of reinforced concrete beam under explosion are in good agreement with the field test results. The damage error is within 3% and the acceleration error is within 10%. It shows that the finite element method can simulate test better. It can be used for more parameter analysis and provide guidance for the design of continuous explosion test project in the future.

Reference:

[1] Wang H M, Liu F and Yan L H. Local Damage Effect of Reinforced Concrete Beams under Contact Explosions *J. Explosion and impact*, 2020,**40** (12):37-45.

[2] Zhang D, YAO S J and Lu F. Experimental Study on Scaling of RC Beams under Close-In Blast Loading *J. Engineering Failure Analysis*,2013,**33**:497-504.

[3] Lin S C, Li D and Yang B. Experimental Study and Numerical Simulation on Damage Assessment of Reinforced Concrete Beams *J. International Journal of Impact Engineering*, 2019, **132**(OCT.):103323.1-103323.15.

[4] Ganchai T, Nicholas H and Priyan M. Behavior of FRP-RC Slabs under Multiple Independent Air Blasts *J. Journal of Performance of Constructed Facilities*, 2011, **25**(5): 433-440.

[5] Kumar S, Petroski H J. Plastic Response to Impact of a Simply Supported Beam with a Stable Crack *J. International Journal of Impact Engineering,* 1985, **3**(1): 27-40.

[6] Hu Z J, Tang X H, and Fang J Q. Analysis of Pressure Field and Response for Concrete Bridge under Close Blast Loading *J. Chinese Journal of Highway*, 2014, **27** (05):141-147+157.

Research on power system risk assessment considering large-scale wind and solar access

* **Wei Pan**

Changzhou Sanxin Power Supply Service Co., LTD.

Liyang Branch

Chang Zhou, China

panweisky@126.com

Abstract**—**The grid connection of new energy will bring some risks to the safe operation of power system. In this paper, the wind-output power model is established according to the wind-output characteristics, and the generator set, line and load probability models are established considering the uncertainty of other components of the system. From the perspective of power grid operation and new energy consumption, the traditional operation risk indexes such as voltage overrun and voltage collapse and the new wind and light abandoning indexes are selected, and the non-sequential Monte Carlo method is used to sample the operation state of the power system after wind and wind connecting to the grid, and the index is calculated. Finally, the influence of wind grid connection on system operation risk is analyzed to provide guidance for large-scale new energy grid operation risk assessment.

1. INTRODUCTION

As carbon neutral strategy is put forward, China will gradually build with new energy as the main body of the new type of power system, because the scenery and other new energy output is random, intermittent and uncertain problems, such as landscape of large-scale power grid tend to cause problems such as insufficient voltage crossing the line, electric power, power imbalances, serious impact on the smooth and safe operation of power system. Therefore, it is an urgent problem to be solved to carry out operation risk assessment research on the power system integrating wind and wind into the power grid and formulate targeted solutions according to the assessment results, so as to reduce the pressure of new energy grid connection on the system and ensure the safe and stable operation of the power grid.

At present, there is a certain research foundation for the system risk assessment after the grid connection of new energy such as scenery. From the perspective of system risk assessment after wind power integration, literature [1] designs progressive three-layer risk assessment indicators to assess system risk, and introduces value-at-risk theory calculation to represent the overall risk level of the power system in the future time interval. Literature [2] evaluated the system operation risk from the three aspects of failure state, economic loss and structural strength, and improved the Monte Carlo method to improve the accuracy of risk assessment. In reference [3], the time cycle characteristics of wind speed are taken into consideration in the system risk assessment after wind power is connected to the grid, so as to reflect the characteristics of the system risk changing with time. An example is made with the help of ieEE-RTS79 system and the actual wind speed data of a wind farm, proving the scientific nature of the algorithm. Literature [4] improved the traditional risk assessment method on the basis of fully considering the error distribution of wind power, and proposed an improved risk assessment method, so as to measure the system operation risk more accurately.

From the perspective of system risk assessment after photovoltaic grid connection, literature [5] introduces third-order Fourier function and random variables subject to three-parameter Weibull distribution to perform fitting and superposition processing on the variation trend of sunshine intensity, which more accurately reflects the time variability of photovoltaic output and system operation risk. Literature [6], on the basis of considering traditional risks, introduces ALARP criterion to grade all risk indicators, and adopts non-sequential Monte Carlo method to carry out risk assessment. When analyzing the operational risk of photovoltaic grid connection, literature [7] introduces semi-invariant series to conduct random power flow research, so as to assist in assessing the risk factors brought by photovoltaic grid connection. In the risk assessment of literature [8], the influence of the combined output of weather and scenery on the line outage rate was considered, and the weight of each indicator was calculated by combining analytic hierarchy Process (AHP) and entropy weight method to make the assessment results more comprehensive and objective.

To sum up, previous studies on the risk of grid connection of new energy sources such as wind power or photovoltaic mostly focused on a single wind power or photovoltaic, while there were few studies on new energy sources including wind power, and the selection of indicators was not fully considered. Therefore, this paper considers the factors of wind power output, system operation state, load uncertainty and so on, and establishes the operation risk assessment model of power system including wind power grid connection. Taking IEEE-RTS79 system as an example, the non-sequential Monte Carlo sampling method is used to calculate the risk index and analyze the operation risk of power system caused by wind power grid connection. It provides reference for making measures to reduce the operation risk of grid connection.

2. Operation Risk Assessment Model of Wind-grid Connected Power System

2.1 *Theory of risk assessment*

The risk research of the traditional coal-fired power system mainly focuses on the transient and static state of the generation and transmission system. With the continuous integration of wind farms and photovoltaic stations into the power grid, the idea of probability assessment needs to be introduced to comprehensively consider the uncertain factors of the new power system for risk assessment, which can be expressed as follows:

$$R\left(X_{t,s}\right)=\sum_{i}P_{r}\left(E_{i}\right)S_{sl}\left(E_{i}\right)\cdot \tag{1}$$

Where: $R\left(X_{t,s}\right)$ is the operation state of the power system at time t, E_i is the system state caused by the disturbance of the i th uncertain risk, $P_r\left(E_i\right)$ is the probability of risk occurrence under the system state, and $S_{sl}\left(E_i\right)$ is the severity of loss caused under the system state.

2.2 *Probabilistic model of output of new solar energy*

2.2.1 *Wind power output probability model*

The output power of wind turbine is related to wind speed, and the uncertainty of wind speed will affect the output of wind turbine. The output probability of wind power generation is as follows:

$$f_{WT}(P_{WT})=\frac{k}{c}\left(\frac{P_{WT}}{c}\right)^{k-1}\cdot\exp\left[-\left(\frac{P_{WT}}{c}\right)^{k}\right] \tag{2}$$

Where; P_{WT} is the active power output of wind; c is the scale coefficient; k is Weibull distribution shape coefficient, which is used to describe the shape of wind speed distribution density function. $k=1$ is the exponential distribution, $k=2$ is the Rayleigh distribution, generally 1.8 to 2.8.

2.2.2 Photovoltaic power generation output probability model

The output power of photovoltaic power station is mainly affected by solar radiation intensity, and Beta distribution is selected to describe its probability density:

$$f_{PV}(P_{PV})=\frac{\Gamma(\alpha+\beta)}{\Gamma(\alpha)\Gamma(\beta)}g\left(\frac{P_{PV}}{P_{PV,\max}}\right)^{\alpha-1}g\left(1-\frac{P_{PV}}{P_{PV,\max}}\right)^{\beta-1} \tag{3}$$

$$\alpha=\mu\left(\frac{\mu(1-\mu)}{\sigma^2}-1\right) \tag{4}$$

$$\beta=(1-\mu)\left(\frac{\mu(1-\mu)}{\sigma^2}-1\right) \tag{5}$$

Where: P_{PV} is the output power of photovoltaic panel; α 、 β is the shape parameter of Beta distribution, which is calculated by the mean and standard deviation of photovoltaic panel output. $P_{PV,\max}$ is the maximum intensity of illumination.

2.2.3 State probability model of power system components

Wind-output characteristics will cause uncertain risks in the operation of power system, but other components in the power system will also have certain risks. The components in the system can be divided into four categories: fan, photovoltaic cell, generator and circuit. In the actual operation of the system, each category of components has probability of failure and shutdown and normal operation respectively. Therefore, Markov two-state model is selected to model various components, whose states can be expressed as:

$$\begin{cases} P(E_n)=\prod\limits_{i\in C_{off}(E_n)} P_{c,i,t}(E_n)\prod\limits_{C_{off}(E_n)}(1-P_{c,i,t}(E_n)) \\ P_{c,i,t}=1-e^{-t\lambda_{c,i}} \end{cases}. \tag{6}$$

Where, $C_{off}(E_n)$ and $C_{off}(E_n)$ are the set of n components that are out of operation and running in the system respectively, and the number of elements in the set is the total number of components in the power system. $P_{c,i,t}$ is the failure rate of component i at selected time period t; $\lambda_{c,i}$ is the failure rate of element i.

Wind farms and photovoltaic power stations are often built in resource-rich regions, so it is necessary to consider the impact of bad weather on lines. When element is line $P_{c,i,t}$, it should be modified as:

$$P_{c,i,t}=1-e^{-\omega\lambda_{c,i}} \quad c_i\in\Omega. \tag{7}$$

Where, ω is the weather correction coefficient under extreme conditions. According to the results of literature [8], c_i is the ith system component, and Ω is the set of circuit components.

2.2.4 Load probability model

For the load of power system, normal distribution is adopted to describe the uncertainty of system load, and its probability density function is:

$$f_{LOAD}(P_L)=\frac{1}{\sqrt{2\pi}\sigma_L}e^{-\frac{(P_L-\mu_L)^2}{2\sigma_L^2}}. \tag{8}$$

Where: P_L is the system load of the selected area; μ_L and σ_L are mean and standard deviation of load respectively.

3. POWER SYSTEM OPERATION RISK INDEX SYSTEM

The selection of power system operation risk indicators should be systematic, representative and comprehensive, and fully measure the risks faced by the system operation.

3.1 Steady-state frequency off-limit index

The steady-state frequency characteristics of the system are considered in risk assessment, and the steady-state frequency off-limit index is introduced, and its severity function is as follows:

$$S_1 = \frac{e^{\frac{\max(|\Delta f| - \Delta f_{\max})}{\Delta f_{\max}}} - 1}{e-1}. \tag{9}$$

Where: $|\Delta f|$ is the steady-state frequency deviation of the system; $\Delta f_{\max}$ is the maximum steady-state frequency deviation acceptable to the system.

$$|\Delta f| = \frac{|\Delta P|}{K_1 + K_2}. \tag{10}$$

Where, $|\Delta P|$ is the total change of system active power; K_1 and K_2 represent the frequency adjustment coefficient of load and generator respectively.

3.2 Line active power overlimit indicator

Considering that the active power on each line exceeds the maximum active power that can be carried, the line active power overlimit index is established based on this, and its severity function is as follows:

$$S_2 = \sum_{n=1}^{N} \frac{e^{\max\left(P_n - P_n^{\max}\right)} - 1}{e-1}. \tag{11}$$

Where: N is the total number of system lines; P_n is the active power of line n; $P_n^{\max}$ is the maximum active power that line n can carry.

3.3 Voltage beyond limit indicator

Consider the voltage amplitude at the node deviating from the normal operating range to set the voltage off-limit index, and its severity function is as follows:

$$S_3 = \sum_{i=1}^{I} \frac{e^{\max\left(\frac{V_i - \overline{V}_i}{\overline{V}_i}, \frac{\underline{V}_i - V_i}{\overline{V}_i}, 0\right)} - 1}{e-1}. \tag{12}$$

Where: I is the total number of system nodes. V_i is the voltage amplitude of the node i.

3.4 Voltage breakdown index

Existing studies usually attribute the consequences of voltage collapse to the loss caused by the loss of load in the system state [9]. The severity function of indicators is as follows:

$$S_4 = kL_h. \tag{13}$$

Where: L_h is the total load under the system state h; k is the conversion coefficient, which is 0.1 in this paper.

3.5 System load loss indicators

The system loss load index is measured by the value of the system loss load, and its index severity function is as follows:

$$S_5 = \sum_{i=1}^{I} P_{l,i} / \sum_{i=1}^{I} P_{r,i} . \tag{14}$$

Where: $P_{l,i}$ is the active power loss of node i; $P_{r,i}$ is the active power load of node i.

3.6 *Abandoned new energy power index*

When the output of new energy such as scenery exceeds the actual acceptable capacity of the power grid, wind and light abandonment risk will occur, and its severity function is as follows:

$$S_6 = Q_h = \alpha P_{h1} + \beta P_{h2} . \tag{15}$$

Where: Q_h is the total electric quantity of wind and light abandoning under system state h; α and β are wind abandoning rate and light abandoning rate respectively. P_{h1} and P_{h2} are respectively the wind and light outputs in system state h.

3.7 *Comprehensive risk indicators*

Assign weights to the above 6 risk indicators to obtain comprehensive risk indicators, as follows:

$$R = \sum_{i=1}^{6} \omega_i S_i . \tag{16}$$

Where: ω_1, ω_2, ω_3, ω_4, ω_5, ω_6 are the weights of the above indicators respectively.

This paper makes comprehensive use of subjective and objective combination assignment method to determine the weight of indicators, namely:

$$\omega_i = \frac{\sqrt{\omega_{1,i}\omega_{2,i}}}{\sum_{i=1}^{6} \sqrt{\omega_{1,i}\omega_{2,i}}} . \tag{17}$$

Where: $\omega_{1,i}$ and $\omega_{2,i}$ are the weights of the ith index determined by analytic hierarchy process and entropy weight method respectively.

4. POWER SYSTEM OPERATION RISK ASSESSMENT PROCESS

4.1 *Non - sequential Monte Carlo method*

This paper defines that there are two states of each component in the power system. In order to evaluate the operation risk of the power system, sampling methods are needed to generate different system operating states. The non-sequential Monte Carlo simulation method is to sample the states. By taking a large number of samples, frequency is taken as the probability, which can be expressed by the following formula:

$$R = \sum_{i=1}^{6} \omega_i S_i . \tag{18}$$

Where:A is the sampling times; A is the total sampling times of period A.

4.2 *AC optimal power flow model*

4.2.1 *Objective function*

The operation risk assessment of new energy grid connected power system is mainly to meet the load demand of users. In this paper, the ac optimal power flow model is selected to calculate the risk index, and the minimum active power loss of the system in one day is regarded as the optimal function, as follows:

$$F = \min\left(\sum_{t=1}^{24}\sum_{i=1}^{N} P_{cut}\right). \tag{19}$$

4.2.2 The constraint

Constraint conditions mainly include system power, system loss load, upper and lower limits of unit output, voltage amplitude, etc. System power is expressed as follows:

$$\begin{gathered} U_i^t \sum_{j=1}^{n} U_j^t (G_{ij} \cos \lambda_{ij} + B_{ij} \sin \lambda_{ij}) + \\ P_{G,i}^t + P_{PV,i}^t + P_{WT,i}^t + P_{cut,i}^t - P_{d,l}^t = 0 \end{gathered}. \tag{20}$$

$$\begin{gathered} U_i^t \sum_{j=1}^{n} U_j^t (G_{ij} \sin \lambda_{ij} - B_{ij} \cos \lambda_{ij}) + \\ Q_{G,i}^t + Q_{PV,i}^t + Q_{WT,i}^t + Q_{cut,i}^t - Q_{d,i}^t = 0 \end{gathered}. \tag{21}$$

Where: $P_{G,i}^t$, $P_{PV,i}^t$ and $P_{WT,i}^t$ respectively represent the active power output of generator, photovoltaic and wind power generation of node i in time period t, and $Q_{G,i}^t$, $Q_{PV,i}^t$ and $Q_{WT,i}^t$ respectively represent the reactive power output of generator, photovoltaic and wind power generation of node i in time period t. G_{ij}, B_{ij} and λ_{ij} represent the mutual conductance, mutual susceptance and voltage phase Angle difference between nodes respectively. $P_{cut,l}^t$ and $P_{d,l}^t$ are the active power and reactive power loss loads of node i respectively. $Q_{cut,i}^t$ and $Q_{d,i}^t$ are the active and reactive load values of node i respectively.

For other constraints, refer to reference [4].

4.3 Risk assessment process

The calculation process of the risk assessment model constructed in this paper is shown in Figure 1, and the specific steps are as follows:

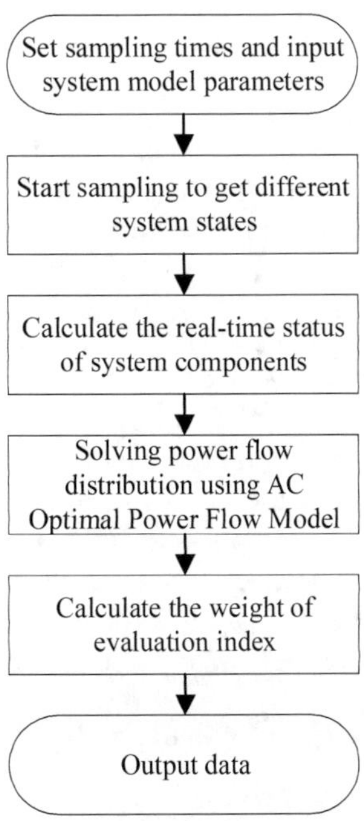

Figure 1. System risk assessment process

Step1: Set the sampling times of sampling method and input risk assessment model parameters, including wind speed, light intensity and system component parameters;

Step2: Start sampling and calculate the real-time state probability of each component in the system under different states;

Step3: Calculate the risk index based on the ac optimal power flow model;

Step4: calculate the index weight, and analyze the system operation risk according to the index value.

4.4 Component risk classification

After calculating the risk index results, it is necessary to evaluate the risk level of components. Therefore, ALARP (As Low As Reasonably Practical) criterion widely used in safety risk management is adopted to determine the risk level of power grid components. The criterion introduces unacceptable risk level line and negligible risk level line, and divides the whole component risk into unacceptable area, acceptable minimum area and negligible area.

Power enterprises can use ALARP criteria to classify the risk of power grid components according to their actual power grid line operation and demand. Using this criterion, the risk negligible level line can be defined as the product of the lowest component failure probability level and the lowest consequence level, and the unacceptable risk level line can be defined as the product of the component failure probability level and the unacceptable consequence level.

Taking the line active power as an example: the relevant risk evaluation criteria can be obtained by calculating the national overhead line data over the years; The lowest attention shall be paid to the level of component failure probability, and refer to the reliability index data issued by China Power Council. The reliability data of overhead lines over the years are shown in the table 1.

TABLE 1. RELIABILITY DATA SHEET OF OVERHEAD LINE OVER THE YEARS

Year	Forced outage rate (Times / Hundred kilometer year)	Non shutdown time (Hour / hundred kilometer year)
2011	0.233	2.75
2012	0.085	3.14
2013	0.059	1.3
2014	0.081	0.556
2015	0.094	0.976
2016	0.067	1.055
2017	0.055	0.453
2018	0.062	2.439

For the data of the above years, the data deviation before 2014 and 2018 is large. Therefore, the data from 2014 to 2017 are taken to calculate the mean value. In recent years, the average forcing rate is 0.07425 times / hundred kilometer year , and the average non-stop time is 0.76 Hour / hundred kilometer year). Therefore, the lowest failure probability level of components that should be paid attention to is obtained:

$$P_{\min} = \frac{0.07425 \times 0.76}{8760} = 6.44 \times 10^{-6} \cdot \tag{22}$$

According to the safety accident investigation regulations of State Grid Corporation of China, when the power supply load is reduced by 4% or less, it does not constitute a power grid event of level 4 or above, and more than 30% of the power supply load is a particularly serious power grid accident. Therefore, the minimum attention to the consequence level is defined as 4% of the power supply load reduction, and the unacceptable consequence level is 30% of the power supply load reduction.

Thus, it can be obtained that the negligible risk level line of component risk is:

$$R_{\min} = 6.44 \times 10^{-6} \times 4\% = 2.58 \times 10^{-7} \cdot \tag{23}$$

The unacceptable risk level of component risk is

$$R_{\max} = 6.44 \times 10^{-6} \times 30\% = 1.93 \times 10^{-6} \cdot \tag{24}$$

Thus, the classification standard of risk level of line active power out of limit is obtained, as shown in the table. The division methods of other indicators are shown in reference [10].

TABLE 2. RISK LEVEL TABLE OF LINE ACTIVE POWER EXCEEDING LIMIT

Risk level	Index risk value
Low	<2.58×10-7
medium	1.93×10-6~2.58×10-7
high	>1.93×10-6

5. EXAMPLE ANALYSIS

5.1 Example design

For the operation risk assessment model and indicators established in this paper, EIEE-RTS 79 system is selected to carry out risk calculation, and the calculation process of relevant indicators is carried out on the Matlab platform. In this paper, the line operation data of the system are referred to the literature [11]. The wind and light rejection rate is calculated based on the wind and light output data of a province in Northwest China. The output of 1MW wind and light unit is shown in the figure 3.

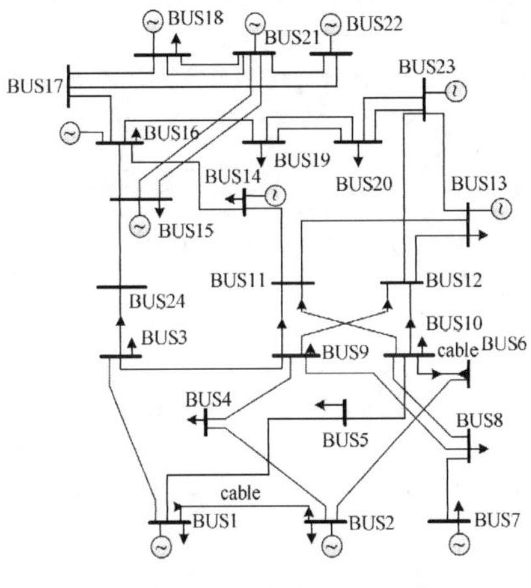

Figure 2. IEEE RTS79-node system

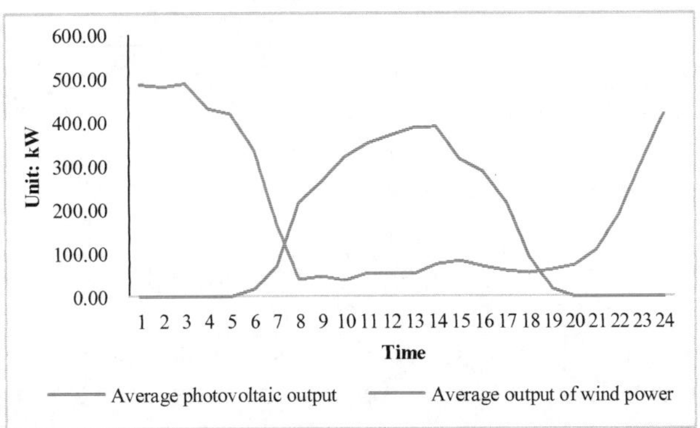

Figure 3. Output of wind turbine units in Northwest China

5.2 *Risk index calculation*

Scenario 1: 200MW wind farm and photovoltaic power station are connected at node 1 and node 2 respectively. The number of Non-Sequential Monte Carlo sampling is 10000, and the outage probability of components in the system is 5%. The calculation results of risk indicators are shown in Table 2.

TABLE 3. CALCULATION RESULTS OF RISK INDICATORS

Risk indicators	Index weight	numerical value	Evaluation level
Steady state frequency out of limit	0.21	1.57×10^{-7}	low
Line active power out of limit	0.04	8.64×10^{-7}	medium
Voltage beyond limit	0.33	9.70×10^{-5}	high
Voltage collapse	0.07	5.54×10^{-6}	high
System loss of load	0.19	2.23×10^{-8}	low
Abandoned new energy power	0.16	2.58×10^{-4}	high
Comprehensive risk assessment		7.39×10^{-5}	high

It can be seen from The results in Table 3 that the risk of Abandoned new energy power, Voltage beyond limit and Voltage collapse are all at a high level after the new energy is connected to the grid. The reason is that the uncertainty of the new energy being connected to the grid increases the operating burden of the power grid system, and the grid structure of some nodes of the system is weak, leading to a large voltage risk.

Scenario 1, the distance between the conventional unit and node 1 and node 2 is far, and the risk of system load loss is low, but the voltage stability of the system grid cannot be maintained. The grid connection distance between wind farm and photovoltaic power station is too close, which will increase the system operation risk due to the output characteristics of scenic resources. Therefore, the access system location of scenic new energy power generation can be changed to reduce the system operation risk.

5.3 *The impact of new energy access location on operational risk*

Scenario 2: Based on scenario 1, change the access position of the wind farm to 15, increase the distance from the photovoltaic power station, and keep other parameters unchanged.

Scenario 3: Based on scenario 2, change the access location of photovoltaic power station to 13, shorten the distance from generator node no. 23, and keep other parameters unchanged.

Scenario 4: Based on scenario 3, change the access position of the wind farm to 18, increase the distance from the photovoltaic power station, and keep other parameters unchanged

The calculated results of risk indicators in each scenario are shown in Table 4.

TABLE 4. CALCULATION RESULTS OF RISK INDICATORS

Risk indicators	Scenario 1	Scenario 2	Scenario 3	Scenario 4
Steady state frequency out of limit	1.46×10^{-6}	1.40×10^{-6}	1.43×10^{-6}	1.37×10^{-6}
Line active power out of limit	8.01×10^{-6}	7.59×10^{-5}	8.11×10^{-5}	6.13×10^{-5}
Voltage beyond limit	9.00×10^{-4}	6.15×10^{-4}	5.67×10^{-4}	5.00×10^{-4}
Voltage collapse	5.14×10^{-5}	5.34×10^{-5}	5.14×10^{-5}	5.14×10^{-5}
System loss of load	2.07×10^{-8}	1.64×10^{-4}	8.07×10^{-5}	1.46×10^{-5}
Abandoned new energy power	2.40×10^{-3}	2.40×10^{-3}	2.40×10^{-3}	2.40×10^{-3}

The main difference between the four scenarios lies in the location of the new energy connection system, including the distance between each other and the distance between the new energy power station and the traditional generator set. By analyzing the change of risk indicators in each scenario in Table 4, the impact of the change of new energy access location on different indicators can be obtained.

In the four scenarios, the two indexes of Abandoned new energy power and Voltage collapse did not change, which may be related to the consumption of new energy and wind-connected grid capacity. The change of access location of wind-connected power station has little influence on the consumption of new energy, so the influence of change of access capacity of wind-connected power station on the risk of system indicators can be further explored.

Steady-state frequency off-limit fluctuation is small, and the change of scenery grid-connected position will increase or decrease the distance from the conventional generator set, resulting in the change of active power output of the generator set and the change of power system frequency. Since voltage mainly depends on the capacity of the system to bear the load, the voltage breakdown index varies little.

The wind access position will have great influence on voltage overlimit, line active power and system load loss. From scenario 1 to scenario 4, the distance between wind farm and photovoltaic power station gradually increases, and the distance between wind farm and conventional unit gradually approaches. Increasing the distance of the scenic power station can avoid the voltage risk brought by the uncertain superposition of the scenery. Meanwhile, the conventional units can also exert the maintenance voltage to reduce the interference of new energy to the system stability. This means that in new energy planning, increasing the distance between solar power stations or close to conventional units can reduce the system voltage risk. In addition, the location change of the new energy access system will change the active power distribution of the original line and affect the off-limit index, and the location selection of the new energy connection in the weak part of the system grid can reduce the load loss event caused by the line.

6. CONCLUSION

This paper studies the operation risk of new energy grid-connected power system. Considering the uncertainty of wind power output fluctuation characteristics, system component state and load, the risk assessment model and index system of power system including wind renewable energy are established, and the following conclusions are drawn.

(1) The access of solar new energy to the power system will lead to the risk of abandoning wind and light, which will have a certain impact on the voltage of the power system. If there are more conventional generating units in the system, the active power fluctuation risk caused by the grid-connected new energy can be reduced. Therefore, large-scale grid connection of new energy requires sufficient unit resources to maintain system voltage and reduce the operation risk of power system.

(2) The grid connection location of solar new energy will affect grid security. The grid connection distance between wind farm and photovoltaic power station should not be too close, and the distance with conventional generator set should not be too far. Therefore, large-scale grid integration of solar new energy requires reasonable planning of grid connection nodes to avoid the superposition of solar uncertainty, which will lead to strong volatility of new energy generation and affect the safe operation of the power system.

REFERENCES

[1] J. MA Yanfeng, YANG Xiaokuan, WANG Zijian et al, Operation Risk Assessment for Power System With Large-scale Wind Power Integration Based on Value at Risk, Power System Technology, vol. 45(03), 2021,pp.849-855.
[2] J. MA Guang, ZHANG Yining, CHEN Zhe, et al, Risk Assessment Method for Hybrid AC/DC System With Large-scale Wind Power Integration, Power System Technology, vol. 43(09), 2019, pp.3241-3252.
[3] J. FU Bingbin, WAN Xiaohua, XIONG Xiaofu, et al. Risk assessment of wind power integrated system based on time-periodic characteristics of wind speed, Power System Protection and Contro, vol. 46(09), 2018, pp.43-50.
[4] J. LI Jinghua, ZUO Junjun Analysis and Assessment of Operation Risk for Power System , Power System Technology, vol. 40(11), 2016, pp.3503-3513.
[5] J. WEI Yong, LI Haoran, FAN Xuefeng, et al. Risk Assessment Method of PV Integrated Power System Considering Time Periodic Characteristics of Solar Irradiance, vol. 42(8), 2018, pp.2562-2569.

[6] J. SHANG Haoyu, LIU Tianqi, BU Tao, et al. Day-ahead operation risk assessment for power system with high penetration of renewable energy based on ALARP criterion, Electric Power Automation Equipment, vol. 41(03), 2021, pp.196-203.
[7] J. ZHANG Wenchao, LIU Liqun, YANG Zhijun, et al. Solar Energy Power Generation Grid Operation Risk Assessment, Journal of Taiyuan University of Science and Technology, vol. 41(05), 2020, pp.344-351.
[8] J. LI Can, ZENG Yuan, QIN Chao, et al. Optimization Method for Power System Dispatching Considering Operation Risk, Proceedings of the CSU-EPSA, vol. 28(06), 2016, pp.73-79.
[9] J. CHEN Wei-hua, JIANG Quan-yuan, CAO Yi-jia, et al. Risk Assessment of Voltage Collapse in Power System, Power System Technology, vol. 29(19), 2005, pp.36-41.
[10] J. SHANG Haoyu, LIU Tianqi, BU Tao, et al. Day-ahead operation risk assessment for power system with high penetration of renewable energy based on ALARP criterion, Electric Power Automation Equipment, vol. 41(03), 2021,pp.196-203.
[11] J. IEEE Subcommittee. IEEE reliability test system. IEEE Transactions on Power Apparatus and Systems, PSA-98 (6),1979, pp. 2047-2054.

Finite Element Analysis and Optimization of the Planar Steel Tubular K-Joints

Ying Chen, Qingtao Yang, Dongchen Wang

State Nuclear Electric Power Planning Design & Research Institute Co., Ltd., 100095, Beijing, China

email：773603548@qq.com

Abstract. Planar steel tubular K-joints feature various geometric forms and intricate load states, whose bearing capacity is influenced by a number of factors. To address the problems above, a research on the bearing capacity of common planar K-joints in practical engineering projects was conducted. Based on numerical simulation, nonlinear finite element analysis on four kinds of planar K-joints was carried out in ANSYS. The bearing capacity and failure modes of those K-joints were figured out. The bearing capacity of each combination was compared. The results provided by finite element analysis was compared with those derived from codes. The design of K-joints was optimized, thereby improving the bearing capacity and satisfy requirements of engineering projects.

1. Introduction

Given the outstanding mechanical properties of steel tubular members, steel tube towers have been widely used in large span transmission lines and high voltage substation gantries in recent years. K-joints play an integral part in trussed steel structures such as transmission line towers, whose mechanical properties and failure modes directly influence the bearing capacity and safety of the whole structure [1].

The research on planar K-joints dates back to a long time ago. In 1950s, the United States commenced the research on steel tubular joints. In 1970, Eastwood and Wood, two scholars from the United Kingdom, recorded the experimental and theoretical research results of welded joints between rectangular and circular tubes at Sheffield University. The design methods they put forward afterwards were adopted by the design manual for steel tubular structural connections which was published as the first one in the world in 1971. In 1981, International Institute of Welding (IIW) published a new set of failure modes of N, T, Y and X shaped tubular joints, including a great number of theoretical and empirical corrections [2][3].

Chinese standard for design of steel structures (GB 50017-2017) presents calculation methods of K-joints [4]. In this paper, four kinds of K-joints are analyzed respectively based on codes and ANSYS, whose stress characteristics are studied. Their stress states and ultimate bearing capacity are compared in order to identify the joint form with the largest bearing capacity and recommend the optimum joint design scheme for real engineering projects. In the meantime, a parametric analysis of the K-joint with the largest bearing capacity is carried out. The effect of each parameter on bearing capacity is investigated. Eventually, the design scheme for the joint is optimized for the purpose of improving its bearing capacity.

2. Analysis model

2.1. Geometric model of K-joint

This paper focuses on the research of planar K-joints, whose geometric model is schematically represented in Figure 1. The typical combination forms of K-joints mainly include:

i) A circular steel pipe serves as the chord member, which is connected with two circular web members

ii) A circular steel pipe serves as the chord member, which is connected with two square web members

iii) A square steel pipe serves as the chord member, which is connected with two circular web members

iv) A square steel pipe serves as the chord member, which is connected with two square web members

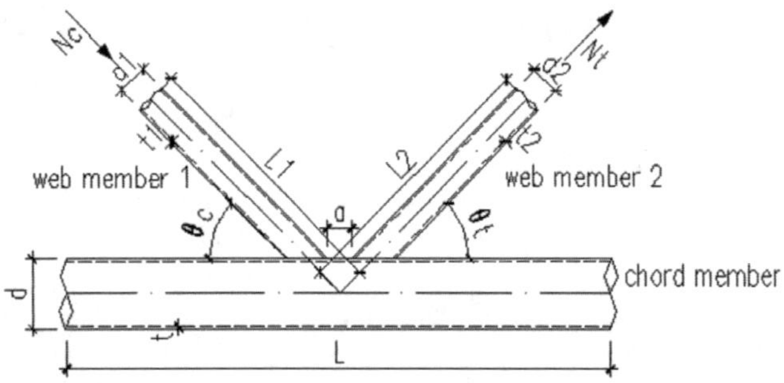

Figure 1 Typical geometric model of K-joints

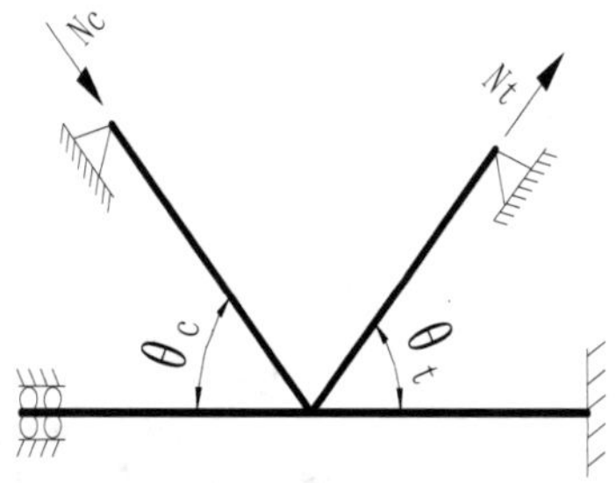

Figure 2 Loading condition of K-joints

The cross-sectional areas of cylindrical members and square members are designed to be equal in order to facilitate the analyses and comparison of the bearing capacity of various K-joints. The length of the chord member (represented by L in Figure 1) is 1200mm. The length of the web members (represented by L1 and L2 in Figure 1) is 500mm. θc equals 45 degrees. θt equals 50 degrees. The thickness of the chord member (represented by t in Figure 1) is 5mm. The thickness of the web members (represented by t1 and t2 in Figure 1) is 4mm. The diameter of the chord member is represented by d in Figure 1. The diameters of the web members are represented by d1 and d2 respectively in Figure 1.

2.2. Boundary conditions and loading modes

Taking into account the requirements of real engineering projects and the stress characteristics of K-joints and referring to extensive domestic data, the model is established combined with the functions of finite element analysis software. The boundary conditions are as follows. The right end of the chord member is fixed. The left end is connected to a sliding support. The ends of two web members are hinged and they are allowed to move along the axes respectively [5].

Two web members are subject to tension and compression respectively. Considering the formulas in design codes, Nc and Nt are applied according to the ratio of sinθc to sinθt until the structure is destructed. The loading condition is shown in Figure 2.

2.3. Element shapes and properties

In the finite element simulation, the elastic-plastic shell element with three dimensions and four nodes is selected in ANSYS. Each element has four nodes and six degrees of freedom, which include displacement in three directions and rotation in three directions. The nonlinear function of Shell 181 shell element are applied, such as large deformation and stress stiffening [6].

2.4. Material properties

Q355 steel with the yield strength f_y of 345mpa, elastic modulus E of 2.06×10^5MPa and Poisson's ratio ν of 0.3 is applied in this model. The material is idealized to be elastic-plastic and comply with Von-

Mises yield criterion. The material is considered as bilinear isotropic strengthening. The Newton Raphson iterative method and the arc length method are combined to establish a nonlinear equilibrium solution method. The influence of initial stress, deformation, residual stress and weld on the ultimate bearing capacity of the joint is not taken into consideration [7].

2.5. Meshing

In the finite element analysis of steel tubular joints, time and accuracy of ANSYS nonlinear calculation depend on the degree of mesh refinement. By controlling the side length of the element, the element size of the chord member close to the intersection area of the chord member and the web member basically equals the thickness of the chord member. The element size of the chord member and web members far away from the intersection area is twice the thickness of the members respectively. Furthermore, the intersection lines of the chord members and web members are refined locally to enhance the density of the mesh [8].

3. Comparison and analyses of joint stress

3.1. Process of joint stress

In the process of loading along the axial direction of the web members, the local bulge occurs to the end of the compressive web member close to the chord member and necking deformation takes place at the end of the tensile web member. The part of the chord member connected with the compressive web member is concave inward and the part connected with the end of the tensile web member is convex outward.

When loads are imposed through the web members, the stress distribution along the axial and circumferential direction of the chord member is apparently uneven due to the fact that the intersection lines of the joint are complex and the radial stiffness of the chord member is significantly different from its axial stiffness. Moreover, local deformation and stress concentration emerges at the intersection lines, while the stress decreases rapidly away from the intersection lines. The circumferential stress at the saddle points or crown points of the joint is the largest, where yield takes place first. However, it does not mean that the joint is destructed immediately at this moment. With the increase of the loads, the plastic zone gradually forms, which results in stress redistribution. Afterwards, the plastic zone constantly expands. The joint eventually fails when significant local plastic deformation occurs [9].

According to the stress cloud chart, when the loads are low, the intersection area between the chord member and the web member starts to yield. With the load increasing, the plastic zone extends outward along the cross section of the chord member. When the joint fails, most of the section of the chord member within the intersection area is in the state of plasticity and the plastic zone has extended to the cross section far away from the intersection area. The final stress result is shown in Figure 3.

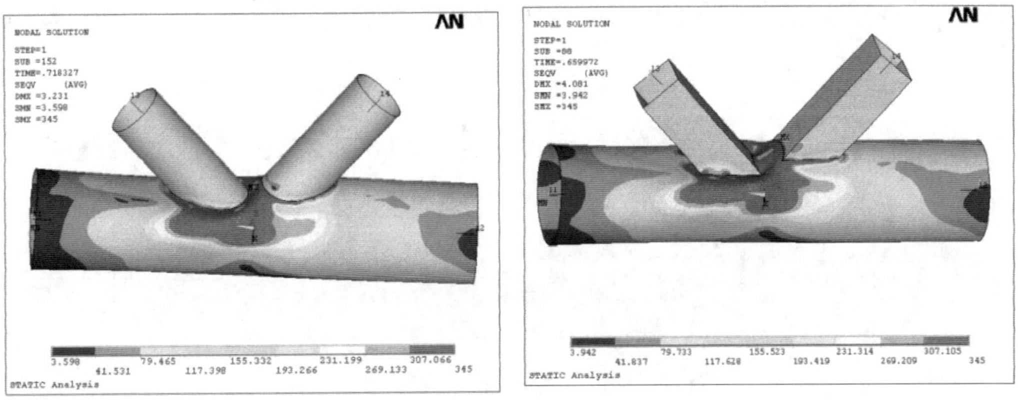

Figure 3 (a) Joint Type A

Figure 3 (b) Joint Type B

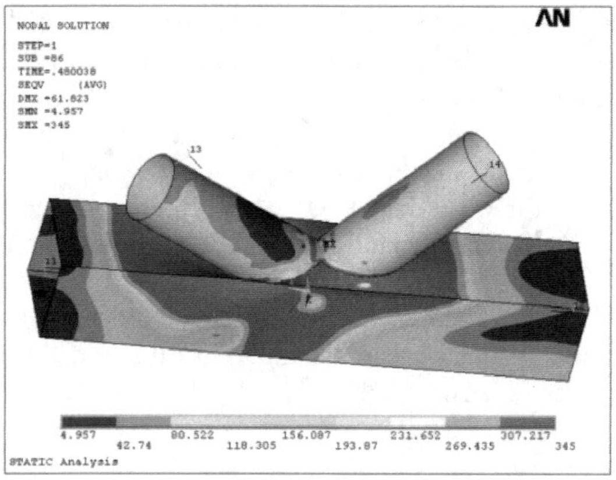

Figure 3 (c) Joint Type C

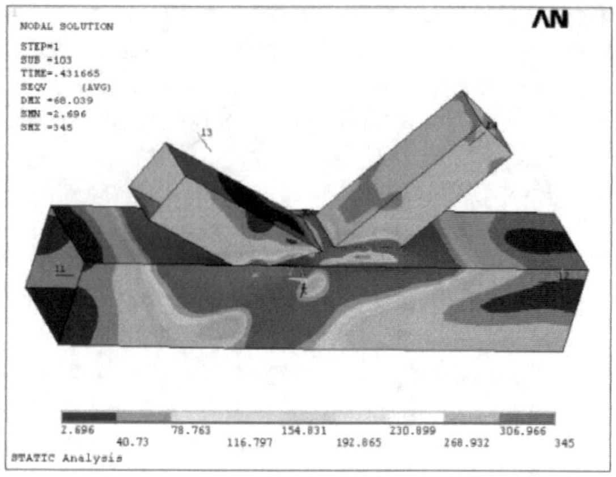

Figure 3 (d) Joint Type D

3.2. Judgment of ultimate bearing capacity

For the joints where members are mainly subjected to axial loads, their ultimate bearing capacity is identified on the basis of three criteria as listed below.

i) the ultimate strength criterion: Load vs. deformation curve is plotted based on the axial force of the web member and the deformation of the chord member pipe wall along the axis of the web member. The crest of the curve is taken as the ultimate bearing capacity.

ii) the ultimate deformation criterion: The deformation of the chord member pipe wall along the chord member direction reaches a certain limit value. In the world, the ultimate bearing capacity is commonly identified on the basis of the load which makes the deformation of the chord member pipe wall in the direction of the web member reach 3% d.

iii) The ultimate bearing capacity is identified on the basis of the load which makes the strain on the surface of tensile web member or in the intersection area of members reaches 20% ε(ε represents the ultimate tensile strain of the material).

In this paper, the ultimate bearing capacity is identified in accordance with the ultimate strength criterion. The crest of the load vs. displacement curve is taken as the ultimate bearing capacity as shown in Figure 4. The results of different joint types are shown in Table 1.

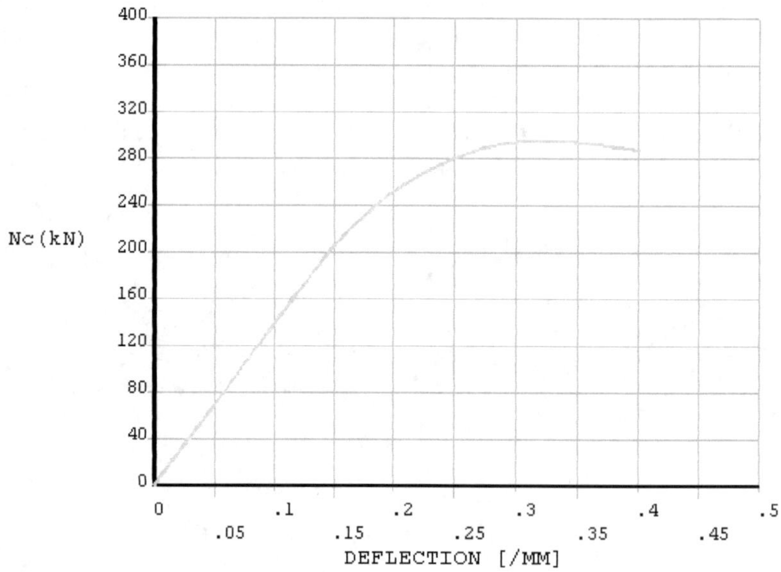

Figure 4 Load vs. Deflection

Table 1 Bearing capacity for different joints

Groups	Joint types	d/ mm	d_1=d_2/ mm	Bearing capacity/kN		N_{C1}/ N_{C2}
				Finite element N_{C1}	National codes N_{C2}	
1	A	273	159	296.1	242	1.22
	B	273	126	271.4	/	/
	C	215	159	192.0	237	0.81
	D	215	126	174.1	239	0.73
2	A	273	180	418.5	310	1.35
	B	273	142	327.0	/	/
	C	215	180	257.1	268	0.96
	D	215	142	216.3	270	0.80
3	A	300	159	286.7	219	1.31
	B	300	126	264.1	/	/
	C	237	159	187.2	226	0.83
	D	237	126	184.7	228	0.81

Note：There is no calculation formula in national codes for the K-joint composed of square chord member and circular web members.

Bearing capacity of three groups of joints are listed in Table 1. On the basis of the premise that the square members have the same cross-sectional area with the circular counterparts, the ultimate bearing capacity of Joint Type A, which is composed of a circular chord member and two circular web members, is the highest and that of Joint Type D, which is composed of a square chord member and two square web members, is the lowest. The difference between these two types of joints is approximately 36% to 45%. The finite element calculation result of Joint Type A is 1.22 to 1.35 times that derived from national codes. The finite element calculation result of Joint Type C is 0.81 to 0.96 times that derived from national codes. The finite element calculation result of Joint Type D is 0.73 to 0.81 times that derived from national codes. It is shown that the result derived from codes is conservative for Joint Type A. On the other hand, the results based on codes are unsafe for Joint Types C and D, which should be adjusted properly.

4. Parametric analyses for K-joints

Parametric analysis is carried out for the joint in between a circular steel chord member and two circular steel web members. Combined with the formulas in codes, the effect of each parameter on bearing capacity is studied.

4.1. The effect of the ratio between the diameters of the web members and the chord member β_1

As shown in Figure 5, with the increase of the ratio β_1 =d1 / d (the outer diameter of the chord member d remains unchanged and that of the web member d1 increases), the extent of stress concentration in the intersection area of the joint dwindles, while the ultimate bearing capacity of the joint rises. The tendencies indicated by the finite element analysis and code calculation are consistent.

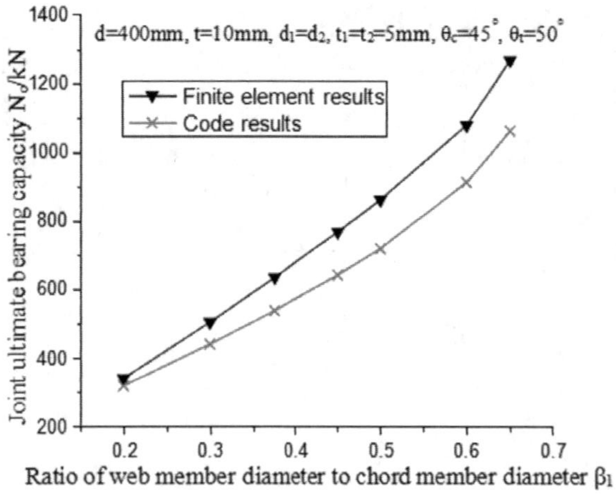

Figure 5 N_c vs. β_1

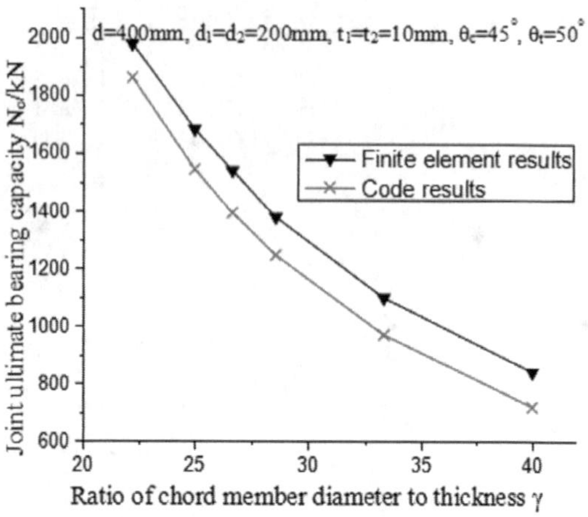

Figure 6 N_c vs. γ

4.2. The effect of the ratio of diameter to thickness of the chord member γ

As shown in Figure 6, with the ratio of diameter to thickness of the chord member γ=d/t increasing (the outer diameter of the chord member d remains unchanged, but the thickness t decreases), the bending rigidity of the chord member decreases and the bearing capacity of the joint diminishes. The finite element analysis suggests the same relationship between N_c and γ with the calculation based on codes.

4.3. The effect of the thickness of the web member t_1

As shown in Figure 7, with the thickness t_1 of the web member increasing, the finite element analysis suggests that the bearing capacity of the joint changes slightly, while code calculation indicates the bearing capacity remains unchanged. The conclusions drawn from two methods are similar.

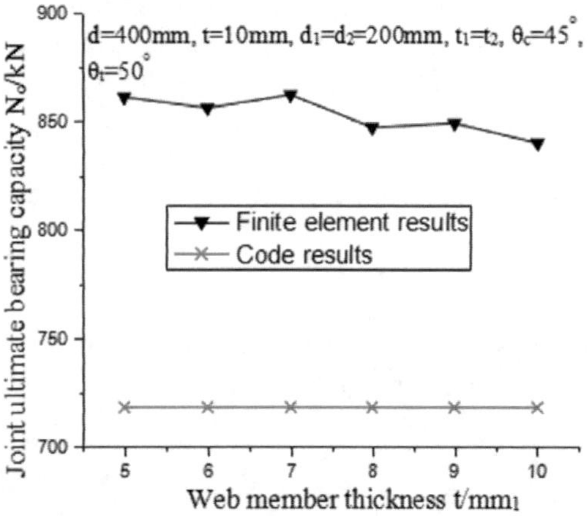

Figure 7 N_c vs. t_1

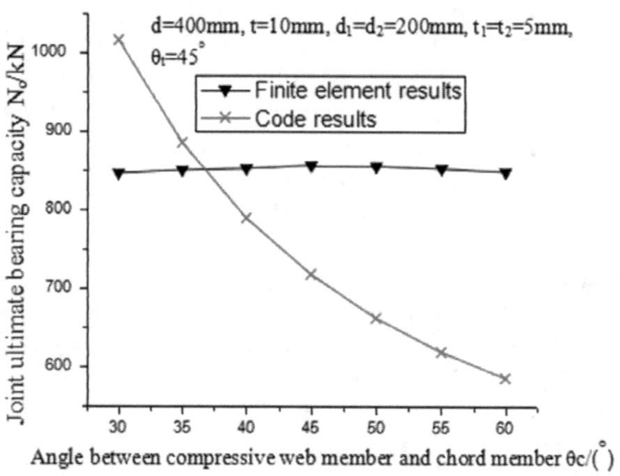

Figure 8 N_c vs. θ_c

4.4. The effect of angle between the web member and the chord member θ_c

As shown in Figure 8, with the increase of θ_c, the finite element analysis suggests that the bearing capacity of the joint rises first and falls thereafter and the change is moderate. However, the results derived from codes denotes that the bearing capacity plunges with the increase of θ_c.

4.5. The effect of the axial compressive stress of the chord member

As shown in Figure 9, finite element analysis and results from codes both suggest that the bearing capacity of the joint diminishes with the axial compressive stress of the chord member σ increasing.

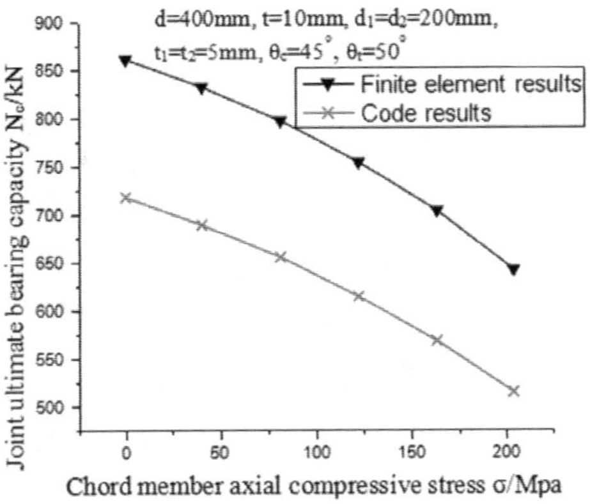

Figure 9 N_c vs. σ

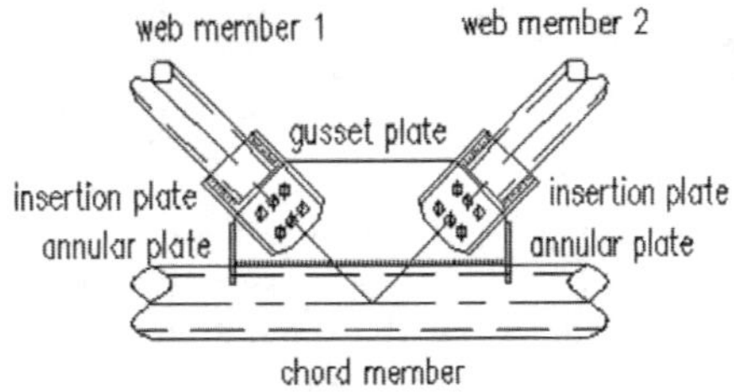

Figure 10 Optimized K-joint

5. Optimization for steel K-joints

The optimization analysis is carried out for the joint whose chord member and web members are all circular steel pipes. The optimized K-joint is shown in Figure 10. A gusset plate, two insertion plates and two annular plates are incorporated into the joint. The gusset plate and annular plates are welded to the chord member. The insertion plates are welded to web members. The gusset plate and insertion plates are connected with bolts. The chord member diameter d equals 273mm and the web member diameters $d_1=d_2=159$mm, which is consistent with the geometric parameters of Joint Type A in Table 1.

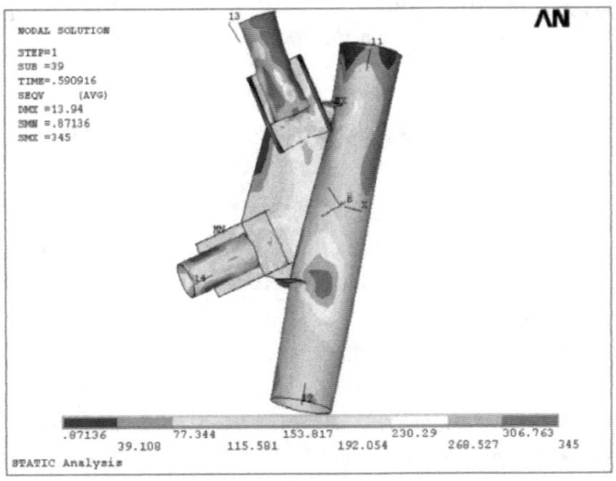

Figure 11 Stress of optimized K-joint

The finite element analysis is conducted for the optimized K-joint. Loads are applied in the same way with Joint Type A. As a result, the stress is shown in Figure 11. With the increase of loading, the web member of the optimized joint reaches its ultimate bearing capacity first, which means that the load cannot be increased any more. The load at this moment is deemed to be the ultimate bearing capacity of the joint, which equals 594.5kN on the basis of computation. The ultimate bearing capacity of the optimized K-joint is twice that of the K-joint of the same specification, which is more competent in the engineering projects with a higher requirement of bearing capacity.

6. Conclusions

On the basis of ANSYS non-linear finite element analysis, this paper brings forward the optimization of planar K-joints. The conclusions drawn from the research are as below.

(1) The majority of the intersection area between the web members and chord member yields when the planar K-joints fail. The failure mode of the joints is primarily plastic failure of the chord member pipe wall.

(2) The joint comprised of a circular chord member and two circular web members boasts the highest bearing capacity. On the other hand, the lowest bearing capacity is present at the joint composed of a square chord member and two square web members. The difference approximates 40%. The finite element analysis result is 1.22 to 1.35 times that derived from codes for the joint whose chord member and web members are all circular. The finite element analysis result is 0.81 to 0.96 times that derived from codes for the joint which is composed of a square chord member and two circular web members. The finite element analysis result is 0.73 to 0.81 times that derived from codes for the joint in between a square chord member and two square web members. It is shown that the bearing capacity derived from codes is conservative for the joint whose chord member and web members are all circular. Nevertheless, the values of bearing capacity based on codes are unsafe for the joints whose chord member is square and web members are square or circular. Correspondingly, adjustments should be made properly.

(3) The ultimate bearing capacity of K-joints primarily depends on the outer diameter ratio of the web member to the chord member β_1, the ratio of the chord member diameter to its thickness γ, the axial compressive stress σ of the chord member. Namely, the factors above play an important part in the destruction of K-joints. On the other hand, the thickness of the web member t_1 and the angle between the web member and the chord member θ_c exert a slight influence on the bearing capacity of the joints. The results derived from codes and the finite element analysis differ with respect to the influence of θ_c on the bearing capacity of K-joints. Therefore, it is necessary to do further research on the relationship between bearing capacity and θ_c and correct the calculation equations provided in codes.

(4) The bearing capacity of the K-joint optimized with a gusset plate and insertion plates is twice that of the K-joint of the same specification. Hence, the optimized K-joint is more adequate to the projects with a higher requirement of bearing capacity.

References

[1] Li Maohua, Xing Haijun, Hu Xiaoguan, *et al.* 2013 *Research on load-carrying capacity of smaller angle k-joints for power transmission steel tubular tower* (Building Structure issue 5 2013) pp 48-53.

[2] Hu Xiaoguang, Yang Jingbo, Li Maohua, *et al.* 2011 *Rib Stiffener Distribution and Ultimate Strength of Steel Tubular Tower's K-joint* (Science & Technology Review vol 29 No. 26) pp 50-56.

[3] Zheng Boxing and Huang Changhua 2008 *Finite Element and Design Method Analysis of Ultimate Strength of Tubular Kt-Joints* (Steel Construction vol 23 No. 7) pp 42-47.

[4] Ministry of Housing and Urban-Rural Development of the People's Republic of China 2017 *Standard for design of steel structures.*

[5] Wan Haiying and Xie Yufang 2006 *Finite element analysis of the ultimate capacity of overlap tubular K-joints* (Journal of Hefei University of Technology: Natural Science vol 29 No. 4) pp 478-481.

[6] Chen Yu and Tang Jumei 2011 *Experimental research and finite element analysis of bearing capacity of K-joints with circular chord and square braces* (Journal of Building Structures vol 32 No. 10) pp 56-64.

[7] Zhang Hong, Pan Ying, Ma Chungang, *et al.* 2013 *Finite Element Analysis of Bearing Capacity of Planar K-Joints* (China Science and Technology Information No. 22) pp 57-59.

[8] You Jun, Li Zhengliang, Bai Qiang, *et al.* 2010 *A Study of Influence of Tubular Axial Force on Ultimate Strength of Steel Tubular K-joints* (J Chongqing Technol Business Univ. Nat Sci Ed vol 27 No. 3) pp 292-297.

[9] Qin Linxiao, Pan Ying and Zhang Pengju 2014 *Bearing Capacity Research of Multi-Planar KKT-Joints Based on ANSYS* (Light Industry Machinery vol 32 No. 4) pp 13-17.

[10] Chen Xin 2014 *Study of Ultimate Strength Experimental and Calculation Method of Steel Tubular K-joints Considering Negative Eccentric* (Beijing Jiaotong University).

2215 (2022) 012026 doi:10.1088/1742-6596/2215/1/012026

Transverse and longitudinal coupled nonlinear vibration of an axially moving Euler beam under variable tension

JunHao Wu[1] , YouQi Tang[2], Ling Chen[3]

1. Mechanical Engineering Institute,Shanghai Institute of Technology,SN: 201400,China

2. Mechanical Engineering Institute,Shanghai Institute of Technology,SN: 201400,China

3. Mechanical Engineering Institute,Shanghai Institute of Technology,SN: 201400,China

yqtang@126.com

Abstract. When describing the vibration characteristics of the transverse parameters of an axially moving beam, the effects of nonlinear factors need to be introduced. Under normal circumstances, it is difficult to solve the nonlinear vibration equations of horizontal and vertical coupling with mixed partial derivative terms of time and space using approximate analytical methods. The longitudinal displacement in the vibration process is very small and much smaller than the lateral displacement, so the equation is simplified to a nonlinear vibration governing equation only about the lateral displacement. However, in these studies, no specific numerical basis for the reason for the simplification is given. Therefore, this paper considers the coupling model of the Euler beam's lateral and longitudinal vibrations, chooses the Galerkin truncation method to decouple and numerically solve it, and obtains the numerical solution of the lateral and longitudinal vibration displacements. Based on the calculation results of the direct multi-scale method and the differential quadrature method of the simplified model, and the solution results of the eight-order Galerkin truncation method of the coupled model, the steady-state response amplitudes of the coupled model and the simplified model are compared to verify the accuracy of the simplified model Therefore, it provides a strong theoretical basis for the simplified model that ignores the longitudinal vibration.

1. Introduction

In the current literature on the continuum of axially variable-speed motion, most of them assume that the axial tension of the continuum is uniform. However, according to Newton's second law, the assumption of uniformity of axial tension is obviously inaccurate because of the acceleration of the axially moving body. Chen and Tang studied the parametric resonance of axially moving structures with longitudinally varying tension and velocity pulsation[1], and soon applied the model to the parametric vibration of Euler beams[2]. Ding et al. studied the parametric resonance and internal resonance of the axially accelerated viscoelastic Euler beam[3], and revealed the bifurcation and chaos of the supercritical axially accelerated viscoelastic beam[4]. Tang et al. analyzed the parametric resonance of an axially accelerated beam with longitudinally varying tension[5], and the steady-state response of an axially accelerated viscoelastic beam under time-varying tension[6]. Tang and Ma analyzed for the first time the

3:1 internal resonance and instability boundary of an axially moving beam under the correlation between perturbation speed and perturbation tension[7,8].

2. The transverse and longitudinal coupling mathematical model of the axially moving Euler beam under variable tension

Considering that the density is ρ, the cross-sectional area is A, the modulus of elasticity is E, the coefficient of viscoelasticity is α, the length between the ends of the support is L, the initial axial tension is P_0 (that is, the tension when the beam is at rest), The time-varying speed $\Gamma(t)$ moves along the axial direction. Applying Euler-Bernoulli beam theory, the coupled plane vibration of the beam consists of longitudinal displacement $u(x,t)$ and lateral displacement $v(x,t)$. **Figure 1** shows the physical model of a viscoelastic Euler beam with axially variable motion.

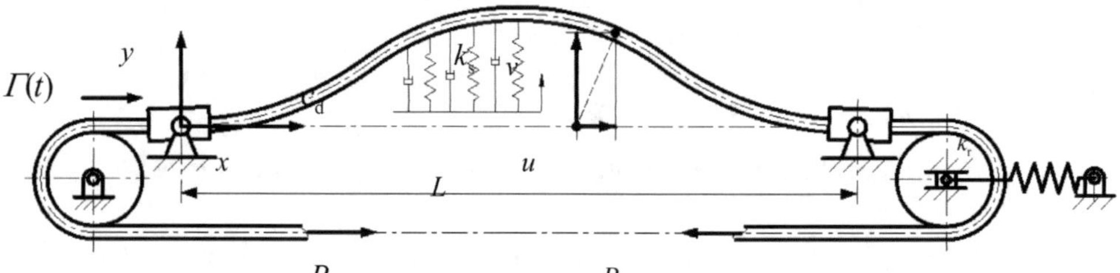

Fig 1 The physical model of an axially accelerating viscoelastic Euler-Bernoulli beam

The governing equation in dimensionless form is

$$u,_{tt} + 2\gamma u,_{xt} + c_{\mathrm{d}}\left(u,_{t} + \gamma u,_{x}\right) + \left(\gamma^2 - k_{\mathrm{N}}^2\right)u,_{xx} + \dot{\gamma}u,_{x} - k_{\mathrm{N}}^2 v,_{x} v,_{xx} = 0 \tag{1}$$

$$\begin{aligned} & v,_{tt} + 2\gamma v,_{xt} + \left[-1 + \left(1-\eta\right)\gamma^2 - \left(x-1\right)\dot{\gamma}\right]v,_{xx} + k_{\mathrm{f}}^2 v,_{xxxx} + k_{\mathrm{s}} v + c_{\mathrm{d}}\left(v,_{t} + \gamma v,_{x}\right) \\ & + \alpha\left(v,_{xxxxt} + \gamma v,_{xxxxx}\right) - k_{\mathrm{N}}^2 u,_{xx} v,_{x} - k_{\mathrm{N}}^2 u,_{x} v,_{xx} - \frac{3}{2}k_{\mathrm{N}}^2 v,_{x}^2 v,_{xx} = 0 \end{aligned} \tag{2}$$

In the above formula, k_f represents the stiffness coefficient of the beam, and k_N represents the nonlinear coefficient of the beam. Since the simply supported boundary is often used as the boundary condition in engineering applications, the simply supported boundary of the smooth sleeve is selected, and the boundary condition form is

$$v\left(0,t\right) = v\left(1,t\right) = v,_{xx}\left(0,t\right) = v,_{xx}\left(1,t\right) = 0 \tag{3}$$

3. Galerkin truncation

For the coupled governing equations (1) and (2) of viscoelastic Euler beams in axial motion under the action of axial variable tension, it is generally difficult to solve by approximate analytical method, so the Galerkin truncation method will be used to separate the variables, combined with the Runge-Kutta algorithm Solve. In the numerical method of solving nonlinear structures, Galerkin truncation method is often used. The principle is based on the variational thought of equations corresponding to fictions, discretizing the separation variables of partial differential equations into ordinary differential equations for numerical solution. Select a finite number of trial functions for superposition, and make the weighted integral of the superposition result in the solution domain and on the boundary satisfy the original equation. The weight function is selected as the trial function itself, which can be simplified into a set of ordinary differential equations that satisfy the boundary conditions and are easy to numerically solve.

Assume that the longitudinal and transverse vibration displacement variables of the beam's motion equation meet the following equations:

$$u\left(x,t\right) = \sum_{m=1}^{M} q_m\left(t\right)\phi_m\left(x\right) \tag{4}$$

$$v(x,t)=\sum_{m=1}^{N}p_m(t)\phi_m(x) \tag{5}$$

Where $\phi_m(x)$ represents the m order characteristic function of the lateral vibration of the simply supported beam at both ends, $p_m(t)$ (m=1,2,…,M) and $p_m(t)$ (m=1,2,…,M) respectively represent the m generalized coordinates of longitudinal and transverse vibration. The characteristic function $\phi_m(x)$ that satisfies the boundary condition equation (3) can be the modal function of the static linear elastic beam, expressed as

$$\phi_m(x)=\sin(m\pi x) \tag{6}$$

Substituting equation (6) into equations (4) and (5), it can be seen that the boundary condition equation (3) is satisfied.

Substituting equations (4), (5), and (6) into the relevant variables in equations (1) and (2), we can get

$$\begin{gathered}
u_{,tt}=\sum_{m=1}^{M}\ddot{q}_m\sin(m\pi x),\quad u_{,xt}=\sum_{m=1}^{M}\dot{q}_m m\pi\cos(m\pi x),\quad v_{,x}=\sum_{m=1}^{N}p_m m\pi\cos(m\pi x),\\
v_{,xx}=\sum_{m=1}^{N}p_m\left(-m^2\pi^2\right)\sin(m\pi x),\quad u_{,t}=\sum_{m=1}^{M}\dot{q}_m\sin(m\pi x),\quad u_{,x}=\sum_{m=1}^{M}q_m\cos(m\pi x),\\
u_{,xx}=\sum_{m=1}^{M}q_m\left(-m^2\pi^2\right)\sin(m\pi x),\quad v_{,tt}=\sum_{m=1}^{N}\ddot{p}_m\sin(m\pi x),\ v_{,t}=\sum_{m=1}^{N}\dot{p}_m\sin(m\pi x),\\
v_{,xxxx}=\sum_{m=1}^{N}p_m\left(m^4\pi^4\right)\sin(m\pi x),\quad v_{,xt}=\sum_{m=1}^{N}\dot{p}_m m\pi\cos(m\pi x),\\
v_{,xxxxt}=\sum_{m=1}^{N}\dot{p}_m\left(m^4\pi^4\right)\sin(m\pi x),\quad v_{,xxxxx}=\sum_{m=1}^{N}p_m\left(m^5\pi^5\right)\cos(m\pi x)
\end{gathered} \tag{7}$$

Second-order Ordinary Differential Equations

$$\begin{aligned}
&\ddot{q}_n-4\gamma\sum_{m=1,m\sim=n}^{M}\frac{mn\left[1-(-1)^{m+n}\right]}{m^2-n^2}\dot{q}_m+c_d\left\{\dot{q}_n-2\gamma\sum_{m=1,m\sim=n}^{M}\frac{mn\left[1-(-1)^{m+n}\right]}{m^2-n^2}q_m\right\}\\
&+k_N^2\pi^3\left[\sum_{m-k-n=0}^{N}\left(-\frac{1}{2}mk^2p_mp_k\right)+\sum_{k-m-n=0}^{N}\frac{1}{2}mk^2p_mp_k+\sum_{n-m-k=0}^{N}\frac{1}{2}mk^2p_mp_k\right]\\
&-2\dot{\gamma}\sum_{m=1,m\sim=n}^{M}\frac{mn\left[1-(-1)^{m+n}\right]}{m^2-n^2}q_m-\left(\gamma^2-k_N^2\right)\pi^2n^2q_n=0\\
&\ddot{p}_n+k_f^2n^4\pi^4p_n-\left[(1-\eta)\gamma^2+\frac{1}{2}\dot{\gamma}-1\right]n^2\pi^4p_n-\\
&\sum_{m=1,\ m\sim=n}^{N}\left\{\frac{4\gamma mn\left[1-(-1)^{m+n}\right]}{m^2-n^2}\dot{p}_m-m^2\left[\frac{(-1)^{m-n}-1}{(m-n)^2}+\frac{1-(-1)^{m+n}}{(m+n)^2}\right]\dot{\gamma}p_m\right\}+\\
&c_d\left\{\dot{p}_n-2\gamma\sum_{m=1,\ m\sim=n}^{N}\frac{mn\left[1-(-1)^{m+n}\right]}{m^2-n^2}p_m\right\}+k_sp_n+\\
&\alpha\left\{n^4\pi^4\dot{p}_n-2\pi^4\gamma\sum_{m=1,\ m\sim=n}^{N}\frac{m^5n\left[1-(-1)^{m+n}\right]}{m^2-n^2}p_m\right\}+
\end{aligned} \tag{8}$$

$$
\begin{aligned}
&k_{\mathrm{N}}^{2}\pi^{3}\left[\sum_{m-k-n=0}^{N,M}\left(-\frac{1}{2}mk^{2}p_{m}q_{k}\right)+\sum_{k-m-n=0}^{N,M}\frac{1}{2}mk^{2}p_{m}q_{k}+\sum_{n-m-k=0}^{N,M}\frac{1}{2}mk^{2}p_{m}q_{k}\right]+\\
&k_{\mathrm{N}}^{2}\pi^{3}\left[\sum_{m-k-n=0}^{N,M}\left(-\frac{1}{2}mk^{2}q_{m}p_{k}\right)+\sum_{k-m-n=0}^{N,M}\frac{1}{2}mk^{2}q_{m}p_{k}+\sum_{n-m-k=0}^{N,M}\frac{1}{2}mk^{2}q_{m}p_{k}\right]+\\
&k_{\mathrm{N}}^{2}\pi^{4}\left[\frac{3}{8}n^{4}p^{3}+\sum_{m=s,k=n,m\sim=n}^{N}\frac{3}{4}m^{2}n^{2}p_{m}^{2}p_{k}+\sum_{s=m+k+n}^{N}\left(-\frac{3}{8}k^{2}msp_{m}p_{s}p_{k}\right)+\right.\\
&\sum_{m=s+k+n}^{N}\left(-\frac{3}{8}k^{2}msp_{m}p_{s}p_{k}\right)+\sum_{k=m+s+n}^{N}\frac{3}{8}k^{2}msp_{m}p_{s}p_{k}+\sum_{n=m+s+k}^{N}\frac{3}{8}k^{2}msp_{m}p_{s}p_{k}+\\
&\sum_{\substack{s+n=m+k,\\ m\sim=s\sim=k\sim=n}}^{N}\frac{3}{8}k^{2}msp_{m}p_{s}p_{k}+\sum_{\substack{k+n=m+s,\\ m\sim=s\sim=k\sim=n}}^{N}\left(-\frac{3}{8}k^{2}msp_{m}p_{s}p_{k}\right)+\\
&\sum_{\substack{s+k=m+n,\\ m\sim=s\sim=k\sim=n}}^{N}\frac{3}{8}k^{2}msp_{m}p_{s}p_{k}+\sum_{\substack{k+n=m+s,\ m=s,\\ k\sim=n,\ m\sim=k,\ m\sim=n}}^{N}\left(-\frac{3}{8}k^{2}msp_{m}p_{s}p_{k}\right)+\\
&\sum_{\substack{s+k=m+n,\ s=k,\\ m\sim=n,\ m\sim=k,\ m\sim=s}}^{N}\frac{3}{8}k^{2}msp_{m}p_{s}p_{k}+\sum_{\substack{s+k=m+n,\ m=n,\\ s\sim=k,\ m\sim=k,\ m\sim=s}}^{N}\frac{3}{8}k^{2}msp_{m}p_{s}p_{k}+\\
&\sum_{\substack{s+n=m+k,\ m=k,\\ s\sim=n,\ m\sim=s,\ m\sim=n}}^{N}\frac{3}{8}k^{2}msp_{m}p_{s}p_{k}+\sum_{\substack{s+n=m+k,\ s=n,\\ m\sim=k,\ m\sim=s,\ m\sim=n}}^{N}\frac{3}{8}k^{2}msp_{m}p_{s}p_{k}+\\
&\left.\sum_{\substack{k+n=m+s,\ k=n,\\ m\sim=s,\ m\sim=k,\ m\sim=n}}^{N}\left(-\frac{3}{8}k^{2}msp_{m}p_{s}p_{k}\right)\right]=0
\end{aligned}
\tag{9}
$$

4. Numerical simulation

In the numerical simulation calculation of this paper, the original parameters of a given beam are L=1.0m、E=210.0×10^{10}Pa、A=0.04×0.03m^2、ρ=7850kg/m and P_0=4.725×10^4N and the corresponding infinite quantity is obtained. Dimension parameters k_N=23.0940, k_f=0.2. Consider other parameters as γ_0=0.86、γ_1=0.1、α=0.0001 and c_d=0.001,. Considering the frequency of velocity perturbation, select the frequency of resonance of the first-order harmonic parameters, that is, ω=2ω_1+σ. Among them, σ represents the detuning parameter, which indicates the degree of change of the velocity perturbation frequency ω around 2ω_1. ω_1 is the first-order natural frequency of the linear free vibration of the beam system, that is, ω_1=2.9171[9].

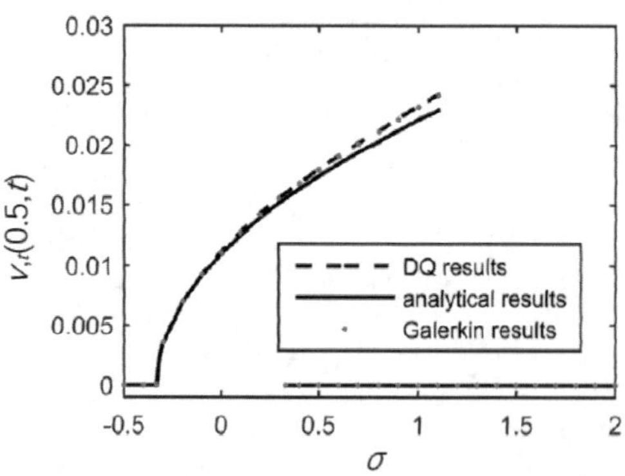

Fig 2 The Comparison of the multiple method and differential quadrature method and the eighth-order Galerkin truncation of the beam on stable steady-state response

Given the parameters $\sigma=0$、$\gamma 0=0.86$、$\gamma 1=0.1$、$\alpha=0.0001$、$cd=0.001$ and $\sigma=0$, the period of the dimensionless velocity perturbation amplitude is $T=2\pi/\omega$. As shown in Figure 3, the time history diagram, time history magnification diagram, phase diagram, spectrum diagram and Poincaré cross section of the first-order harmonic parameter resonance under the horizontal and vertical coupled vibration model are given. Observing Figures 3(a) and Figures 3(b), it can be seen that after the vibration of the axially moving beam enters a steady state, the longitudinal vibration amplitude and the transverse vibration amplitude differ by 10-3 in dimension, that is, the longitudinal amplitude is much smaller than the transverse amplitude. It can be seen from Figures 3(c) and Figures 3(d) that the phase diagram of longitudinal vibration exhibits irregular attenuation motion, and the phase diagram of lateral vibration is periodic motion.

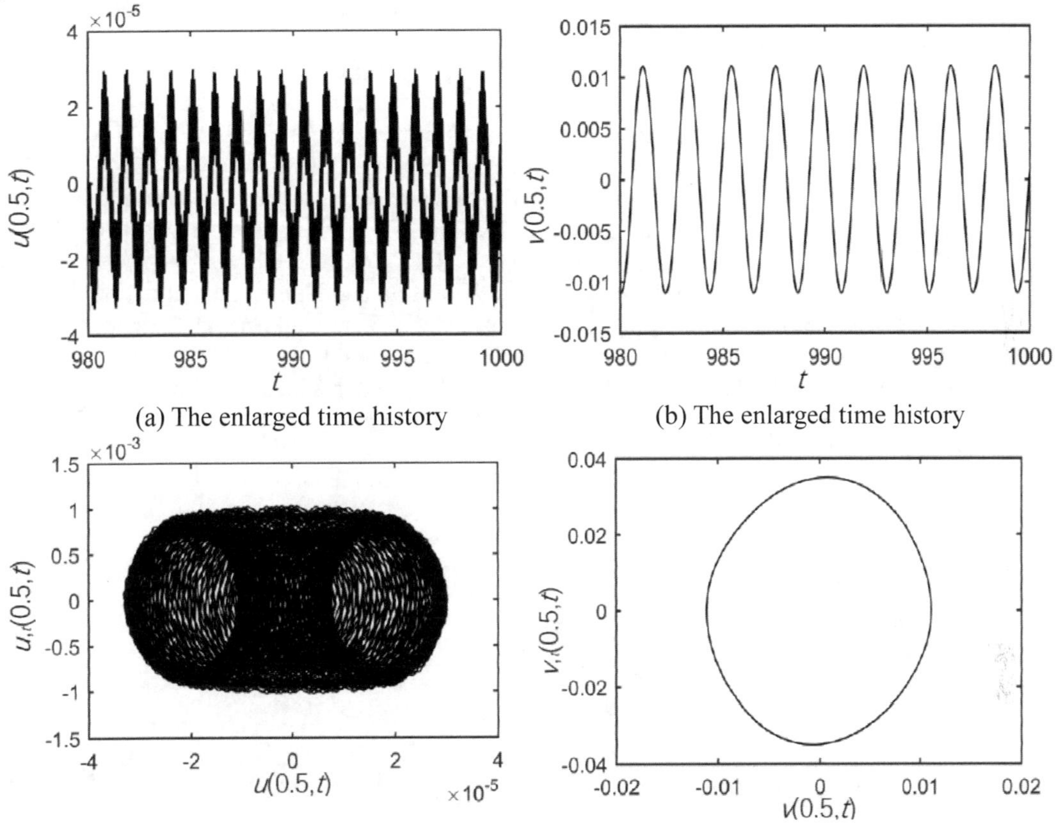

(a) The enlarged time history (b) The enlarged time history

(c) The phase diagram (d) The phase diagram

Fig 3 The longitudinal and transverse vibrations of the midpoint of beam

5. Conclusion

This paper describes the steady-state vibration response of a viscoelastic Euler beam under axially variable tension. First, the radially varying tension is introduced, the viscous damping coefficient and support stiffness are introduced, the transverse and longitudinal coupling mathematical model of the viscoelastic Euler beam with axially variable motion is established, and the corresponding homogeneous boundary conditions are established. Based on the eighth-order Galerkin truncation method and the fourth-order Runge-Kutta method, the nonlinear partial differential governing equations are numerically solved. The vibration amplitude results of the coupled model and the simplified model under the approximate analytical method and different numerical methods are compared and analyzed. The specific numerical basis for ignoring the influence of longitudinal vibration displacement is given. Finally, given parameter values, the horizontal and longitudinal vibrations of the beam system are shown.

Reference

[1] Chen L Q, Tang Y Q. Combination and principal parametric resonances of axially accelerating viscoelastic beams: Recognition of longitudinally varying tensions [J]. Journal of Sound and Vibration. 2011, 330: 5598-5614.

[2] Chen L Q, Tang Y Q. Parametric stability of axially accelerating viscoelastic beams with the recognition of longitudinally varying tensions [J]. Journal of Vibration and Acoustics. 2012, 134: 1-11.

[3] Ding H, Yan Q Y, Chen L Q. Chaotic dynamics in forced vibration of an axially accelerating viscoelastic beam [J]. Acta Physica Sinica. 2013, 62 (20): 1-7.

[4] Ding H, Yan Q Y, Zu J W. Chaotic dynamics of an axially accelerating viscoelastic beam in the supercritical regime [J]. International Journal of Bifurcation and Chaos. 2014, 24 (05): 1-19.

[5] Tang Y Q, Zhang D B, Gao, J. M. Parametric and internal resonance of axially accelerating viscoelastic beams with the recognition of longitudinally varying tensions [J]. Nonlinear Dynamics. 2016, 83 (1-2): 401-418.

[6] Tang Y Q, Zhang Y Y, Yang X D. On Parametric Instability Boundaries of Axially Moving Beams with Internal Resonance [J]. Acta Mechanica Solida Sinica. 2018, 31: 470-483.

[7] Tang Y Q, Ma Z G. Nonlinear vibration of axially moving beams with internal resonance, speed-dependent tension, and tension-dependent speed [J]. Nonlinear Dynamics. 2019, 98: 2475-2490.

[8] Tang Y Q, Ma Z G, Liu S, Zhang L Y. Parametric Vibration and Numerical Validation of Axially Moving Viscoelastic Beams with Internal Resonance, Time and Spatial Dependent Tension and Tension Dependent Speed [J]. Journal of Vibration and Acoustics. 2019, 141 (6): 061011.

[9] Tang Y Q, Zhang D B, Gao J M. Parametric and internal resonance of axially accelerating viscoelastic beams with the recognition of longitudinally varying tensions [J]. Nonlinear Dynamics, 2016, 83 (1-2): 401-418.

Journal of Physics: Conference Series **2215** (2022) 012027 doi:10.1088/1742-6596/2215/1/012027

Research on ship detection of optical remote sensing image based on Yolo V5

X M Sun and Y J Zhang*, H Wang, Y X Du

School of Computer Science and Software Engineering, University of Science and Technology, Liao Ning, China

Corresponding author. E-mail: zhangyujun@ustl.edu.cn

Abstract. In order to achieve ship target detection based on remote sensing images in complex backgrounds, this paper proposes a ship target detection algorithm based on YOLO V5. First, the improved Kmeans is used to cluster the data set, DIOU_loss is used for the candidate frame selection strategy, and finally the channel attention mechanism is introduced. The experimental results show that under the same conditions, the improved network has a slight decrease in detection speed compared with the original YOLOv5 network, and the effect of images with densely distributed small targets is significantly improved. On the same test set, the accuracy rate is reduced by 3.84%, the recall rate is increased by 32.73%, and mAP is increased by 22.48%.

1. Introduction

With the continuous development of satellite technology, it has become very common to use satellites to monitor the conditions of various sea areas in real time. The use of remote sensing images for ship target detection has extremely broad application prospects in civil and military fields [1]. In civilian use, it can be used to find missing ships at sea; it can also be used to monitor marine fisheries; it can also monitor illegal ships, such as pirate ships, ferry boats, and so on. In military use, it can be used to detect the location of enemy ships, guide long-range missiles to accurately strike; it can also monitor the sea during wartime, illegal intrusion of warships, and so on. Due to the long-distance shooting, most of the objects to be detected in the obtained remote sensing images show the characteristics of small targets [2]. The small target in the image loses a lot of feature information after the model changes. Therefore, the recognition of small targets in remote sensing images is a hot topic with high research significance.

At present, the detection methods of ships in remote sensing images can be divided into two categories: traditional methods and deep learning-based methods. The current deep learning method is the mainstream method of optical remote sensing image detection. In the field of two-stage target detection algorithms, Gu Jiaojiao et al. used Faster RCNN to redesign the number and size of its anchor frames, which effectively improved the problem of easy repeated detection of objects to be detected in remote sensing images[3]. In the field of single-stage detection algorithms, Ma Junjie et al. used the YOLO network to establish the nonlinear observation equation of the remote sensing ship target, which was transformed into a linear equation, and then solved the specific position coordinates of the ship, which improved the positioning accuracy of the ship[4].

In summary, based on the YOLO v5s network, this article improves the YOLO V5s network to solve the problem of poor recognition of dense small targets in remote sensing images. First, use the improved Kmeans to cluster the used data set and improve the selection strategy of the target frame.

Introduce channel attention mechanism. The experimental results show that compared with the original YOLOv5 network, the improved YOLOv5s network has an improved effect in the task of remote sensing image ship target recognition.

2. Ship remote sensing detection model based on YOLO V5

2.1. Yolo V5 introduction

YOLO v5 currently has four versions: YOLO V5s, YOLO V5m, YOLO V5l, YOLO V5x.Among them, YOLO v5s is the smallest and fastest version of the model. This article uses the YOLO v5s version. YOLOv5 can be divided into 4 modules in structure: input terminal, backbone network Backbone, Neck and prediction terminal Prediction. The network structure is shown in Figure 1.

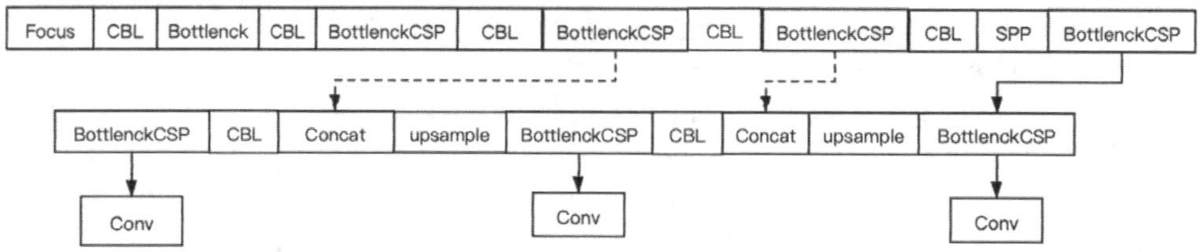

Figure 1. Figure with yolov5 structure

The image preprocessing stage of YOLOv5 introduces image preprocessing operations such as Mosaic data enhancement, adaptive image scaling, and adaptive anchor point calculation[5]. This is conducive to the image proportion not deformed after the image is zoomed. The Backbone stage of YOLOv5 adopts the Focus and CSPDarknet53 structure, which reduces the amount of calculation for the model while ensuring accuracy. The Neck structure is mainly to generate feature pyramids. This structure generates pooled feature vectors of different fixed sizes, and also strengthens the expression ability of features, which is conducive to detecting targets with large changes in size.

2.2. Improved yolo v5 model

2.2.1. NMS improvements

Non-Maximum Suppression (NMS) is mainly used to eliminate redundant candidate frames. The target detection algorithm model will generate a lot of target frames during operation, and one target will be selected by many target frames . The role of NMS is to eliminate redundant target frames, so that a target retains only one target frame with higher confidence. However, for small targets that are too close to each other in the original NMS algorithm, the neighboring target frame may be eliminated. The original NMS algorithm will greatly reduce the detection effect of moored ships in the port terminal. This article introduces DIOU-NMS. Unlike the original NMS, DIoU-NMS not only considers the value of IoU, but also considers two boxes. The distance between the center points. A new formula is used to determine whether a box is deleted:

$$S_i = \begin{cases} s_i, IoU - R_{DIoU}(M, B_i) < \varepsilon, \\ 0, IoU - R_{DIoU}(M, B_i) \geq \varepsilon, \end{cases} \quad (1)$$

$$R_{DIoU} = \frac{\rho^2(b, b^{gt})}{c^2} \quad (2)$$

Among them, R_{DIoU} is the distance between the center points of the two Boxes, Expressed by the above formula (2):Where p(.) is the distance, b and bgt represent two boxes, and c is the diagonal length of the smallest box containing two boxes.

2.2.2. *Channel attention mechanism*

Convolutional neural networks use the idea of local receptive fields to fuse spatial information and channel information to extract image information features. The channel attention mechanism (Squeeze-and-Excitation (SE), SENet) improves its representation ability by modeling the dependence of each channel, and it can adjust the features channel by channel, so that the network can learn Global information is used to selectively enhance features that contain useful information and suppress useless features[6].

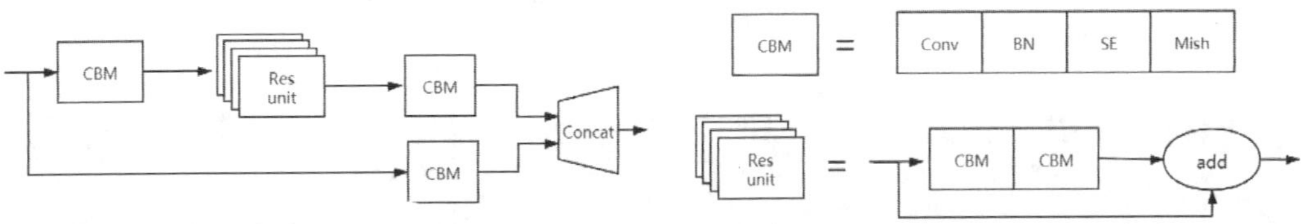

Figure 2. SENet is introduced to YOLO v5s structure diagram

The algorithm structure is composed of three parts: Squeeze, Excitaion and Reweight, and explicitly constructs the interdependence between feature channels [7]. Specifically, it uses a method of recalibrating feature channels.In the process of learning features, the network learns the contribution value of each channel to the overall feature at the same time, and then according to this contribution value to improve useful features and Suppress features that do not contribute much to the task [8].

3. Experiment

3.1. *Data set*

The data set of this experiment comes from the public remote sensing image data set and the image data intercepted from the Internet. Among them, 2432 pictures in the training set and 57 pictures in the validation set are selected from the public data set HRRSD. The test set selected the aerospace remote sensing target detection data set annotated by NWPU Northwestern Polytechnical University and some of the image data made by myself.

3.2. *Experimental platform*

The deep learning framework used in this experiment is the torch framework, and the operating platform is the ubuntu 18 system.The running memory is 16g, the video memory is 6g, the graphics card model is GTX1660ti, GPU accelerated environment 10.0, and the framework of the training model is pytorch.

3.3. *Evaluation Index*

Because the test set used in this article contains a part of data that only contains a single target. As a result, when the map value is calculated, the maximum accurate value corresponding to each recall stage is 1, so the effect of this experiment cannot be effectively evaluated. Therefore, the evaluation indicators of this experimental model mainly use precision (Precision, P), recall (Recal, R) and F-score, and the specific calculations are shown in equations (3), (4).

$$recall = \frac{T_p}{T_p + F_n} \tag{3}$$

$$precision = \frac{T_p}{T_p + F_p} \tag{4}$$

In the formula, TP means that the ship is correctly detected; TN means that the target is mistakenly detected as a ship; FP means that the target is not a ship, and the detection result is not a ship; FN means that the ship is mistakenly detected as another target. The mean precision AP is the area value of the curve enclosed by the precision rate P and the recall rate R, and the mean precision mean mAP is the average of the mean precision values of all classes of learning.

3.4. Analysis of experimental results

After optimizing the loss function of YOLO V5 and the screening function of the prediction box, introducing a series of operations such as the channel attention mechanism, the improved detection result is shown in 11. After using CIOU_LOSS as the bounding box loss function of the predicted target and DIOU_NMS processing, it is obvious from Figure 11 that the ships in densely arranged ports and occluded locations can be effectively identified, and the confidence score has also been improved.

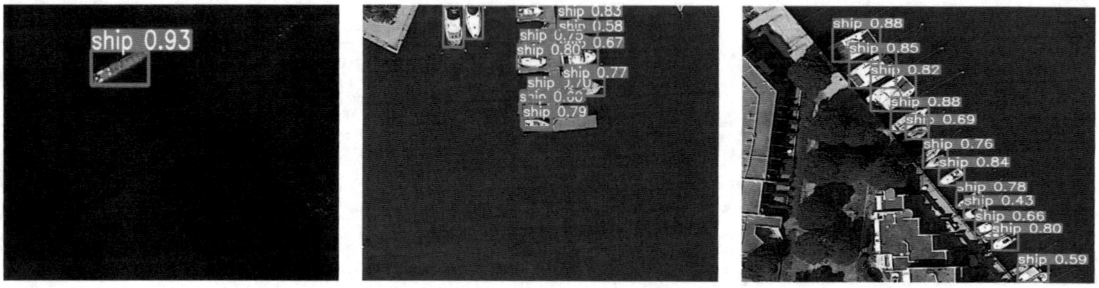

Figure 3: Improved YOLO V5 test results

There are a total of 30 pictures in the test set, which contain a total of 1437 boats. Using the improved YOLO V5 for detection, a total of 1 ships were detected, 1320 ships were correctly detected, 13 ships were falsely detected, and 122 ships were missed. They are in the same test set as Faster RCNN, SSD, YOLO v4, and YOLOv5s models. For experimental comparison, the detection performance of different models is shown in Table 1.

Table 1. Comparison of detection performance

Network	Precision rate	recall rate	mAP
Improved YOLO V5s	89.34%	88.95%	92.6%
Original YOLO V5s	93.19%	56.22%	70.12%

It can be seen from Table 1 that the performance of the improved algorithm has been greatly improved compared with the original YOLOv5s, more small target detection samples have been detected, and the number of false detections has also increased. The recall rate increased by 32.73%, and the accuracy rate decreased by 3.84%. mAP increased by 22.48%. Compared with some other algorithms, there are similar changes. From this point of view, the performance of the improved model is significantly improved.

4. concluding remarks

Aiming at the problem of poor detection of dense small targets in remote sensing image ship target detection, this paper makes some improvements to the YOLO V5s network based on the original YOLO V5s network, and proposes a remote sensing image ship target recognition based on the improved YOLOv5 network. method. The experimental results show that the F-score of the improved target recognition network reaches 95.17%, and the accuracy and recall rate are significantly improved, and the deficiencies of the original model are improved. The next step is to improve the remote sensing image data set and improve the recognition under complex conditions. Research on accuracy and other aspects.

References

[1] Jin Wenchao. Optical remote sensing image ship target detection technology based on complex

background[D]. Harbin Engineering University, 2013.
[2] Guo Liqiao, Lin Jianyuan, Wang Wenying. The application of refined urban management based on high-scoring remote sensing data[J]. Urban Development Research, 2012, 19(11): 57-63+69.
[3] Gu Jiaojiao, Li Bingzhen, Liu Ke, et al. Infrared ship target detection algorithm based on improved Faster R-CNN [J]. Infrared Technology, 2021, 43 (2): · 92 · 170-178.
[4] Ma Junjie, Huang Daqing, Liu Yaohui, et al. Ship target positioning based on improved ship y-olo [J]. Electronic Measurement Technology, 2020, 43(5): 112-115.
[5] Lin Qingping,Zhang Qilin,Xiao Lei.Remote sensing image target recognition method using improved YOLOv5 network[J].Journal of Air Force Early Warning Academy,2021,35(02):117-120.
[6] Wang Dongfei. Application of Convolutional Neural Network Based on Channel Attention in Image Super-resolution Reconstruction[J]. Radio and Television Technology, 2018, 45(06): 63-66.
[7] HU Jie, SHEN Li, ALBANIE S,et al.Squeez-e and excitation networks [EB/OL].[2018-09-30].https1://ai-xiv.org/abs/1709.01507.
[8] Li Kewen,Li Xinyu.An improved Faster R-CNN pedestrian detection model based on SENet[J].Computer Systems & Applications,2020,29(04):266-271.

Low power optimization method for contactless micro intelligent voltage sensor communication network

Hongdi Sun[1,a], Renze Chen[1,b], Ke Zhou[2,c], Xiaoming Wang[2,d]

[1]Technology Research and Development Center, Digital Grid Research Institute, China Southern Power Grid, Guangzhou, China

[2]Electric Power Research Institute , Guangxi Power Grid Co., Ltd., Nanning, China

[a]sunhd@csg.cn

[b]chenrz1@csg.cn

[c]zhou_k.sy@gx.csg.cn

[d]wang_xm.sy@gx.csg.cn

Abstract—Aiming at the problems of high power consumption and low success rate in contactless micro voltage sensor communication network, a low power consumption optimization method of contactless micro voltage sensor communication network is proposed. By calculating the average power consumption of micro voltage sensor nodes in the communication network area, the energy balance degree of micro voltage sensor nodes is obtained. When the micro voltage sensor nodes meet the sleep probability of communication network area coverage factor, the power consumption of voltage sensor communication network is calculated. By defining the information entropy of communication network data set, the conditional entropy of data feature classification of communication network is obtained, and the high-power data is eliminated by using the validity kernel function of data features. Combined with the low-power optimization algorithm of micro voltage sensor communication network, the low-power optimization of micro voltage sensor communication network is realized. The experimental results show that the optimization method in this paper can not only reduce the power consumption of data transmission in contactless micro voltage sensor communication network, but also improve the success rate of data transmission, and has better optimization effect.

1. Introduction

Voltage measurement, often applied in the engineering field, is of great significance in the academic circle. However, in electrical system, voltage measurement is usually realized with micro voltage sensors [1]. The main function of micro voltage sensors is isolating the electricity at the high-voltage side from that at the low-voltage side. When the electrical system is operating normally, micro voltage sensors could reduce its pressure, transforming high voltage into low voltage that is more supportive for the stable operation of the electrical system. However, micro voltage sensors have their limits in application. When the electrical system malfunctions, especially under extreme weather, it is very difficult to find the accurate malfunctioning area. To the contrary, in contactless micro voltage sensors, the communication network could accurately measure the breakdown signal. The principle is maintaining the smoothness of the communication network by taking advantage of the relationship between the intense magnetic field surrounding the semiconductor and the electric conductor, featuring simple structure and rapid response. By studying the method for low power consumption optimization of communication network of micro voltage sensors, this paper aims to provide strong theoretical support for future development of electric grid and sensor intelligence.

Li Bin et.al. [2] proposed a MEC-based method for low power consumption optimization of high-speed railway communication network, using traditional MEC technology to analyze wireless network of high-speed railway. The study offered an optimization scheme for wireless communication network, verified the performance of the optimization method with real cases under high-speed railway scenario and obtained the conclusion that the method would consume a relatively large amount of power, which provided strong theoretical support for the optimization of wireless network. Li Bin et.al. [3] proposed an electric power communication network optimization algorithm based on risk joint equalization. In his study, data about features of electric power communication network were extracted and existing resources of each network nodes were analyzed to build electric power communication network optimization model. The electric power communication network optimization algorithm based on risk joint equalization could effectively improve the operation efficiency of the network, with the power consumption still being relatively high.

To improve the optimization effect, this paper, focusing on the communication network of contactless micro voltage sensors, designed a low power consumption optimization method based on the aforesaid studies.

2. Design of Low Power Consumption Optimization Method for Voltage Sensor Communication Network

2.1. Calculation of Power Consumption of Voltage Sensor Communication Network

First, set the energy consumption equilibrium index of each transmission node to obtain the average power consumption of different transmission modes. Then, calculate the power consumption of communication network at each node according to the coverage rate of transmission nodes in the communication network of micro voltage sensors based on the power consumption equilibrium index of voltage sensors. In this way, the calculation of power consumption of voltage sensor communication network is completed. Detailed steps are as follows:

It is assumed that the power consumption value of all network transmission nodes against neighbor nodes is l_{sd} , $D(E_r)$ is the variance of residual power consumption of each network transmission nodes, E_s is the power consumption value of each transmission node where communication network data transmission does not occur,and E_w is the power consumption of each transmission nodes where data transmission is performed through voltage sensors. Assuming that there are N transmission nodes in any area of the voltage sensor communication network and K transmission nodes are under operating mode, formula (1) can be used to calculate the average power consumption of each transmission node in voltage sensor communication network every X period. The formula is:

$$\bar{E} = \frac{E_s \times E_w}{D(E_r)} \times \frac{\{1_{sd}\} \times K}{X \pm N} \tag{1}$$

Within X periods, K transmission nodes are under the operating mode. df_{xg} is the coverage rate of transmission nodes in the voltage sensor. I_{ef} refers to one transmission node in voltage sensor communication network. p_{sd} is the communication network transmission period of each transmission node. σ_{klj} is the power consumption per bit of information of voltage sensor communication network. Formula (2) can be used to indicate the degree of balance of voltage sensor communication network at each transmission node:

$$d_{fgjk} = \frac{X \times K}{df_{xg} * I_{ef}} \otimes \frac{\{p_{sd} \times \sigma_{klj}\}\}}{g_{dfh}\,\mathrm{m}u_{djh}} \pm h_{hjk} \tag{2}$$

Where g_{dfh} indicates the working state of each transmission node in the voltage sensor; u_{djh} refers to the power consumption of each transmission node under operation; h_{hjk} is the probability of any area being covered by transmission nodes in the communication network.

If the power consumption of each transmission node is μ_{sf} , ω_{hj} is specific location of transmission node in the voltage sensor, p_{ij} is all the power consumed to transmit communication information from node i to node j, t_{ij} is the time taken to transmit communication information from node i to node j. Formula (3) can be used to calculate the power balance of voltage sensor communication network:

$$W_{asgg} = \frac{\sum t_{ij} p_{ij}}{\mu_{sf} \text{ m } \omega_{hj}} \oplus \frac{(i, j)}{g_{fh} \times r_{sg}} \quad (3)$$

In the formula, g_{fh} is the gross power consumed during the transmission process, r_{sg} is the total transmission time of all transmission nodes.

In voltage sensor communication network, the number of bits transmitted between transmission node i and j by each signal is θ_{pop} . The number of bits of communication information is d_{ag} . Formula (4) can be used to calculate the coverage rate of all transmission nodes of voltage sensor communication network:

$$p_{cbij} = \frac{d_{fgjk} \times \bar{E}}{\theta_{pop} \times d_{ag} \text{ m} W_{asgg}} \times f_{dh} \quad (4)$$

In the formula, f_{dh} is the number of two neighbor transmission nodes. The power consumed to send information from voltage sensor communication network from level i to level j is k_{sh}. Formula (5) can be used to calculate the power consumption of voltage sensor communication network:

$$W_{dfpp} = \frac{\{i, j\}}{k_{sh}} \times e_{dg} \oplus p_{cbij} \times W_{asgg} \quad (5)$$

The calculation of power consumption of communication network of contactless micro voltage sensor is hereby completed.

2.2.Elimination of High Power Consumption Data in Voltage Sensor Communication Network

In the communication network of micro voltage sensor, the data sample can be mainly classified into m categories c_i (i=1, 2,L, m), and $|c_1|$+ L+$|c_m|$=S. The probability of data sample transmitted through the communication network of any voltage sensor belonging to category c_i is $P_i = c_i / S$. The information entropy of data sample set S transmitted through voltage sensor communication network is hereby defined as:

$$\text{Info}(S) = -\sum_{i=1}^{m} P_i \log_2 (P_i) \quad (6)$$

Assuming that the features value of data transmission of voltage sensor communication network is t_i, the training sample set of data transmission of voltage sensor communication network is divided as $S = \{S_1, S_2 \text{L}, S_q\}$, where the number of data sample contained in S_i is $|S_i|$. After classifying the features of data transmitted through voltage sensor communication network [4],the condition entropy obtained can be expressed by formula (7):

$$\text{Info}_{t_i}(S) = \sum_{i=1}^{m} |S_i| / |S| \, \text{Info}(S) \quad (7)$$

The above result is used as the information gain $e^{\mathrm{Gain}(t_i)}$ of features t_i of data transmission of voltage sensor communication network:

$$\mathrm{Gain}(t_i) = \mathrm{Info}(S) - \mathrm{Info}_{t_i}(S) \tag{8}$$

In the formula, the feature vector of data transmission of voltage sensor communication network can be indicated as the vector value of each feature in the sample [5]. The feature vector of data transmission of voltage sensor communication network can be indicated with formula (9):

$$f_i = \left(\mathrm{Gain}(t_i) / \sum_{i=1}^{n} \mathrm{Gain}(t_i) \right) e^{\mathrm{Gain}(t_i)} \tag{9}$$

In the formula, $e^{\mathrm{Gain}(t_i)}$ is the feature vector of a certain transmission node sample. If the feature vector matrix in feature t_i of data transmission of voltage sensor communication network can describe n features, the validity of voltage sensor communication network can be obtained with the following formula:

$$P = \left(f_t^1 \mathrm{L}\ f_t^n \right) \tag{10}$$

In the formula, f_t^n is the feature validity of the data transmission of the ith communication network. Set K as transmission data matrix and P as the validity matrix of features in data transmission of n-tier voltage sensor communication network, and the validity function is:

$$K\left(x_i, x_j\right) = K\left(x_i^T P, x_j^T P\right) \tag{11}$$

In the function, $x_i^T P$ is the feature vector matrix of the ith data transmission of the voltage communication network; $x_j^T P$ is the feature vector matrix of the jth data transmission of the voltage communication network; r is the power consumption value of the validity function. The power consumption function K' of voltage sensor communication network can be expressed with formula (12):

$$K\left(x_i, x_j\right) = \exp\left\{-\gamma[\left(x_i - x_j\right) P \cdot f_t^i\right\} \tag{12}$$

The smaller power consumption of data transmission of voltage sensor communication network f_t^i is, the smaller the influence the ith feature in the data sample transmitted through voltage sensor communication network has on the calculation of power consumption of data transmission would be. [6] In this light, power consumption function of data transmission of voltage sensor communication network can be used to distinguish high power consumption data and low power consumption data and thereby eliminate the high power consumption data in voltage sensor communication network.

2.3. Design of Low Power Consumption Optimization Algorithm for Micro Voltage Sensor Communication Network

Set the transmission range of communication data of network transmission node m as [0,T], the probability distribution function of transmission node can be expressed as:

$$P(m,a) = \begin{cases} 0 & a < 0 \\ \dfrac{a}{T} & 0 < a < T \\ 1 & a > T \end{cases} \tag{13}$$

In the formula, the threshold of the probability distribution of transmission node m in micro voltage sensor communication network is a. When the information between two nodes is being transmitted in voltage sensor communication network, assume the probability of communication network information transmission is $1-P(T_X,0)$, the transmission model of voltage sensor communication network can be expressed as:

$$F(t) = P\left(T_f,\ t\right) = \frac{P(T_x,\ t) - P(T_x,0)}{P(T_x,0)} \tag{14}$$

In the formula, the power consumption is e_r when the coverage probability of voltage sensor communication network is P_r. a refers to the constraint function of actual data transmission of communication network. E_k refers to the residual power consumption after the transmission nodes complete transmission of communication information.

This paper has optimized the power consumption of the communication network of micro voltage sensors with residual power consumption optimization method,and obtained the final constraint equation of low power consumption optimization of voltage sensor communication network:

$$c(L) = \frac{E_k + \alpha e_r}{e_r} \tag{15}$$

In the equation, $e_t(A)$ and $e_i(B)$ refers to the power consumption value of transmission node A and B under micro voltage sensor communication network respectively. T_{ab} represents the time node A and B have to take to transmit communication data. r indicates the size of communication network data. p refers to the transmission rate of communication network data. T_c is the transmission time after communication network data is compressed.

To this end, a low power consumption optimization algorithm for micro voltage sensor communication network is designed to realize the low power consumption optimization of communication network.

$$G\left(\frac{e_t(A)}{e_{rf}(A)} - \frac{e_i(B)}{e_{rf}(B)}\right) - T_{ab} - \frac{r}{p} = T_c \tag{16}$$

In the formula, the time interval of communication network information transmission at node x is T_x . t is the time constraint threshold of the communication network. Calculate the above formula, obtain the duration T_f [7] of the data transmission of micro voltage sensor communication network, and thereby construct a data transmission model for voltage sensor communication network. Compress the low power consumption data of voltage sensor communication network with data compression algorithm [8],the following formula is obtained:

3. Test and Analysis

3.1.Preperation

Verify the effectiveness of the optimization method proposed in the paper through simulation analysis. Select an area of 150m*150m in the communication network of contactless micro voltage sensor, and put 150 voltage sensor nodes in the area randomly, with the initial power consumption of each node being 10W (as shown in Figure 1).

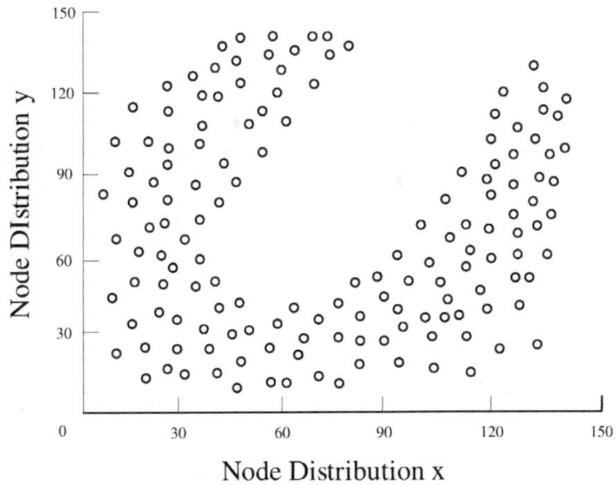

FIG. 1 DISTRIBUTION OF NODES OF MICRO VOLTAGE SENSORS

In Figure 1, the optimization method based on MEC (the method in Literature [2]) is compared with the optimization method based on load and risk joint equilibrium (the method in Literature [3]). The two methods are both applied to lower the power consumption of micro voltage sensor communication network. The following test results are obtained.

3.2.Analysis of Test Results

For the communication network of micro voltage sensors, low power consumption optimization aims to lower the transmission power consumption of the communication network. In the study, three optimization methods were applied to lower the power consumption of voltage sensor communication network with data transmission time being the independent variable. The power consumption of data transmission of voltage sensor communication network was as shown in Figure 2.

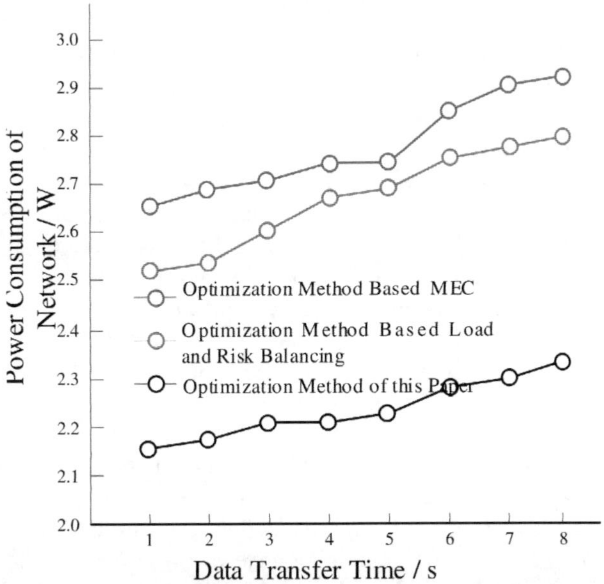

FIGURE 2 COMPARISON OF POWER CONSUMPTION OF COMMUNICATION NETWORK

From Figure 2, it can be concluded that the longer the transmission time is, the power consumption of communication network of the three optimization methods would all increase, with the power consumption of MEC-based optimization method and the optimization method based on load and risk joint equilibrium being relatively high within specific transmission time. The MEC-based optimization method can keep the power consumption of communication network within the range of 2.65W~2.95W, while the optimization method based on load and risk joint equilibrium can keep the power consumption of communication network within the range of 2.5W~2.8W, almost the same. In contrast, when optimizing the communication network of micro voltage sensors with the method proposed in the paper, the transmission power consumption is relatively stable, between 2.15W~2.35W, smaller than the power consumption range of the other two methods, which indicates that the method proposed in the paper performs better in lowering the power consumption of micro voltage sensor communication network.

In order to further verify the effect of the optimization method proposed in the paper, success rate of data transmission is used as an indicator of the optimization effect of micro voltage sensor communication network. Success rate of data transmission refers to the proportion of data successfully transmitted in the network. The result is shown in Figure 3.

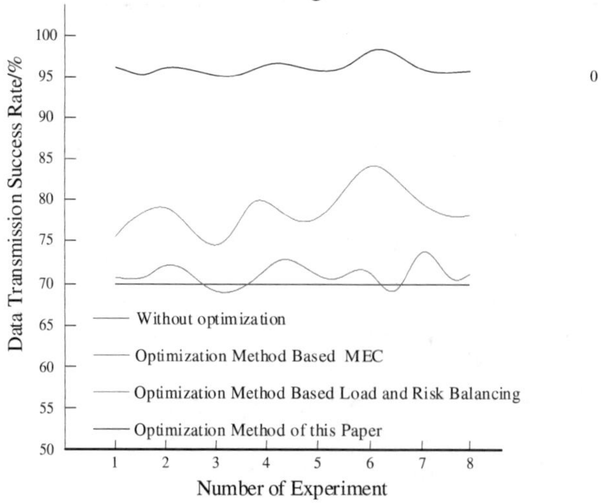

FIGURE 3 TEST RESULT OF DATA TRANSMISSION SUCCESS RATE

From Figure 3, it can be concluded that when MEC-based optimization method is applied, the success rate of data transmission in micro voltage sensor communication network is close to that before optimization; before the 2nd -3rd test and 6th -7th test, the success rate of data transmission after the optimization method is applied is lower than that before optimization. When the optimization method based on load and risk joint equilibrium, the success rate of data transmission is higher than that before optimization (fluctuation between 75% and 85%), fluctuating quite significantly, which is not as stable as the data transmission in network. When the optimization method proposed in the paper is adopted, the success rate of data transmission can be raised to over 95%, and the fluctuation of the success rate of data transmission is relatively stable, realizing better low power consumption optimization effect for micro voltage sensor communication network.

4. Conclusion

The paper proposed a low power consumption optimization method for contactless micro voltage sensor communication network. Tests have shown that the optimization method could achieve better effect in the communication network of micro voltage sensors and thereby lower the power consumption of traditional method in communication network. However, the method proposed still needs to be improved. In the future, studies should, while ensuring the optimization effect of the power consumption and data transmission success rate of the communication network, increase the speed of data transmission in micro voltage sensor communication network to avoid network congestion.

Acknowledgement
Supported by science and technology Project of Guangxi Power Grid Co., Ltd (GXKJXM20200119)

REFERENCES

[1]. Jiang Taoran, Liu Xizhe. "Non-Contact Voltage Sensor Based on Topology Transformation". Transactions Of China Electrotechnical Society, Vol.34 No. 1, pp 153-159 ,Jan, 2019.
[2]. Li Bin. "Wireless communication network optimization scheme for high-speed railway based on MEC ".Vol.11, pp 88-95 , Nov, 2019.
[3]. li bin, lu chao, jing dongsheng, et al. "An Optimized Routing Algorithm With Load and Risk Joint Balance in Electric Communication Network".Proceedings of the CSEE, Vol.39 No.9, pp 2713-2722, May, 2019.
[4]. Xie Wenwang,Sun Yunlian."Optimization algorithm for de-mapping module of broadband power line communication based on OS-ELM". Electrical Measurement & Instrumentation, Vol.56 No.13, pp 44-50, Jul, 2019.
[5]. He Hao, Yi Weidong, Chen Yongrui, et al. "WSN Timer Resolution Adjustment Based on UKF Approach". Journal of Electronics & Information Technology, Vol. 41 No. 3, pp 687-693,Mar, 2019.
[6]. Li Bin, Jia Bincheng, Sun Yi, et al. "Cross-layer Protection Algorithm Based on Network Coding in Power Communication Network". Automation of Electric Power Systems, Vol. 44 No. 3, pp 132-138, Feb, 2020.
[7]. Yang Jiaying, Li Ting, Xie Peizhong. "Energy Efficiency Optimization Algorithm for Heterogeneous Cellular Network Based on Stackelberg Game".Journal of Signal Processing, Vol. 36 No. 11, pp 1923-1930, Nov, 2020.
[8]. Feng Wei, Wang Feng, Xu Dan, et al. "Jointly Security R outing and Power Allocation in Wireless Sensor Networks". Chinese Jou r nal Of Sensors And Actuators, Vol. 32 No. 4, pp 610-617, Apr, 2019.

AUTHOR INDEX